PICTURE CONTROL

WRITING SCIENCE

EDITORS   Timothy Lenoir and Hans Ulrich Gumbrecht

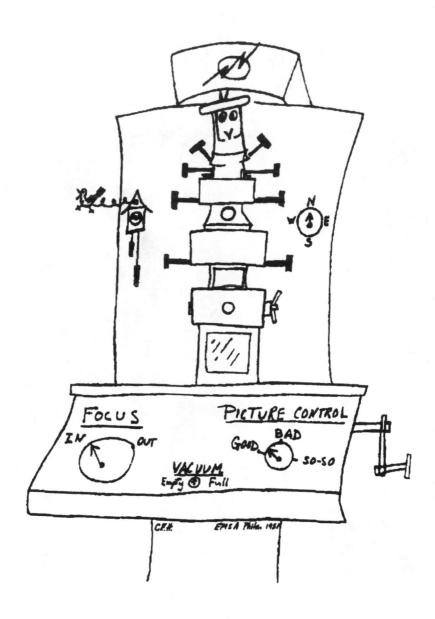

PICTURE CONTROL

The Electron Microscope and

the Transformation of Biology

in America, 1940–1960

Nicolas Rasmussen

STANFORD UNIVERSITY PRESS

STANFORD, CALIFORNIA

Stanford University Press
Stanford, California
© 1997 by the Board of Trustees of the
Leland Stanford Junior University
Printed in the United States of America

CIP data are at the end of the book

Frontispiece: Cartoon of an electron microscope, signed by
Cecil Hall ("CEH") and inscribed "EMSA Phila. 1951," presumably
circulated at the EMSA (Electron Microscope Society of America)
meeting that year. Courtesy of the Microscopy Society of
America (formerly EMSA).

To Harry Sootin:
chemist, historian, experimenter,
and grandfather beyond compare

Preface

This book is in a sense the culmination of the ten years or so I have spent working (half-time or more), at a variety of levels from media-preparer to technician to doctoral student, in biology laboratories. My experience has covered a range of the life sciences: in addition to short stints in labs of half a dozen subdisciplines, including neuroscience, yeast genetics, and fly genetics, I have spent substantial time laboring in the ranks of clinical microbiologists, cancer physiologists, bacteriophage molecular geneticists, and finally plant developmental biologists. I was in excellent labs and had good learning experiences.

At one point I began studying the philosophy of science, and discovered that for all its elegance, this branch of thinking seemed to have nothing to do with what I knew of science from the laboratory. The main reasons are not hard to find: I was a biologist, whereas the philosophy of science has since its beginning dealt almost exclusively with physics; and I (like almost all life scientists) was an experimenter, whereas the philosophy of science has, for most of this century at least, concentrated on theory at the expense of experiment. Perhaps also there is something to Lukács's thoughts on the privileged position of the worker for understanding the economic system that applies here; certainly, my point of view as a technician was quite different from the much more remote one of most philosophers of science (though there were a few exceptions with whom I discovered considerable sympathy, e.g., Ludwik Fleck, Michael Polanyi, Thomas Kuhn, Stephen Toulmin, and my supervisor at the University of Chicago, William Wimsatt). In the decade and a half since I began studying the philosophy of science, alert philosophers have been attending increasingly to the epistemology actually used in experiment, while the ever-increasing status of molecular biology seems to have drawn more attention to the experimental life sciences, and thus the philosophy of science has begun to make amends for its two large oversights. I expect that someone making a similar transition today might find the discrepancy between the familiar science of the biology lab and the science of the philosophers less jarring than I did.

I moved into the history of science and found that although experiment had been rediscovered and was, thankfully, a thoroughly fashionable topic, nobody seemed to be answering a question that had been bothering me since the first few times I had been trained to perform experimental protocols and routines. I was wondering where these practical rituals of research came from; for there were seldom any publications one could turn to for detailed advice, and when asked about the source, those responsible for my training, the technical personnel or scientists proper, usually referred to people who had given them their training. Sometimes it was "we have always done it that way," sometimes it was "I went to the X lab for Y months and learned how to do that," and more rarely it was "I spent Z months [or years] improvising protocols to make that work." But never was it sufficient to look at a standard reference publication—even though many procedures are highly standardized between laboratories. So I came to understand, as all experimenters must, that there is much tacit information you need to know if you want to repeat, or even carefully read and interpret, another scientist's published experimental work. I suspected that this massive body of practical lore had an interesting and for the most part sensible and well-justified historical underpinning—though I have encountered more than one odd superstition with the authority of tradition behind it—but I could not think how to unearth this history.

Eventually I received training as an electron microscopist: primarily as a scanning electron microscopist, but I learned the basics of transmission techniques for biology as well. I had good teachers. I encountered a great deal of interesting lore about how to identify and avoid artifacts. Looking for the sources of this lore, I discovered that the electron microscope had been in use for only two generations and moreover, most of the important biological electron microscopists had trained in one of very few laboratories. This looked like an opportune area in which to research the history that would answer my nagging question about where the traditional practices came from. Digging a little, I discovered that the subject was even more opportune than I had imagined: historical circumstances had kept the circle of early biological electron microscopists small enough to study thoroughly, archival sources were just becoming available, and many people from the early days were still living and friendly and available to talk with me. The electron microscopists wanted their story told, and they as well as others I am about to name helped and encouraged me. I doubt I have satisfied all of them, but this book represents, I hope, a start at giving them an accurate chronicle and at answering my nagging question.

I owe great thanks to all the friends and colleagues who have read chapters, commented on talks, or gone over this material with me during the five years of this book's gestation, including (but certainly not limited to) Rachel Ankeny, Alan Chalmers, Hasok Chang, Guy Cox, Evelyn Fox Keller, Jean-Paul Gaudillere, Gerald Holton, Christophe Lécuyer, Wendy Lynch, Ted Porter, Hans-Jörg Rheinberger, Jim Strick, Jacqueline Taylor, and anonymous reviewers. Thanks also to Paul Green, Fran Thomas, and the other good people of the Stanford biological sciences department for their fine training and tolerance of my extracurricular activities. And thanks to the historians and philosophers I have learned from over the years, notably Arthur Adkins, Peter Galison, Nick Jardine, Tim Lenoir, Joel Snyder, and Bill Wimsatt; though I am not sure all will recognize their contributions, I assure them that the book would have been worse without their instruction. Most of all, I am grateful for the criticism, collegiality, and friendship of Tim Lenoir, without whose encouragement at many stages I might never have undertaken this book, and Felicia McCarren, without whose support I might never have finished it in my lifetime.

I would also like to express my appreciation to colleagues who have so generously shared with me their background knowledge, unpublished works, or research materials: Mary Bonneville, Angela Creager, Peter Hawkes, Sterling Newberry, Henry Slayter, Susan Spath, and Jim Strick. In addition, I am very grateful to the electron microscopists who have allowed me access to their papers or spent the time to talk with me about their early work: Mary Bonneville, George Chapman, James Hillier, Marie Jakus, Keith Porter, Francis Schmitt, Carlton Schwerdt, and Fritiof Sjöstrand.

This book could never have become a reality without the kind help and skillful assistance of the librarians and archivists at the American Philosophical Society, the Marine Biological Laboratories, the Massachusetts Institute of Technology, the National Museum of American History, the Rockefeller Archive Center, Stanford University, the University of

California's Bancroft Library, the University of Colorado at Boulder, and the University of Pennsylvania. The research was accomplished with the support of a professional development fellowship from the National Science Foundation, and the final stages have been assisted by a University Research Grant from the University of Sydney.

Earlier versions of portions of several chapters have previously been published. Much of Chapter 3 appeared in the *Journal of the History of Biology* 28 (1995): 381–429, © 1995 Kluwer Academic Publishers. All rights reserved. Chapter 1 appeared in *Studies in the History and Philosophy of Science* 27 (1996): 311–49. Some portions of Chapters 4 and 5 appeared in *History of Science* 35 (1997): 245–99.

N.R.

Contents

✥

Plates follow pages 152 and 196

Tables and Figures

Tables

Figures

Note on Usage of Technical Terms

In this book I use certain common terms concerning experiment in specific ways. *Experimental practice* is divided into *technical practice* (or *technique*), the set of actions and equipment and materials employed in an experiment, and *interpretive practice* (or *epistemological practice*), the set of operations on the outcomes of technical practice needed to make sense of the results. When necessary, particular elements within technical practice may be distinguished, such as *experimental objects*, materials that provide the specimen being investigated (as well as more self-explanatory elements like instruments, preparative techniques, reagents, etc.). An *experimental system* is a combination of technical and interpretive practices, described at a medium to high level of generality, that can be used to perform an indefinite number of varied experiments all sharing the same basic experimental design. *Method* bears the broad meaning it usually does when used by biologists: techniques or some combination of techniques with interpretive practices as fitted together in an experimental design. I believe that all of these usages are consistent with those common among experimental biologists, even if they may conflict with usages among physical scientists or philosophers (e.g., *method*). However, where I use *methodology* I do so in something close to the standard manner of philosophers, meaning intellectual methods including but not restricted to interpretive practice; this usage is also consistent with one of the senses of the term among biologists, so no confusion should result. Finally, where I use *epistemology* I mean the theory of scientific knowledge and how best to obtain it that is implicit in the interpretive practices and other elements of method employed by scientists. This limited definition of their subject matter is only one of several occupying epistemologists, but it is the only one that need concern us here.

Introduction: Scientific Knowledge and Its Means of Production

❖❖❖

The title of this book is multiply appropriate because early biological electron microscopy involved a struggle for picture control on a number of levels. As the reader will discover, a picture control figured in a biologist's subjective experience of the electron microscope as one of the three relevant readouts and, along with focus, one of the two open to intervention. Of course, there was no such thing among the seven indicators and nineteen switches and knobs on the console of the Radio Corporation of America (RCA) EMU microscope, the instrument depicted in Cecil Hall's 1951 cartoon (see Frontispiece). All the more reason to strive for an understanding of what that knob was for. Control of who could make pictures with the electron microscope, how pictures should be made, what pictures would be printed, and how those pictures ought to be used in establishing biological facts were the dominant issues when the new instrument was introduced to biologists at the outset of the Second World War, as Chapter 1 discusses. By the end of the war, a community of scientists in whom expertise was vested—authorized microscope users who for the most part agreed about who should use the instrument and in what manner for which purposes (a group roughly coextensive with the biological membership of EMSA, the Electron Microscope Society of America)—was established, and assumed a basic level of regulatory control. But for individual microscopists, control of the characteristics and interpretation of pictures remained a problem, and one that was divergently addressed in different biological subfields, even in different research programs within them. The middle chapters of this book examine several manners in which the electron microscope was integrated into new experimental ways, and how these ways of experimenting were related to changing boundaries of, and control over, the disciplines adopting electron microscopy. Aesthetic factors (e.g., picture style and quality), epistemological factors (e.g., picture interpretation and artifact identification), and social factors (e.g., collective decisions on goals and standard practices) all play roles in determining the meaning of an electron microscope picture. None of these could be controlled by any knob made by electrical engineers.

Historical circumstances conspire to make the case of biological electron microscopy in America especially favorable for studying the origins of a tradition of experimental practice (see Note on Usage of Technical Terms) with a scientific instrument. As noted, the initial serious steps were being taken to make the electron microscope useful to biomedical researchers just at the onset of the war. The war first isolated the circle of early American and Canadian electron microscope users from those overseas, then limited their number because of wartime production priorities, thus circumscribing an unusually small and well defined (thus, easy to study) set of scientists responsible for making the rare and expensive machine into a useful research instrument. For five years and more, the use of the electron microscope in North American biomedicine was almost totally confined to the handful of characters in this book. After the war, with Europe and Japan a shambles, the biologists of this charmed circle were so far ahead of other scientists (foreigners and native upstarts alike) with whom they might otherwise have had to compete or reconcile their ways, that the circle's wartime experience effectively made it the world's source of expertise. Thus the contingencies of war gave the microscopists dealt with in this study disproportional influence; to a large extent the practices developed by North America's handful of wartime biological electron microscopists became the foundation of all subsequent biological practice with the instrument. Moreover, in the initial period the requirements for instrument domestication were brought to the surface and made especially obvious for reasons having to do not only with peculiar wartime circumstances, but also with the peculiar power of electron microscope images to create public sensation. Nonetheless, one can reasonably suppose that for all new instruments there are basic similarities, in terms of the difficulties that must be overcome to win acceptance among scientists, as well as the subsequent changes wrought by acceptance, even if the story of each instrument is in some respect unique. Therefore this compact study of biological electron microscopy is suited to shedding new light on general questions about the introduction of novel scientific instrumentation, and its impact.

How then do new devices come to be taken up as experimental instruments by scientists, and how does the adoption of new instruments affect scientific knowledge? These are the major questions motivating this study. Though the pair of questions may seem at first glance simple, so many ramifications soon emerge that no one book could possibly do justice to the full set. Among these can be counted, for example, historical questions about how, by whom, and why new instruments are introduced, or about how another, different set of instruments might belong to science given alternative, counterfactual social and cultural circumstances; philosophical questions about the extent to which and manner in which the content of scientific understanding of the world depends on the technique and methodology scientists happen to use; and sociological questions about how the organization of work within disciplines and laboratories and other scientific institutions may depend on the equipment employed. Although I will not be able to follow enough strands in this study to satisfy every stripe of historian, philosopher, and sociologist of science—let alone scientists personally involved with the events and the individuals discussed—there should be something for everyone. The simple pair of questions about innovative instrumentation with which this study begins has so many implications because it opens onto one of the largest puzzles about science: science must conserve and accumulate, yet science must continually progress and thus overthrow its own past. Science grows by radical conservation, or perhaps by conservative revolutions. The nature of scientific change is a difficult subject precisely because of this basic oxymoron or, better, antinomy.

Interest of the Study for Historians

Big Science and the War

In the brief period around the Second World War, most Western countries saw the rapid growth of the modern administrative state and a concomitant transformation of the scale of science together with an elevation of its prestige. In wartime America in particular, massive support suddenly flowed from the central government into basic research, and many scientific fields found themselves suddenly metamorphosed from the intellectual equivalent of cottage industries to heavy industrial plants. The era of "big science" had come. Results like the atom bomb, penicillin, and radar were presented as marvels proving that public patronage of such large-scale research projects was a wise cultural and military investment for America. Beginning with Vannevar Bush's 1944 broadside, *Science: The Endless Frontier*, there were widespread calls to carry the government funding of scientific research on into the postwar era, and these soon came to fruition.[1] The frenzied pace of wartime physical science was maintained after Hiroshima and Nagasaki through funding both from the defense budget and from the "civilian" Atomic Energy Commission (AEC), so the era of big-budget high-energy physics launched by the Manhattan Project continued on a similar footing (only now, with the Cold War, coming to an end).[2] In the life sciences, postwar support quickly out-stripped wartime levels, first with funding from the Office of Naval Research (ONR) and then—after the 1950 legislation finally enacting it—the National Science Foundation (NSF), but most of all with the burgeoning research budget of the National Institutes of Health (NIH).[3] The NIH budget grew from $3 million in fiscal year 1946 to $8 million in 1947 (the year it assumed remaining wartime medical research contracts), to $26 million in 1948, to $52 million in 1949. Thus in these four years, government patronage of biomedical research through the NIH alone swelled from about half of what had been provided by private philanthropies to an order of magnitude greater. After a brief plateau from 1950 to 1952, related to the Korean war, the NIH budget continued its dizzying climb to $211 million in 1958, and beyond. The reasons why Congress was "force-feeding" the NIH in this period are complex, and are related to the American Medical Association's blockage of Truman's national health plan, a sort of diversion of the flow of obstructed resources within the medical sector of the budding American welfare state (and perhaps important in its ultimately stunted development).[4] The magnitude of new financial support for biology is a brute fact that must be taken into account.

This mounting flood of monies brought profound material and cultural consequences for the life sciences. Before the war, grant applications for large sums had been rare and the peculiar province of philanthropic foundations, but afterward the writing and evaluation of applications became the endless occupation of the mature American biologist, who had perforce to become something of an executive (or a "principal investigator," in the new parlance of grantsmanship). The numbers of workers in the biomedical research laboratory shot upward as grants permitted the hiring of extra technicians, and as the numbers of graduate students and postdoctoral researchers training in the most active fields rose far beyond the replacement levels that had prevailed in the Depression. And pricey, complex technology—ultracentrifuges, electrophoresis apparatus, isotopes and isotope counters—suddenly moved to center stage in the post-war laboratory of many fields of life science. The electron microscope was another such expensive instrument, avidly taken up by biologists in the first flush of government support of the late 1940s and 1950s. Larger-scale historical problems potentially illuminated by exploring the introduction of this instrument include: How did America's de facto science policy, in the behavior of granting agencies, influence the adoption of new bio-medical techniques and instruments in general, and were certain technologies especially favored? Which sort of scientific institution promoted and which resisted the new government-funded experimental technologies? What effect did the quickly introduced expensive technology have on the status of biological disciplines in American research institutions? Fine-scale history of science yields the right sort of data to answer such large general questions about biology and the war. Here, for instance, we shall see that of four groups pursuing electron microscopy after the war, three swelled quickly with people and machines and gave rise to flourishing research schools while the fourth, viable in the 1940s, withered partly because of institutional obstacles to exploiting new funding sources.

Experiment and Technique in the History of Science

After a generation's hiatus, such questions about the role of technique have only recently come back into fashion among practitioners of the specialty of history of science, and of the overlapping specialty, philosophy of science. Traditional positivist thinking once held that science progresses continuously, as solid facts accumulate in an ever-increasing stock. To the Comtean positivist historians and successors, new instruments seemed important because they allowed scientists to invade previously inaccessible niches of nature and thus bring new facts to the storehouse of knowledge. But there is also a long tradition more idealist than empiricist, in which the important advances in science are superior theories about the world, and in which observable facts (e.g., experimental findings) tend to be relegated to secondary status as the raw material that theory must explain. After Thomas Kuhn's 1962 *Structure of Scientific Revolutions*, theory change in science went from being one historical issue among many (one that happened to be important for epistemologists) to becoming the preeminent question for historians of science. Kuhn argued brilliantly that the important changes in scientific knowledge are discontinuous, as comparatively static periods of routine or "normal science," wherein knowledge accumulates under one guiding theoretical worldview (or "paradigm"), give way occasionally to "paradigm shifts," theoretical revolutions that sweep the slate clean and remake the world picture. Old scientific knowledge became "incommensurable" with the new, merely a relic of lost worlds where extinct conceptual schemes informed all scientific data and interpretation.[5] For more than two decades after Kuhn, historians of science turned their attention to conceptual schemes, striving to reconstruct the dead paradigms that grounded past scientific knowledge. Revolutionary paradigm shifts were the key to explaining scientific change. In the new focus on the discontinuous, and especially on the conceptually discontinuous, the material and the continuous in science—elements like instrumentation and traditions of experimental practice—tended to fade into the background.

Recently a sea change seems to have reversed this trend, as limitations of the theory-dominated view of science have come to be appreciated, and attention has returned to instruments and schools of research practice. Historians of physics are elaborating in careful detail the stories of the complex, expensive, and gigantic apparatus used, ironically, to study the smallest and simplest components of matter.[6] In the history and sociology of the life sciences, new attention has been paid to the careers of experimental systems, which, as all practicing biologists are well aware, are often associated with long histories and strong traditions that can profoundly shape experimental practice.[7] At the outset of the present book, which draws both on the study of instrument use as the historians of physics especially have been doing it, and on the study of biomedical experimental practice affected by and even built around instruments, it is worth considering what this general historiographic trend has to offer. The answer, I think, has to do with the small-scale historical evidence one needs to study in order to tell a story about evolving laboratory practices—details about short-lived prototypes, failed attempts, training programs, calibration procedures, specimen preparation, and data-gathering methodology—and also with the midscale information concerning laboratory administration and financing one needs in order to understand local trajectories of experimental work. With the emphasis on the local, the quotidian, and the uncertain in the early stages of what may already be understood by practitioners as a new kind of science, contingency and material constraints are bound to come into focus. With such material we can ask: What motivated the particular life scientists who after the Second World War so eagerly adopted the expensive instruments they were suddenly able to buy? How did local personalities, politics, and resources figure in the design of experimental activities involving the electron microscope? Who benefited by the standardization of one kind of microscopical procedure rather than another? How did particular research agendas reflect the institutional contexts of electron microscope labs? Such fine- and medium-scale detail about any technical enterprise's early stages is interesting because it suggests the multiple determinants of history's actual pathway, and the alternate pathways that history might have taken.[8]

Still, one must not lose sight of the fact that there is a big picture that cannot be surveyed using the high-power immersion lens of microhistory (to borrow terminology from the light microscopist). But even though a full account of important relevant issues, like the impacts of Hiroshima and Sputnik on American science, lie beyond the scope of any detailed study on a field of scientific inquiry, that does not mean that all effort is futile. In order to gain some modicum of purchase on issues of this larger sort, I have constructed this book as a set of microstudies of roughly simultaneous research programs in biological electron microscopy at different sites, all sharing certain common wartime roots. The big picture emerges through the linkages, resonances, and contrasts among the parallel microstudies. In a sense, the historians of particle physics experimentation have had an easier task because the expense and scale of instrumentation for this science means that major laboratories are few; thus the microstudy of one or two active sites covers a large proportion of the scientific field. Because the experimental life sciences are institutionally (and intellectually) much less unified than the physical sciences, the difficulty of generalizing from a highly variegated set of objects cannot be bypassed by the historian of biomedical knowledge.

The Rockefellers and the Origins of Molecular Biology

There was a time when *molecular biology* and *biophysics* were synonymous, or at least when one field was seen as a subfield of the other. That era includes essentially the whole period covered in this study, and for most of it biological electron microscopy was considered a form of biophysics. Only at the end of the 1950s, when cell biology took firm shape as a discipline that encompassed much of biological electron microscopy (see Chap. 3), did it become incorrect to classify the microscopists dealt with in this book as "molecular biologists" or "biophysicists." Shortly thereafter, the term *molecular biology* became the definitive label for a smallish subset of the fields that had counted in the 1940s and 1950s as biophysics (which originally included not only molecular genetics, and ultraviolet and electron microscopy, but also physical chemistry of macromolecules, radiation biology, nerve electrophysiology, mathematical biology, and "bioenergetic" metabolism studies—especially when isotopes were employed). Elsewhere I describe in detail the dramatic postwar blossoming of biophysics, and how it was related to large-scale cultural compensation for the terror of the atomic bomb.[9] Especially in Chapters 4 and 5, we shall see how the work of a number of biological electron microscopists was linked with the postatomic enthusiasm for a new physics of life. But what matters most for this book as a whole is that the branching research traditions of biological electron microscopy whose development I chronicle here all share common roots in that admittedly ill defined and variegated interwar discipline known as "general physiology" or, alternatively, "biophysics" and "molecular biology."[10] To be sure, resources were drawn from sciences other than biophysics in the electron microscopical research traditions I describe, especially from bacteriology (see Chap. 2) and histology (see Chap. 3). Nevertheless, the history of biological electron microscopy must be considered in the context of the history of "molecular biology," even though it lies almost entirely outside the field(s) denoted by the latter term today.

Therefore this study touches on one of the most celebrated episodes in the history of twentieth-century science, wherein America's most powerful industrialists nurtured the infant molecular biology cum biophysics through the lean Depression years. On April 11, 1933, the *New York Times* headlines told of a world in crisis: the continuing Depression and drastic economic measures from the new president, Franklin Roosevelt, a massive Japanese offensive in China, persecution of Jews and political opponents under Germany's new führer, and blacks rioting in New York over the Scottsboro case. (The sense of crisis over the beer shortage that followed Prohibition's recent repeal was not small, either.) At a meeting of the Rockefeller Foundation that day, the new director of natural sciences, Warren Weaver, presented a novel program to introduce methods and instruments from physics into the biomedical sciences he considered backward. Weaver said:

The welfare of mankind depends in a vital way upon man's understanding of himself and of his physical environment. . . . Science has made magnificent progress in the analysis and control of inanimate forces, but it has not made equal advances in the more delicate, difficult, and important problem of the analysis and control of animate forces. . . .

Important questions are: Can we obtain enough knowledge of the physiology and psychobiology of sex so that man can bring this aspect of his life under rational control? Can we unravel the tangled problem of the endocrine glands and develop a therapy for the whole hideous range of mental and physical disorders which result from glandular disturbance? Can we develop so sound and extensive a genetics that we can hope to breed in the future superior men? Can we solve the mysteries of the various vitamines [sic], so that we can nurture a race sufficiently healthy and resistant? Can psychology be shaped into a tool effective for man's everyday use? In short, can we rationalize human behavior and create a new science of man?[11]

Weaver's presentation suitably impressed the board of trustees, who were university administrators, medical men, J. D. Rockefellers Jr. and III, Chicago meat tycoon Harold Swift, 1924 Democratic presidential nominee J. W. Davis, former Federal Reserve economist Walter Stewart, financier and onetime New York City police commissioner Arthur Woods, midwestern news publisher W. A. White, and Owen D. Young, board member of General Motors, General Electric, and RCA.[12] As one would expect, these were not just any pillars of society interested in political stability, but ones especially friendly to the Rockefeller financial interests and the family's Democratic Party alignments.[13]

Over the next two decades Weaver channeled substantial money to carefully selected and closely managed scientists (especially biophysicists) who studied biological problems in terms of molecular mechanisms, funding techniques such as electrophoresis, ultracentrifugation, spectroscopy, and X-ray crystallography. He also funded biological electron microscopy during the war years. But what does this Rockefeller patronage mean? Weaver has claimed for himself a prime role in fostering what became "molecular biology," a term he seems to have coined himself in 1938.[14] Weaver's program was certainly an early example of modern science policy, and possibly a key step toward the postwar "big science" style of biomedicine. Historians of science have given widely ranging interpretations of the nature and magnitude of Weaver's impact. From a Marxist perspective, the Rockefeller program has been regarded as a diabolically effective scheme to naturalize the power relations of the late industrial revolution, which not only helped save capitalism from itself in its hour of crisis but subordinated academic biology to industrial management and thereby laid the foundations of today's genetic engineering establishment.[15] At another extreme, Weaver's program has been seen as an ill-informed effort at technology transfer from physical to life science, resulting only in temporary colonization of biology by opportunists who ultimately contributed nothing to the triumphant molecular genetics of the 1960s.[16] Intermediate positions are plentiful.[17] But the funding of new technologies for science can be credited with influence only to the extent that technique itself can be credited with power over the development of science. Instead of offering my own direct answer to the question of Weaver's significance, I point out powerful conservative forces conditioning the acceptance of the electron microscope, forces that in general would seem to limit the influence that technique and its patrons can ever have. That is, the biologists were hardly passive partners in their retooling. The knotty question of the extent to which technology should be *explanans*, and to what extent *explanandum*, in accounting for historical change—otherwise known as the issue of technological determinism—is another that cannot be altogether avoided in the history of novel experimental technique.[18]

From History to Philosophy of History

As noted, only recently have historians of science returned to the study of techniques and instruments, after a generation's hiatus spent attending to the importance of paradigms, conceptual schemes, and the like. Where can one anticipate the new focus on practice and technique to improve on the recent focus on theory, where can one expect it to disappoint (even where efforts are made to overcome the aforementioned localism needed to begin any study of technique)? In short, what is the relationship between concept and material technique, mental and bodily work, in experimental science? The answer to these historiographic puzzles depends fundamentally on the extent to which technical change is important in shaping science. This big issue—which overlaps a related basic problem in the philosophy of history about how much our worldview depends on our material culture—is as crucial for science policy as for the history of science, because technical innovations can be major sites of influence on science by elements of the greater social sphere (as in the case of the Rockefeller Foundation, noted above). Despite their difference, the traditional types of history of science have something in common, both the positivistic strain emphasizing the material diligence, openness to serendipity, and ingenious handiwork of science's great crafts-men, and also the theory-driven strain emphasizing conceptual resources and their visionary transmutation in the minds of geniuses. This some-thing in common may well contribute to the seeming intractability of the theory-practice relation and similar problems. That is, both tell the his-tory of science as a progressive series of *solutions* (inventions, discoveries) or *answers* (theories, models) to questions about the nature of the universe and how best to master it. With few exceptions, the history of the moti-vating *questions* has been told only incidentally, if at all.[19]

In *Scenes of Inquiry*, Nicholas Jardine has taken an innovative step toward resolving these difficulties by recasting them as problems concern-ing the history of scientific questions. A scene of inquiry, for Jardine, is the milieu in which investigators regard a question as real and deem it worth pursuing. Questions are considered real when a community of inquirers finds them intelligible (the questions do not entail unacceptable premises or conclusions, etc.), and when evidence relevant to their resolution is obtainable. Conceptual developments can alter a scene of inquiry, for example by reorganizing premises, making formerly absurd questions newly real. But innovations in experimental technique can also radically alter the scene of inquiry in a science. New techniques make new forms of evidence available, and thus can lead to redefinition of standards of evidence and explanation in already established problem domains. New techniques can also introduce whole new problem domains by making previously unsuspected phenomena accessible. Thus technical innovation not only alters scientific doctrines by introducing new evidence and un-dermining old, but also changes the questions to which doctrines are the answers by suddenly making whole sets of questions real, which may alter the perceived adequacy of existing doctrines.[20] In turn, changes in ques-tions can motivate innovation in technique. Examples from the history of science could easily be enumerated to illustrate all these kinds of change, and they frequently would indicate conflict between the advocates of new questions, based in new concepts or new techniques, and advocates of the traditional ways. In studying these conflicts the historian, whether attend-ing primarily to theories or to practices, ought to be mindful of how the scientific questions change, not just the scientific answers. In conclusion I make some suggestions about the nature of the mechanism linking changes in question reciprocally to changes in technique. Having brought this introduction from historical issues to the brink of the philosophy of history, I turn now to issues of philosophy and sociology.

Interest of the Study for Philosophers and Sociologists

Since the years around the Second World War, philosophy of science has relegated experiment to a marginal role as little more than that which helps scientists pass judgment on fully explicit, logically coherent theories (as, for instance, in the work of Karl Popper). No doubt this trend has been compounded by a continuing selective focus on physics—the science in which theory is most fully developed and separated from experiment—as the most important science, and the one best exemplifying scientific method. The Kuhnian revolution that brought discontinuous "paradigm shifts" into the historical limelight did nothing to divert philosophical attention away from conceptual change as the driving force of scientific growth. But finally, in the last decade, the abandonment of efforts to define one canonical scientific method in favor of empirical forays into the methods specific to the several sciences has brought, among philosophers of science, a significant and growing reorientation toward the laboratory, with the slogan "experiment has a life of its own" as its watchword.[21]

This return to experiment has often taken the form of detailed studies of scattered and usually quite recent episodes of experimentation. Scientific publications and some interviews provide the material. The technical and interpretive practices of the experimenters are analyzed to extract generalizable rational principles, thus making more explicit the reasoning behind the conclusions reached in the described experiments. Ian Hacking visited some biology and some physics laboratories in this spirit, Ronald Giere followed the work of one particle physics lab for a period, Peter Kosso has analyzed the logical structure of experiments in physical sciences, and Allan Franklin has described a variety of recent physics experiments.[22] But this return to experiment has typically neglected diachronic studies that would reveal how these guiding principles came to be applied, how the principles may have mutated in character over time, or how the principles may have changed in their mix or hierarchy relative to one another amongst the set deployed by experimenters in a field. Rather, even at its best this approach can be likened to taking snapshots of assorted scientific experiments, looking carefully for features common to all, and identifying these as the essential (because universal) elements of sound experiental reasoning (worse, sometimes this new empiricism seems to yield no more than a paraphrase of the publications of the scientists under study, in abstract but not more felicitous language).

Several candidates for the status of fundamental principle have received broad support among analysts applying this kind of approach. Agreement or convergence between one instrument or technique and another that functions on a different basis is one such candidate, and interestingly enough, both Ian Hacking and Allan Franklin discovered it in the biological electron microscopist's practices of examining the same specimen with another type of microscope, or the same specimen prepared differently (though they also find analogs to this practice in other sciences).[23] In this book we will look into the origins of this strategy and its later evolution along several pathways. Another very popular putative "principle" is the validation of an instrument's (or a technique's) output by a rigorous account in terms of the physical theory of its operation and interaction with the object of knowledge, and Kosso in particular has stressed that this theory of the instrument ought in addition to be independent of the theory being investigated by the experiment so that "nepotism" can pose no threat.[24] Elsewhere I have questioned the applicability of this sort of vindication in life science, because biological specimens usually are excessively complex and insufficiently well characterized for thorough calculations according to physical theory; in this study we will encounter a number of further examples where physical theory has disappointed the biological electron microscopist.[25] Calibration of a new instrument or technique, either against prior results with related techniques or older similar instruments, is yet another commonly invoked general "principle."[26] There is no question that scientists do use this strategy a great deal, but it is not clear how far this goes in solving the reliability question. As Percy Bridgman pointed out long ago, calibrating a new device can raise large puzzles and problems, because to the degree that the device is new and can do what no other yet has done, it cannot be calibrated in the interesting part of its performance range.[27] The experimenter must make a novel device behave itself in the range of phenomena accessible to established techniques, and develop theory of the technique to assess its reliability in the new range of phenomena, but this operation requires reference to established techniques and theories that are open to reinterpretation in the light of the new device and the novel phenomena it brings to light. Thus validation is only secured by reference to historically grounded standards embodied in a web of techniques that are ultimately grounded on one another—a web the novel device and its associated techniques must now be made to join.[28]

Many cases of experimental reasoning classifiable under these three heads—convergence, calculation, and calibration—will emerge in the pages to follow; however, I point to the ways in which the instantiation of these "principles" (especially the first) varies greatly over time, and from biological subfield to subfield, and even among practitioners in a subfield at any one time. Just because such general principles can be identified ubiquitously does not mean that their particular variations are uninteresting and that the job of understanding experiment ceases with their mere identification. Indeed, to me it is not so much ubiquity as patterns of *variation* within it that call for explanation, and the explanation will be in terms of the history of practice. There are a few philosophers of science currently studying experiment in ways that more thoroughly take into account changing experimental epistemologies, and with their work my diachronic approach is in harmony.[29]

The drive in recent philosophy of experiment to uncover the universal general principles guiding experiment, discussed above, appears largely to be aimed at countering the "strong program" in sociology of knowledge, which proposes that scientists reach conclusions in both theoretical and experimental matters on grounds of social expedience, rather than on the rational scientific grounds they offer after the fact. Thus, for some philosophers the main goal has simply been to show that rationality drives scientific decisions; in the context of meeting this goal, nuances such as origins and development of these rational grounds for reaching conclusions are of little importance. For their part, the sociologists of knowledge have mostly taken a similar "snapshot" or synchronic approach to experiment, aiming to show that the rational grounds offered by scientists, for example in their collective conclusion that a certain kind of gravity wave does not exist or that the sun emits neutrinos in a certain way, lack full internal coherence and rigor, and that conclusions reached can be re-explained politically.[30] Thus there seem to be many on both sides that have not considered that grounds for belief can be negotiable, provisional, and changing according to collective decisions, but still rational (in some less than absolute sense) nonetheless—with a validity that partly depends on a history that is left out of the account.

There is another current school of sociology of knowledge, loosely groupable under the heading of "ethnological," that differs in remaining agnostic on the big epistemological questions that preoccupy philosophy of science; instead, the project is simply to observe and explain the behavior of scientists in the same empirical fashion that one describes participants in any cultural group.[31] The goal of describing the behavior of scientists in their own everyday terms (rather than in the received, "official" terms distorted by the role into which society—and all too often, philosophy—casts its scientific sages), and of relating this to general principles of psychology and social organization, has much to recommend it, even if it is unsatisfying (at least to someone who already knows what it is like to be a scientist) as an exclusive and final goal. In particular, Michael Lynch's description of how electron microscopists are trained to look at micrographs captures very nicely the ordinary practices of such a biology lab; and although I have come to the present project independently (see Preface), I can find little to fault in Lynch's study and recommend it to anyone curious about what everyday life in biological electron microscopy is like.[32] However, because ethnology depends on living amongst practitioners for a limited period, the ethnological approach to experimental practice is also of the synchronic, "snapshot" variety (and must be, unless an ethnographer should be so fantastically lucky as to stumble upon an episode of dramatic change). Therefore, ethnology is in no better position to provide evidence about *why* traditional practices are what they are. Thus, all my earlier comments about the need for a historical, diachronic study of experimental practices back to their origins apply here, too. For only in the history of a field do the patterns of, for example, indoctrination of graduate students in certain ways of thinking, seeing, and acting—indoctrination so carefully documented by the ethnomethodologists—emerge not as arbitrary or demanding of explanation in terms of functionalist sociology,[33] but as grounded in collective memory of practical experience. Although I do not attempt any detailed sociology, throughout this book I explore the historical reasons for many of the basic practices into which budding electron microscopists are indoctrinated today.

Recently a few American philosophers in the phenomenological tradition, notably Patrick Heelan, Don Ihde, and Robert Crease, have addressed some of the issues surrounding scientific experiment from a different perspective. Setting aside Husserl's own not altogether friendly attitude toward modern science, and drawing on Ihde's work on bodily interaction with technology, Heelan has developed a Husserlian "first-person" approach to scientific experimentation based on the primacy of perception, and on a fundamental homology between a reader's interpretation of a literary text and a scientist's interpretation of experimental data. Experimenters can be described as "perceiving" natural phenomena with their familiar techniques and practices—not as a passive spectator but in the same strong sense as everyday perceiving, because the skilled ex-

perimenter can be described as "embodied" in the experimental apparatus and engaged with the phenomenon in the same direct way that one
is engaged with one's material surroundings in other forms of activity
(as the shoemaker is embodied in his hammer, to cite Heidegger's favorite example).[34] The scientific experimenter seeks to grasp the essence
or "inner horizon" of a phenomenon by revealing its different perceptual profiles in multiple and varied experimental settings. The essence is
grasped when, through an interpretive struggle with the data, the different profiles can be understood as manifestations of the same invariant,
manifestations transformable into one another by means of a theory mod-
eling or otherwise accounting for the phenomena. Theory is thus secondary to experiment, and results from an interpretive process with marked
similarity to the hermeneutic circle describing how literature and other
texts are grasped by readers. This essence as grasped is not the absolute
essence of a natural thing, any more than a reader can ever grasp the single
and final essence of a book; rather, the essence is always relative to the
given background theories and practices and cognitive capacities of the
experimenters. Moreover, this essence or "inner horizon" never appears
except against an "outer horizon," which is the place the phenomenon
holds in the historically specific scientific community and the larger society that embeds it.[35] This seems to be the beginning of a philosophy of
science capable of taking seriously the experience and personality of the
individual scientist—and the cultural context also—without collapsing to
a totalizing psychology or sociology of knowledge wherein political clout
is supposed to decide matters of scientific fact. I apply Heelan's hermeneutical approach at the end of this work, in order to address the particularly intractable problems about the role of novel instrumentation in new
knowledge raised at the outset.

Robert Crease has recently extended Heelan's approach to experimentation by developing an analogy between the way experimenters
produce various profiles of natural phenomena and the way actors and
production staff stage theatrical performances, and drawing a distinction
between the hermeneutics of production and of reception of experimental data. Although I do not follow Crease in the present study (largely
because many of the enlightening parallels between theater and science
apply best to grand-scale particle physics experiments in which rehearsals,
stage sets, and the like have much clearer counterparts than they do in
smaller-scale biology experiments), I appreciate Crease's emphasis on the
perceived *problematic situation* among scientists that not only motivates

experimentation on a given phenomenon but delineates the outer horizon in which it manifests itself. That is, experiments are only capable of giving a limited range of answers about nature, and this range is determined by an experimental design that reflects a particular problem or question about the phenomenon as understood by a community of inquirers. And as Crease points out, the role of a community's perceived problems in constituting the objects of knowledge they investigate invites treatment in the terms of American pragmatist philosophy from the early twentieth century.[36]

A pragmatist treatment might help fill the lacunae left by answers to the now traditional philosophical problem of how theory-laden observation may vitiate conclusions from empirical evidence. After all, there is more to a question, experimental or otherwise, than theoretical premises and background beliefs; in addition there must be a special motive for addressing one particular difficulty for a coherent view of the world, because an always incomplete understanding of nature offers limitless other potential foci for investigation. Moreover, there must also be some sort of predisposition about what constitutes a serious answer, informed by the motive and by available methods. Thus observation is also laden with motive and question (and therefore method); theory-ladenness need not bear such a heavy explanatory burden. Jardine's anatomy of question change noted above, also rooted in pragmatism, is a beginning in unraveling these issues. Here, where I subscribe to a philosophical position, it is based on John Dewey's pragmatism. Increasingly, there has been optimism about the relevance of pragmatism for science studies,[37] but few efforts to apply it explicitly in any detail, Jardine's work being a notable exception. This book is such an effort, one that builds a bridge from Dewey to phenomenology, rather than toward logical positivism—the direction favored by American pragmatists of the 1930s and 1940s.[38] Although the main goal is to show (in proper pragmatist fashion) the fruitfulness of the approach through use rather than to develop or discuss it at a high level of abstraction, there is one area, apart from my unusual but theoretically unrevolutionary adventure into the lived experience of experimenters, in which I endeavor to go beyond pragmatist trails already blazed. This is on the issue of standardization of experimental practices in science. It seems to me that a strict standardization must take place whenever new methods (and the knowledge they generate) are added to the existing structure and pattern of scientific work, just as for the effective growth of a rail network, standardization of tracks and signals is essential, not to mention the landscape, and even time.[39] Standardization is needed for the building of new knowledge onto the old in a way that maintains the coherence of the whole; and this is crucial for the whole edifice of scientific knowledge, for what was once contingent when new may become the increasingly unshakable substance of scientific truth when it is built crucially into subsequent practice. I do not claim great originality, for the point clearly follows from basic pragmatist principles as understood from the beginning: as Sidney Hook put it so nicely in 1939, "problems are settled and stay settled only when common agreement has been secured by the use of a common method of inquiry."[40] However, the original generation of Deweyan pragmatists applied the point less to science than to matters of ethics and social policy, their dominant interests. In what follows I illustrate concretely and in detail not only how changing methods shape scientific questions, but also how matters of scientific fact (and also matters of the social organization of science) are only settled through agreement on standardized methods of inquiry. Whether different methodologies would ultimately converge on the same conclusions over a sufficiently "long run" is a further question, which I do not attempt to address.

I might here add a few words on the notion of "mapping" employed throughout this book, for although it is consistent with everyday intuitive connotations of the term, my own usage does have a distinct pragmatist color. According to Dewey, every idea, even the most abstract theoretical proposition, in essence represents a plan or record of action. So, a fortiori, must be any representation of the world that, like a map, is literally a record of the measuring activities of explorers and surveyors; indeed the aspect of maps as plans becomes quite obvious when they are used to guide military intervention or the mobilization of material resources in commerce. The electron micrograph certainly can be regarded as closely analogous to a map (whether or not *all* forms of scientific knowledge can):[41] electron micrographs are automatically recorded maps of the distribution of electron-scattering matter in a specimen, each point on the micrograph corresponding to a point in the imaged field. Maps are made in different ways for different uses. One need only consider road maps, topographical survey maps used by hikers and timber companies, and aircraft navigation charts showing only coastlines and major city lights and invisible radio beacons, to grasp both the range dependent on purpose and mapping method, and the nonsense of the notion of one "true map"—unless it be, as Borges imagined, a perfect duplicate of the territory itself, identical in scale and composition.[42] Just so, very different electron micrographs of the same starting material are made by different specimen preparation procedures (and to a limited extent, also by different instrument configurations), depending on the purpose of the inquiry. Thus objects of knowledge such as biological entities, when investigated by electron microscopy, yield a variety of representations, or maps, all of which are the result of some particular mediating methods used in imaging. Sometimes it is easy to move between two representations, locating all the features of interest in one kind of micrograph that correspond to the features of interest in another; sometimes the different sorts of image made by different methods may look so unlike one another that it is difficult to identify *any* feature in one micrograph with its counterpart in the other. In either case, the work of finding correspondences between diverse types of representations is what I call "mapping" between the two electron micrographic media, or put more generally, representational spaces.[43] Mapping is important work because it coordinates and generates links between diverse fields of activity, and it can be terribly puzzling work, too. How many qualities must two appearances have in common to justify identifying them as the same thing? Are some shared qualities (shape, size, distribution, internal arrangement) especially definitive of identity? When should differences in appearance be attributed to differences in technique and when to differences in the object? (It is not usually possible to view the self-same specimen by different methods;[44] typically, one compares views of originally similar specimens prepared differently, making this problem inevitable.) And it must be kept in mind that whenever maps are made, they can be used to move physically as well as mentally; electron microscopy has been used effectively to guide the extraction of pure cell components by means of centrifuges, for instance.

Science, as Crease noted, is antinomic in being both fully a social product and also, equally, knowledge of what is not just a human creation. This seems the same insight captured by Bruno Latour with his image of science's Janus face.[45] It strikes me as a fruitless effort to attempt to decide empirically between "strong program" sociological relativism and traditional realism when the same evidence can be used to support both—by the differing standards that each side sets for itself. The *terms* of the increasingly tiresome realist-relativist debate demand attention before there can be any hope for resolution, and even then I suspect prospects are dim.[46] Authoritative science and the truth about nature, that is, who is right and what is right, are established simultaneously, so it is unclear how one could ever prove whether the causal arrow points from truth to power or from power to truth.

Even though it features no direct engagement with the realist-relativist debate, I hope this book will be of interest to philosophers concerned with experiment. It details not just the history of laboratory techniques used by electron microscopists and the scientific findings gleaned thereby, but the history of the ways they designed experiments and interpreted their results. Thus the book is largely a *history of epistemology* as implemented by scientists in these particular fields of biology, in the sense of a *history of questions* and the conditions under which they come to be posed (for no experimental question is posed without reference to the means expected to answer it), together with the more typical history of theories given as answers. This kind of diachronic story is rather uncommon, compared with the type of study that gives a "snapshot" glimpse of experimental epistemologies for purposes of philosophical (and sociological) illustration noted above, and although my diachronic or "longitudinal section" is only twenty years deep, it includes the origin of the cluster of research traditions that are collectively known as biological electron microscopy. That many "snapshot" synchronic analyses have been taken from the practice of biological electron microscopists will, I hope, enhance the interest of this book for readers familiar with this philosophical literature, because here they can read about where those epistemological practices came from and why they were accepted as valid by the microscopists.

Moreover, the material this book provides bears on another large issue. With the slow demise of grand projects aimed at defining a single overarching and universally valid scientific method, the conclusion that there are scientific methods that vary from science to science—and change over time within a science—has come to seem increasingly inevitable.[47] With this, the serious question for both believers and skeptics of science then becomes: in what way can the changing epistemologies employed by a science be said to *improve* over time? That is, by what yardstick can one establish that, as Dudley Shapere put it, science "learns how to learn" better? This issue has received much more thorough attention in French scholarship, presumably through the influence of Bachelard and successors, than among anglophone science scholars, for whom Bachelardian "epistemic breaks" became Kuhnian "conceptual revolutions," thereby shifting attention from changes in scientific questions and methods for answering them to changes in theories—that is, answers.[48] (Kuhn did, of course, include methods among the highly heterogeneous items standing under his "paradigm" rubric, but with no special emphasis.) To put the problem differently, does one need to specify a meta-standard by which one can judge conflicting but internally consistent standards of scientific knowledge in order to claim progress in the intellectual methods of the sciences? Imagine, for example, two conflicting construals of the world both judged to be best for understanding and predicting phenomena—by two different methods and notions of adequacy going with them. Such conflicts are, I believe, not hard to find in the history of science, even though the transnational structure of scientific disciplines makes such heterogeneity unstable (except where portions of the scientific community are isolated, for instance by war) so that they should appear deceptively rarely in brief "snapshot" glimpses. As already intimated, post-Kuhnian treatments of such situations in terms of "theory incommensurability," and concomitant analyses of the extent to which change to a victorious theory can be rationally justified by those who accept it, fail to engage what is at issue here, which can roughly be described as "problem incommensurability" (taking problems to include both questions and methods). To begin to address this sort of issue, we need more histories showing in detail how epistemological practices actually come to be established in particular sciences, for in these accounts one can expose the standards applied over time to epistemology itself, as utilized by scientists. In this study I concentrate on showing simply *that* standards change, and do not venture more than an occasional suggestion in accounting for that change. These few suggestions about the cause of change are provisional, and I do not doubt that they might not withstand well-considered challenges such as I hope they may spark.[49]

There is one more philosophical issue to which I seek to draw attention in this work. In biological electron microscopy, and I suspect in many if not all other fields of experimental science, aesthetic considerations play a powerful role—alongside but not opposing the more formal devices of epistemology—in the production and evaluation of experimental evidence. That aesthetics figure indispensably and constitutively in experiment is not news to the scientists themselves, and I will at many points in the story highlight both conscious and unconscious aesthetic decisions made by the biological electron microscopists. Too often, however, when the role of aesthetics has come to the fore in recent studies of experiment, this has been assumed to strike a blow against science's rationality.[50] But this reaction trades on outmoded notions of an inflexible, logically specifiable, and universal scientific method, I would suggest; thus, denial may not be necessary for the defender of scientific objectivity. Aesthetic standards can be publicly specified, learned and refined through practice, and (at least conceivably) made conducive to reliable interaction with the world in varying degrees.[51] The constitutive role of aesthetics in experiment needs much more study, and should not simply be rejected as scandalous by philosophers—even by the analytic philosophers most implacably devoted to slaying their "antiscientific" dragons. If their goal is to defend science as it actually is, and not as they imagine *per impossible* it ought to be, then vindication of aesthetic standards is called for. Within Continental philosophy the harmony of reason and aesthetics perhaps need not strike such a discordant note, except that the scientific view of things has been presumed to lack such subjective participation (a lack that makes science appear suspect). As noted above, only recently have English-language philosophers drawing on the phenomenological tradition begun to explore how aesthetic sense might feature constitutively and constructively in experimental science. I see no a priori reason, in an intellectual climate that has, one hopes, outgrown both crude Cartesian rationalism and overindulgent romanticism, that a reconciliation between human tastes and a nonarbitrary grasp of nature cannot obtain.

Interest of the Study for Scientists

Scientists well versed in the areas of research discussed in this narrative, and particularly those personally involved in the events described, may find my account of the science oversimplified or somewhat imprecise. They may also object that credit for important advances has not been carefully enough distributed. To the second of these objections let me plead guilty immediately—for it is not among the purposes of this book to say who deserves credit for what. That the main characters discussed herein are all very important contributors to the development of biological electron microscopy is beyond question, and no doubt they do deserve much credit; by concentrating on them in particular, though, I do not mean also to concentrate all the credit in them. These characters are chosen both because of their influential roles in the community of electron microscopists and because they represent clear examples of different styles in the use of their instrument for biological experimentation. They were indeed stylistic leaders, who promoted their manner of using the electron microscope and who educated many in that style. But they were not the only leaders; some have been left out, and to these and those close to them I apologize. This study is primarily a history of the science of biological electron microscopy, not of the microscopists per se. To the first objection, that the science has been oversimplified, I must plead guilty but invoke extenuating circumstances. The growth of scientific knowledge has been described only in the degree of detail necessary to show the electron microscope's role in various biological research programs, and to give examples of the kind of impact it has made. And my account does not attempt to bring the scientific matters discussed up to date, either—except occasionally where present knowledge seems important for understanding the way matters appeared to the scientists in the period being described. Communicating the present state of scientific knowledge is the proper role of science textbooks, not histories of science. Though some scientist readers may feel deprived of a meatier review of the scientific knowledge gained by electron microscopy, there are compensations in the broader perspective gained from a sketchy overview, in the understanding of past knowledge in its own terms, and in the accessibility to nonspecialists of a nontechnical history.

Indeed, scientists themselves may gain from this overview of biological electron microscopy, however much they may be experts immersed in some aspect of the field themselves. Too often elements of the big picture elude those intensely involved in day-to-day work. Electron microscopists will likely find interesting the origins of some of their familiar practices, handed down through apprenticeship for so long that the origins of this inherited wisdom have grown obscure. Scientists working with newer experimental technologies will, I hope, find interesting the challenges that confronted the first biological electron microscopists, and they may find useful suggestions in how these problems were dealt with. Scientists curious about the origins and life cycles of research programs, and the characteristics of the more successful programs, will find grist for their mills. And scientists concerned with the big questions that I find so interesting, about how our knowledge depends on our technology, should find stimulating material for reflection herein. Even if we never get clear answers to these last questions, asking them may help us appreciate both how wonderful it is that we know what we do, and how very much more awaits us through future techniques.

I should perhaps address an issue related to biological electron microscopy's appropriateness as a basis for generalizations on experimental practice. All scientific fields have their dissenters and would-be revolutionaries, and biological electron microscopy is no exception. Physiologist

Harold Hillman in recent years has attracted considerable attention to his unorthodox views on cell structure, views at odds with those based in cell biology research using electron microscopic and ultracentrifuge techniques.[52] Although some of the interpretive uncertainties that Hillman points to are indeed real difficulties, and although on the other hand electron microscopists can readily offer cogent ripostes to many of his particular arguments (for instance, to the claim that far too many narrow "unit" membrane profiles appear in electron microscopic thin-section views as normal to the section plane, it can be retorted that microscopists select normal cuts for publication—among many other views in their sections—because these yield the most accurate measurements), there is a larger issue. In a field already noteworthy for the diversity of its experimental methods and manners of reasoning from data (although, as I will argue, there are certainly standard epistemologies, these allow considerable leeway), Hillman assesses cell biological evidence in an extremely idiosyncratic style of his own, which he has most helpfully spelled out.[53] He differs from the majority both in the weighting and mix of approaches (or principles) he thinks ought to be applied in experiment, and perhaps also in the formulation of some of these approaches. Particularly striking is Hillman's stark claim that "the constancy of appearance is not evidence that it is not an artifact"—in contradistinction to most cytologists (and philosophers following them; see above) who consider that although this may not constitute proof, the more different techniques can reproduce an appearance, the less plausible its interpretation as a technique-dependent artifact becomes.[54] Another marked difference is Hillman's higher estimate of the reliability of mathematical geometry and physical theory, which most cytologists have found quite unreliable in the past, given the uncertainties about specimen composition and other factors that must be assumed (because not known) in order to make the calculations. Practicing biologists are understandably too busy to engage much in the tricky business of metaepistemology, but philosophers ought to be taking up the slack more than they are. A great deal more work could be done to analyze the tried-and-true standard epistemologies of experiment, and to explore measures by which these measures of truth might themselves be compared. As I urge above, this is an interesting and important topic.

To conclude, it seems to me that consistent with the existence of such vociferous dissidents (albeit contrary to their protestations), biological electron microscopy is a suitable field for the study of sound experimental practice. Where the field is atypical is in ways that, if anything, make it

especially appropriate and favorable for the study of epistemology as practiced: the research is done in small laboratory units in a variety of institutional settings from hospitals to drug companies to universities to basic
research institutes; it is engaged in by practitioners in a variety of biological disciplines; it has a tradition of users groups in which methods and
interpretations are openly debated; and because the microscopes are not
so expensive as to be rare these days, it is a relatively democratic form of
big science—at least compared with fields like particle physics or space
science, which can only be pursued at a handful of sites worldwide. All of
these pluralism-enhancing factors have tended to encourage open disagreement on epistemological matters, making the topic of heterogeneous and evolving epistemologies easier to analyze. By the same token,
these factors must—if it is true that debate and competition have an
improving effect—have fashioned a superior set of epistemological tools
and a sound body of knowledge derived therefrom.

The Electron Microscope and Its Earliest Origins

Even though this book deals not with the electron microscope per se, but
its use as an instrument by biologists, an introduction would not be
complete without a few words on the remarkable machine itself, and on
the engineers and physicists who brought it into the world. Resolution of
a microscope—the minimum separation between points distinguishable
by observation—is limited by the wavelength of radiation used for imaging. Thus electron microscopes are capable of much greater resolving
power than light microscopes because of the shorter wavelength of electrons: whereas the best light microscopes cannot do better than about
one-fifth of a micron (symbolized as μ, one-thousandth of a millimeter)
and seldom actually approach that limit, electron microscopes quickly
obtained a resolution several hundred times better, in the range of one
millimicron (or $m\mu$, one-thousandth of a micron, equivalent to ten Angstrom units, or Å), without even nearing their theoretical limit. Many
good sources are available to explain the principles and construction of
electron microscopes.[55] Basically, a transmission electron microscope is
made by directing a high-voltage electron beam from a cathode through
an evacuated column, where the beam passes through a specimen and several magnetic or electric fields that act as lenses, to a plate or luminescent
screen at the far end. (Electrostatic lenses were tried on a number of early
microscopes.)[56] The ultimate magnification obtained is a function of the

focal lengths and positions of all lenses along the beam path, just as in a light microscope. With an electron lens that is an electromagnet solenoid, the focal length varies with current in the coil. In electron microscopes, focal lengths are generally varied electronically and the lens positions are fixed, whereas in a light microscope the opposite is the case, but the consequences are equivalent, and as noted can be treated by the same optical theory. An image forms on the screen of a transmission electron microscope through the scattering (without much absorption) of incident electrons by atoms inside the specimen, creating a shadow pattern of greater and lesser electron transmission. Apertures installed along the beam path may increase contrast and decrease spherical aberration by limiting the beam to the central areas of lenses.

The invention was something of a natural outgrowth from cathode-ray technology, particularly the cathode-ray oscillograph. The theory and technique of this device, improved electronic tubes and vacuum technology, and the matter-wave theory of de Broglie, which implied that optical concepts could be applied to electrons, all made the time ripe to develop an electron microscope around 1930. That year, electrical engineers Max Knoll and Ernst Ruska began building one in Berlin, and by the mid 1930s independent microscope projects had also sprung up in America, Britain, France, Holland, Sweden, Belgium, and Canada, where in the physics department of the University of Toronto a series of talented graduate students applied themselves to the technology under Eli F. Burton. In 1937 Siemens hired Ruska to build a commercial microscope, the first of which was delivered in late 1939. Before the war turned decisively against Germany, Siemens promoted the use of its microscope among a variety of scientific and medical researchers in a manner similar to that of RCA (see Chap. 1).[57] However, scientific communication was curtailed, so electron microscopical developments in Germany and on the Allied side proceeded for the most part independently.

Although biomedical applications were hoped for from the outset, the difficulties presented by biological specimens stymied early electron microscopists. The electron beam requires a vacuum inside the microscope so—because life is a wet condition—specimens cannot be alive, and have somehow to be dehydrated in a way that minimizes distortion. Because the electron beam interacts strongly with matter, the specimen has to be thoroughly preserved or "fixed" so as not to change too much during further preparation and under the beam. Indeed, electron radiation experienced by a specimen in the microscope beam is said to be about the same as would be received from a ten megaton H-bomb blast 30 yards away.[58] Biological specimens not chemically altered (and thus made of the low-atomic-weight substances native to living things) lose around one-third of their dry mass in the first minutes of viewing, as their substance is stripped away under the intense bombardment.[59] Thickness of the specimen presents another main problem area with living things; the penetrating power of the beam is insufficient to give good pictures of any but the thinnest cells, and even where sufficient, too many layers of superimposed structures within a cell make interpretation difficult. Contrast also is particularly problematic in the case of biological specimens, the different parts of which usually vary little in opacity to electrons. Specimen preparation techniques were required to overcome these problems: ways to dry, preserve, slice thinly, and enhance contrast chemically. As the story of these techniques unfolds, it will become clear that in some cases specimen preparation for electron microscopy would evolve from techniques developed in the late nineteenth and early twentieth centuries for light microscopy,[60] while in other cases completely novel protocols had to be devised. By the late 1930s simply building an electron microscope was a more obvious matter than making one useful for life science. That was a matter for biologists and the "Picture Control" Hall referred to, not for instrument developers.

CHAPTER 1

RCA and the War Years

❖❖❖ A new opinion counts as "true" just in proportion as it gratifies the individual's desire to assimilate the novel in his experience to his beliefs in stock. . . . It makes itself true, gets itself classed as true, by the way it works; grafting itself then upon the ancient body of truth, which thus grows much as a tree grows by the activity of a new layer of cambium. . . .

The trail of the human serpent is thus over everything. Truth independent; truth we *find* merely; truth no longer malleable to human need; truth incorrigible, in a word; such truth exists superabundantly—or is supposed to exist by rationalistically minded thinkers; but then it means only the dead heart of the living tree, and its being there means only that truth also has its paleontology, and its "prescription," and may grow stiff with years of veteran service.

—William James, *Pragmatism*

William James raises an issue of central importance for anyone seeking acceptance of a new scientific idea, experimental finding, or indeed any claim to knowledge. The scientific community (in the role of James's "individual") only accepts novel or controversial claims if they do not conflict too much with the body of established knowledge. Moreover, that concretion of secure knowledge consists of elements that were themselves once controversial or potentially so, and were adopted when assimilated successfully to what had earlier been accepted. So scientific knowledge grows by a historically contingent, almost organic process. Some elements of the body of accepted facts with which new knowledge must fit are indisputable, having grown sturdy with use—essential and indispensable because routine practices for dealing with our world depend on them. Thus anyone promoting the acceptance of a new scientific claim faces the problem of fitting or "calibrating" the novelty with the world as previously understood by science, and the new must not clash with anything essential in the old, lest it be counted unacceptable. The same goes for anyone promoting acceptance of a new technique for *making* scientific claims or discoveries; indeed, from a pragmatist perspective no sharp distinction can be drawn between tools of knowledge and products of those tools, because nothing counts as knowledge that is not proved useful. Particularly in the case of an instrument or other technique, the task of minimizing cognitive dissonance is paralleled by a task on the level of social engineering, that is, the task of nondisruptively introducing new work practices and the social relations that go with them to established laboratories. Except in the odd circumstance of an entire scientific discipline being founded around one instrument or technique all at once (and those joining such a new discipline might still bring an established commitment to that one central technique from another context),[1] a new device generally must be provided with instructions for its use that integrate it with the techniques and other practices already established in the laboratories that might adopt it.

This then was the dual task, at once epistemological and sociological, of those wishing to introduce the electron microscope to biological research. They needed to make pictures that conformed to and yet extended accepted knowledge about living things. And they simultaneously needed to make the process of producing and interpreting pictures with the new machine convenient for biologists. That is, they needed to show biologists how the new machine could be used constructively in their research. In a word, they needed to domesticate the electron microscope by establishing a set of experimental customs around the new machine, taming it and making it safe to bring into the lab and the discipline as an instrument.[2] This chapter argues empirically the case just sketched theoretically: that domestication of the electron microscope (and by extension, instruments in general) was achieved when a set of practices was developed that fitted and integrated, *in minimally challenging and disruptive ways*, the new imaging machine and its pictures with the existing knowledge and ways of life prevailing in certain fields of biological science. These new adaptive practices that *channeled* or *restricted* the manner of microscope usage quickly "stiffened" and became traditional, thereby further extending through time the established ways of life to which the device had been domesticated. The implication of my case is that conservative forces govern the acceptance of novel technique in scientific practice, and that these set sharp limits on the capacity of innovations to bring about the overnight technological revolutions or theoretically revolutionary discoveries sometimes invoked in romantic accounts of science. These conservative forces (whose operation I can adumbrate historically here, but which demand sustained inquiry elsewhere in their own right with the help of additional social theory), ensure a certain continuity within experimental sciences, both culturally in terms of the patterns of practice and authority within a field, and intellectually in terms of the scientific knowledge held to be true by it.

RCA Casts the Die

One of the pioneers in both microscope construction and observation of biological specimens was Ladislaus Marton, a physical chemist at the Free University of Brussels, who had previously worked on spectroscopy at the Tungsram Lamp company.[3] In 1932, only a year after the first microscope had been built in Berlin by Knoll and Ruska, Marton built a simple electron microscope to study filament emissions and the photoelectric effect. By early 1934 he had built an improved machine, independently keeping up with the Germans despite a shoestring budget. Marton tried out his microscope on sections of plant tissue fixed with osmium tetroxide, which he obtained from biologist colleagues in Brussels. These electron micrographs, published in *Nature*, were regarded as exciting largely because they disproved the common belief that biological specimens would be incinerated in the electron beam.[4] Marton worked on the theory of specimen-beam interaction, and seems to be widely credited (at least outside of Germany) with recognizing that with very thin specimens such as tissue sections, image formation depends more on scattering than on absorption of electrons.[5] In 1935 Marton finished a third microscope, which incorporated practical improvements for photography and specimen manipulation. Although he was generally unable to find biologist collaborators, over the next three years Marton became known for his work with biological specimens. The electron micrographs he produced, however, were more curiosities holding promise of future biological electron microscopy than experimental findings with any real meaning to life science. Their lack of significance to biologists is an indication that domestication had yet to begin.

Well before the Rockefeller Foundation took an active interest, RCA gave impetus to the adaptation of the electron microscope for biomedical research. By 1938 Vladimir Zworykin, then RCA's head of electronics research and later famous as the "father of TV" in America, had decided to start work on a salable electron microscope, and had already convinced the company's president, David Sarnoff, to bankroll the effort under his direction.[6] A few other American firms, such as Kodak and GE, built electron microscopes in the late 1930s, but none were very serious about commercializing them, believing that even a handful might saturate the market.[7] It seems doubtful that Sarnoff and Zworykin had a dramatically higher estimate of the electron microscope's profitability than did the executives of the other firms that decided not to manufacture them. Rather, it is worth considering that the RCA decision may have had much to do with the firm's famous $50 million investment in television. By 1938 RCA had a system of transmitters and receivers, based on Zworykin's patents, ready for production, and its NBC network (the National Broadcasting Company, RCA's principal profitable subsidiary during the Depression) was poised to move into this new field of broadcasting. But RCA's competitors in radio-receiver manufacture argued that RCA's transmission technology was inadequate; moreover, others including archrival Columbia Broadcasting were making impressive progress with alternative television systems. RCA launched an intense publicity campaign for its television business in early 1939, and put on a spectacular demonstration beside the Trylon and Perisphere of New York's World Fair, but they were rebuffed for a monopolistic approach to the new medium by the Federal Communications Commission (FCC), which decided that permanent frequency allocation was premature. Negotiations among manufacturers eventually led to adoption of provisional standards, but the state of national emergency declared with the outbreak of war in France and Britain had already postponed commercial TV for the duration, delaying any chance for RCA to recoup its investment and giving competitors extra time.[8] Given that technical inferiority was one of the charges against Zworykin's television system, it is possible that the prestige to be won during the war for RCA's cathode-ray technique via the electron microscope may have been one reason that RCA embarked on what must at first have seemed an unremunerative program in scientific patronage. Certainly RCA used its enhanced technical image after the war to sell its TV system, as we shall see.

In autumn 1938 Zworykin brought Marton, who had just fled to England, to his RCA lab in Camden, New Jersey, in order to build an electron microscope for possible commercial production. The choice of Marton suggests that RCA was especially interested in making a microscope useful for biological work, though he was certainly prominent among microscope designers as well as users. Marton's machine, dubbed the "model A," was working by the spring of 1939 (Plate 1.1). Basically the model A was the column of Marton's previous microscope design enclosed in a second metal jar, the entirety of which was evacuated. Airlocks allowed access to specimen holders and photographic plates. Advantages of the bell-jar concept included better magnetic shielding of the column and stability against vibration. Drawbacks of the concept far outweighed the advantages, according to John Reisner, who was an engineer at RCA in the 1940s, and doomed the model A. The machine was large, heavy, and expensive. Because such a large volume had to be evacuated, it took a very long time to pump down, and the vacuum ultimately attained was not good. Because air got in, the column contaminated rapidly with charging substances, causing loss of resolving power until it was cleaned out. This frequently required cleaning was difficult because the jar restricted access to the column. Reisner recalls that it was immediately obvious to Zworykin that the model A would never be commercial.[9] The frequent expert attention demanded by the microscope was perhaps the principal obstacle to its marketability.

Nonetheless, the model A did function when carefully attended, and Zworykin wanted it used for biology. Through their mutual friend Katherine Polevitzky, a bacteriologist on the faculty of the University of Pennsylvania (Penn) dentistry school (who became Zworykin's second wife in 1951), Zworykin invited Stuart Mudd, head of medical microbiology at Penn, to try the microscope in his research.[10] Polevitzky, Mudd, and Mudd's junior colleagues from Penn were soon enthusiastically imaging many common microbes. The bacteriologists were quickly joined by Mudd's friend Wendell Stanley of the Rockefeller Institute's Princeton department of plant and animal pathology, who brought some of his plant viruses to RCA for imaging. Though only in his mid thirties, Stanley was already famous as the first to crystallize a virus, the tobacco mosaic virus (TMV). Conceived as giant enzymes, viruses were at the time widely considered "naked genes" and/or the basic units of life; they were often represented as giant molecules capable of assembling themselves into higher units of biological structure. They were fascinating entities on the border between the animate and inanimate, as Stanley often proclaimed.[11]

Zworykin was unable to induce Marton to undertake the radical changes in his microscope design that would be required for marketability. In February 1940, with little to show Sarnoff for all his efforts, Zworykin hired James Hillier, who was building an electron microscope for his doctoral thesis in physics under Burton at the University of Toronto. Zworykin installed Hillier in a lab nearby Marton's, and understandably, a rivalry soon developed between the two. Hillier worked quickly, designing easily made parts by virtue of prior experience as a machinist, and by July 1940 had his microscope, to be known as the model B, running. Thanks to a compact, feedback-regulated 60 kv power supply just developed by an engineer in the electronics research group, the whole machine including vacuum pumps fit, as Zworykin had desired, into one self-contained unit about the size of a refrigerator. And already Zworykin was hatching plans to introduce the RCA electron microscope to biomedical research on a larger scale. Beginning late that spring and summer, Hillier has recalled, there was mounting tension with Marton, no doubt fueled by Zworykin's favoring the younger man's model B design for production. Indeed, Marton's duties even included demonstrating the model B's virtues to visitors (Plates 1.2, 1.3). To cover the budget shortfall caused by hiring Hillier, at the end of 1940 Zworykin sold Hillier's prototype to the American Cyanamid firm of Connecticut, where the research head, R. Bowling Barnes, was a friend of Zworykin. The model B was priced at $10,000, ten times the cost of a luxury car and approaching that of a cyclotron at the time.[12] Very few life scientists then could have been expected to afford one.

The Rockefeller Foundation and Electron Microscopy

By the time the first model B was delivered in December 1940, Marton was communicating with the Rockefeller Foundation, asking for help in finding a university post. He was told, in conformity with foundation policy, that he should make institutional arrangements himself and later apply for funding.[13] Nonetheless, Warren Weaver was keenly interested in encouraging electron microscopy, and behind the scenes was trying to place Marton at the Massachusetts Institute of Technology (MIT). There Francis Schmitt, a zealous young neurophysiologist and "general physiologist" (in the manner of the previous generation's flamboyant Jacques Loeb, iconoclastic founder of the discipline in America) long cultivated by Weaver, was about to take over a biology division newly reorganized with massive Rockefeller funding.[14] Marton had sent the foundation a plan for applying the microscope to biology, discussed below, and Weaver passed it on to Schmitt and the MIT administration. Marton sent out other feelers, and by early January 1941 he had interested Stanford chemistry professor James W. McBain, who had been collaborating with Marton in research on the microstructure of soap.[15] McBain in turn recruited some Stanford chemists and biologists, and approached Ray Wilbur (a Rockefeller Foundation trustee and longtime Stanford president) about bringing Marton and his microscope to campus with Rockefeller help. McBain at this point considered the RCA model B to suffer excessively "high cost and shortcomings and limited applicability," sharing what must have been Marton's opinion—that Marton could do better.[16] Certainly, the foundation was improving his job prospects. In February 1941 Hillier tired of the Marton situation at RCA and, Hillier recalls, gave Zworykin an ultimatum: "either he goes or I go!"[17] Soon Marton was on the phone to the Rockefeller Foundation announcing his availability and pressing for "news." Marton obviously understood the active part Weaver took in promoting physical instrumentation for biology, and was trying to capitalize on that Rockefeller patronage system to move back into academia.[18] Thus the Rockefeller influence was felt even beyond academic biology, in an industrial research and development lab.

Although Marton fumbled MIT by coming across as an "arrogant European" (in the words of graduate school dean John Bunker) when he gave a trial talk there in January 1941,[19] a job for him at Stanford did eventually emerge after plenty of collusion between Weaver and McBain. In mid-February 1941, McBain and Stanford bacteriologist Edwin Schultz visited Weaver, who encouraged them to draft a proposal to the foundation, emphasizing the electron microscope's use for a wide variety of life science and chemical research. The proposal Weaver invited—a report claiming that 38 Stanford scientists were interested in the microscope, and listing numerous interdisciplinary, biological, and chemical projects in which to employ it—appeared on his desk in short order. By late February 1941 Weaver had already decided to fund electron microscopy at both MIT and Stanford; MIT would concentrate on physiological applications, and Stanford was a "first rate opportunity to supplement the MIT program" by microscope improvement and ventures into bacteriology and colloid chemistry.[20] It was not unusual for Weaver to advise researchers whose projects he favored far in advance, and sometimes he would virtually dictate what sort of proposal they should make. Many of the scientific clientele, for their part, knew very well how to appeal to the Rockefeller audience: the ethos of cooperative individualism, which had penetrated to all political and cultural spheres from corporate America during the later teens and the Republican 1920s, still reigned in Rockefeller circles, and throughout the interwar years scientific projects por-trayed as cooperative efforts in undeveloped fields between disciplines were in vogue with foundations and other funding bodies such as the National Research Council (NRC).[21] The NRC, a board of prominent scientists organized during the First World War to facilitate military re-search by providing a link between the government and the scientific community, afterwards became a central clearinghouse for distributing philanthropic aid for research in America. It has aptly been likened to a trade association for science.[22]

Marton's initial pair of informal proposals, written near the end of 1940, before his detailed consultation with Stanford introduced the dis-tinctly American rhetoric of interdisciplinary cooperation, show most clearly what he had in mind.[23] The first is entitled "Project of a 'Sub-Microscopical' Research Center." Clumsily recognizing the value of ap-pearing to till an undeveloped interdisciplinary field, it proposes a labora-tory to explore "the gap between the limits of light microscopy and of the atomic dimensions," which would pursue, along with teaching and mi-croscope development, studies in polymer chemistry, bacteriology, and virology with electron microscopes and a number of other physical in-struments, including "electron and X-ray diffraction cameras, ultraviolet microscope, ultracentrifuge, etc." The second proposal details the order in which Marton wished to proceed. First, he would establish himself in a chair of "electron optics and electron-microscopy," and start teaching in order to train potential laboratory helpers. Next, Marton would build two microscopes, a service model roughly similar to the RCA machines, and an experimental model with easily interchangeable lenses and suited for much increased accelerating voltages, which would be far superior when complete. At the same time he would install the X-ray and electron diffraction apparatus. Then he would start work on viruses and bacteria, investigating the "cell under different conditions (variations with age, medium, etc., chemical reactions in the cell—action of disinfectants, drugs, etc.)." Marton wanted a "bacteriologist attached to the research center with full equipment," and similarly for virology, "the full time of one specialist . . . with ultracentrifuge, Tiselius apparatus, viscosity appa-ratus, etc." For macromolecular chemistry he envisioned using much of the same virology equipment and working with a "skilled physical chem-ist." Marton imagined a grand, centralized laboratory, in which he would command a brigade of underlings more knowledgeable about the biolog-ical specimens than himself (and who could fill in the "etceteras" for him), experts armed with an expensive battery of the latest instruments—but his microscope would be first among them.

The grant proposal submitted by Stanford in May 1941, more modest and attuned to the cooperative American ethos, was skillfully crafted to play on the sympathies of Weaver and the foundation.[24] Stating that the electron microscope in 1941 "is as far short of its probable possibilities as was Galileo's telescope in comparison with the two hundred inch telescope now building" (at Mount Palomar, with Rockefeller funding through the International Education Board), the proposal flatters the foundation while aiming to exploit an apparent trend favorable to gigantic scientific instruments. The Stanford electron microscope was being positioned as a flagship project for Weaver to rival the spectacular Palomar instrument—the same strategy successfully employed by Ernest Lawrence to win Weaver's support for that "mighty symbol," his 184-inch Berkeley cyclotron.[25] The quest for pure knowledge (such as provided by telescopes), not practical power over phenomena, was implicit as the grounds for these prewar appeals for funds. The Stanford grant proposal casts doubt on the quality and real availability, given defense priorities, of the RCA model B. It then promises that Marton would not only work on improving microscope technology, but also "within a few months" build a service machine for the use of biologists and chemists who "would be responsible for . . . all specimens and their interpretation as regards their special fields, while Dr. Marton would be responsible for the interpretation insofar as electron optical features" are concerned. Cast in fashionable terms of interdisciplinary collaboration, this ambiguity about who should take charge of micrograph interpretation glossed over a fundamental obstacle: Marton did not know enough about biological experiments to invent ways to make his microscope useful, so he needed to negotiate an arrangement with Stanford life scientists.

Taking the place of Marton's earlier vague notions about interesting biological specimens, the Stanford grant proposal gives a list of fourteen biological and chemical collaborative projects with faculty members to employ the microscope in studies on cancer, various cell types, bacteria, and a host of currently hot topics including protein structure and antibody production,[26] viruses, "chromosome and gene structure," and more. Stanford's physics and electrical engineering departments would cooperate and provide facilities, and involved departments would cover the costs of their own projects. These last two provisions were supposed to meet the foundation's requirement that host institutions show financial commitment; minimal though they were, they would suffice for approval. One has the impression of a hastily prepared proposal half written by an eager Weaver. The $65,000, five-year grant was given as a lump sum on July 1, 1941.[27]

Meanwhile, Schmitt's efforts to obtain Rockefeller support for electron microscopy at MIT also went well. Although Schmitt was reportedly "enthusiastic" about taking on Marton when he heard Marton's dreams of "submicroscopical" research in December 1940, he lost no enthusiasm for the dreams when Marton himself alienated Dean Bunker. At the beginning of February 1941, Bunker visited the RCA microscope lab in Camden, immediately followed by Francis Schmitt and his brother Otto (a biophysicist at the University of Minnesota), who spent a couple of days looking at red blood cell membrane samples that Francis had brought with him from St. Louis. Bunker and Schmitt both concluded that the RCA model B was adequate, that Marton was not essential, and that they would like to hire Otto for the MIT project instead.[28] Schmitt, who like Bunker considered Marton someone he would prefer to avoid as a colleague (for social rather than intellectual reasons), was now "so enthusiastic" about electron microscopy that he said he was going ahead with it "Hell and High Water, money or no money."[29] Weaver balked at taking Otto away from the Rockefeller-funded program at Minnesota, and after trying Hillier (Zworykin blocked a job offer behind the scenes), and Albert Prebus at Ohio State, Schmitt decided to hire Cecil Hall from Kodak's electron microscope lab (Plate 1.4). Hall and Prebus were both, like Hillier, trained in Toronto.[30] Because Burton's group was the principal source of practical expertise with electron microscopes in North America at a time when both theoretical and craft knowledge about their construction was considered crucial for working with them, his students represented the most viable alternative to Marton.

In late February 1941, Schmitt and Bunker submitted a draft electron microscopy grant proposal for Weaver's perusal and comment. Schmitt's lengthy appendix to the draft lists numerous potential investigations in the "molecular structure" of biological specimens, including muscle fibers, cilia, mitotic spindles, collagen, blood clots, membranes, nerve fibers, chromosomes, mitochondria, and the then-fashionable "protein megamolecules" (i.e., viruses and enzymes).[31] All of these were macromolecular assemblies extractable in relatively pure form from the organism, like red blood cell membranes and the other favored specimens of general physiologists; however, what exactly would constitute the most useful application of the new microscope was evidently no more obvious in advance to Schmitt than it was to Marton. Like Marton, Schmitt planned to use the electron microscope on these biological specimens in conjunction with various physical methods such as ultracentrifugation, X-ray crystallography, and spectroscopy. But unlike Marton, Schmitt had prior experience working with his proposed biophysical specimens and methods. Moreover, Schmitt had specialists in these methods on his staff, including X-ray diffraction expert Richard Bear and ultracentrifugist David Waugh, former junior colleagues for whom Schmitt obtained MIT faculty jobs when he took charge.[32]

Schmitt's proposed goal was to visualize and measure the chemical architecture of cell components, so that phenomena such as nerve conduction, cell division, and muscle motion would be explicable by automatic mechanisms like molecular self-assembly. This represented the kind of life science of which Weaver wholeheartedly approved; there was no need to correct Schmitt, who by now was proficient at appealing for Rockefeller money, and his plans in the official grant application of May were unaltered.[33] The $70,000 MIT grant was officially approved in June, but Schmitt had evidently placed an order with RCA long before, because he took delivery of a model B directly on receipt of his Rockefeller money in July. By August 1941 Schmitt said he was already getting good protein fiber pictures with his microscope,[34] although his model B—one of the first batch serially produced—suffered from electronic problems that afflicted the MIT group with plentiful downtime and frequently required service visits.[35]

Wartime Biological Electron Microscopy at MIT and Stanford

Aside from putting Hall in charge of the microscope itself, Schmitt attached to him a lab assistant, and assigned graduate student Marie Jakus, who had also come with Schmitt from St. Louis, to the microscopy project.[36] Among the earliest projects Schmitt, Hall, and Jakus undertook was with the protein collagen, a major component of skin. This material had the habit of spontaneous reassembly from solution that interested Schmitt (because such processes in which higher-order organic structures emerge automatically point the way toward physical and chemical accounts of all vital phenomena), and was thought from X-ray and polarized light work to have structural features large enough to see with the model B. After

Pearl Harbor, Schmitt decided to concentrate on collagen because of its potential medical applications. In January 1942, Schmitt was already producing a so-called artificial skin from collagen sheets, and then went on to develop artificial sutures by spinning the collagen with equipment borrowed from the American Viscose textile firm.[37] He arranged a contract from the Office of Scientific Research and Development (OSRD), a body of leading scientists closely akin to the NRC, which during the Second World War distributed all government research work and also guaranteed resources and sheltered personnel from the draft. The contract allowed Schmitt to put together a sizable group to study wound healing, with some local doctors. A similar contract protected George Beadle's biochemical genetics work at Stanford; to get one, it helped to have friends in Washington.[38] By the summer of 1942, Schmitt wanted another microscope for the collagen project, and was able to buy one with help from industry connections, mainly leather and rubber manufacturers.[39] The second instrument was reserved for the military work on collagen and rubber, with a special security room and separate operator. Significantly, the only student Schmitt committed primarily to basic biological research with the microscope was Jakus, who as a woman was exempt from the draft; draftable scientists had to work on defense projects. Electron microscopy has always had a relatively large proportion of prominent women, and one reason must be that it was founded during wartime, giving female scientists an enhanced opportunity to enter at least the biological part of the field.[40] Like Rosie the Riveter, they had a harder time when men returned to work after the war, but the precedent seems to have made some difference.

Jakus and Hall worked together very closely at first in exploring the microscope's potential, Hall sharing his knowledge of how to use the machine to best effect and Jakus sharing her knowledge of biology and of specimen preparation procedures for light microscopy. For her doctoral research project, Jakus studied a structure in paramecium called a trichocyst, a capsule that rests under the cell membrane until, when some stimulus threatens the organism, it suddenly bursts out of the cell as a defensive spike several times its original length. She found that the shaft of this spike consists of fine protein fibrils exhibiting regular molecular structure when fully extended, which were probably tightly folded in the trichocyst capsule before extrusion. In line with the theory that changes in affinity for water molecules could explain how protein molecules in solution produce motion (favored at the time by Schmitt among others),

Jakus concluded that the fibrillar molecules of the shaft undergo a chemical change that causes them to take up water and expand into a long, hollow tube supporting the sharp tip. The MIT group also made some exciting new findings on the molecular architecture of nerve, Schmitt's chief research topic before the war, and on muscle, both of which will be discussed in Chapter 4 together with the MIT postwar research programs on these fronts. The group's productivity was impressive, given the troubles of wartime; Jakus, Hall, and Schmitt produced some dozen publications on basic biological problems under their Rockefeller grant. What all of the MIT wartime research projects had in common was a focus on protein fiber structure and function, a general interest of Schmitt's before the war that the new microscope moved to the fore.[41]

The MIT group worked out how to precipitate pure collagen in differing forms, and also myosin, an abundant protein in muscle, by tinkering with salt solutions along lines loosely guided by contemporary colloid chemistry theory. Adapting the traditional cytological stain phosphotungstic acid, they learned how to stain fibers of these and other protein molecules with the reactive metal to bring out details. (Here is one simple example of how the crossing over of standard light microscopy procedures—continuities of practice—played a role in domesticating the electron microscope.)[42] With differential density staining, the MIT lab was able to measure structural patterns directly from micrographs, and in Schmitt's hands the electron microscope became a tool to gauge dimensions of the same sort of macromolecules he had begun studying before the war. Plate 1.5 shows a type of myosin (from clams obtained at Woods Hole, where Schmitt enjoyed summer research) analyzed in this way, the spacing figures checked by X-ray crystallography by Bear.[43] The continuity between the problems and objects of knowledge in Schmitt's earlier research and his war work indicates how smoothly Schmitt adapted the new microscope and his established research programs in general physiology to one another. After the war, when Schmitt built up and diversified "molecular biology" at MIT, electron microscopy remained the centerpiece.[44] And as will become plain in later chapters, microscopists trained at MIT have carried on his style of direct measurement of molecular assemblies from micrographs.

The war story at Stanford was not one of equivalent productivity. Marton was put in charge of a freestanding Division of Electron Optics, under the executive supervision of a committee of the biologists, chemists, and physicists who had brought him there. In his division Marton attempted to realize, in miniature, the grand "Center for Sub-Microscopical Research" he had originally envisioned, but such a realization would have required a degree of authority that he lacked. The electron microscope committee chairman, McBain, had to countersign all expenditures, and most decisions about staff, space, and other resources had to be made at committee meetings. When he arrived in mid-1941, Marton was set up in a corner of the aging chemistry building, with a view to moving him later to the new chemistry extension then being planned, and given limited access to the chemistry machine shop. He worked slowly with one part-time machinist, possibly awaiting his promised new quarters. However, after the December outbreak of war, building plans were curtailed and materials became harder to obtain. Not much material help was forthcoming from the departments involved in Marton's grant proposal, however much they may have hoped to benefit from his microscope. Marton's space problem would not be solved until December 1943, when Samuel Morris, the engineering school dean, volunteered his roomy offices for conversion to a microscope laboratory.[45]

During his initial slow period Marton pressed to popularize his electron microscope research. His most prominent effort in this regard was a whimsical piece entitled "Alice in Electronland," published in the high-profile journal *American Scientist* and written around the end of 1942. Marton distributed copies to Stanford's president and chancellor, Ray Lyman Wilbur and Donald Tresidder, evidently in the hope of communicating his industriousness and the importance of his work.

The premise of this piece is that Marton discovered an unpublished manuscript of Lewis Carroll's that treated, with remarkable prescience, electron microscopy.[46] It opens:

Alice was beginning to get very tired of sitting by her brother on the bank and having nothing to do. She idly picked up some papers which were lying beside him and gazed questioningly at some extra-ordinary photographs.

"Curiouser and curiouser," cried Alice (she was so surprised that for the moment she quite forgot how to speak good English), "now I am getting smaller and smaller than the largest microscope ever showed."

The world was singularly changed. All kinds of objects, some of them considerably larger than herself, were floating in the air. She had some difficulty clinging to the bank, because she was being kicked all the time and could not discover from where the kicks came.[47]

After explaining that Alice was a colloid experiencing Brownian motion, and that she could not see her fingers because they were smaller than the wavelength of visible light, Marton introduces a pedantic *Bacillus* (who resembles a "child's sausage-shaped balloon") to narrate a rather lackluster lesson in electron optics and the utility of electron microscopy for bacteriology.

Gradually the didactic content of the piece gives way to some imaginative passages that are revealing of Marton's situation and character. Especially noteworthy are the veiled jabs at his old colleagues at RCA, in a section in which the *Bacillus* guide describes for Alice some other microbial denizens of Electronland: " 'You might have seen the picture of cousin Joe Strep. Here he is. He says his picture was taken, and he did not mind it at all, but'—the creature came very close and whispered in Alice's ear—'I don't believe him. He is such a liar and tells tales to build up a reputation.' "[48] Apart from explaining that electron "photography" is lethal for microbes, this segment appears to be describing a boastful, egotistic character who claims authority on matters about which his testimony is unreliable. Given that Marton illustrated Joe Strep with the *Streptococcus* electron micrographs he had made with Stuart Mudd, and which Mudd had used for his first electron microscopy publication (with Lackman),[49] it seems likely that Mudd—never shy to blow his own horn—is parodied in this loud bacterial reputation-seeker. Later, after Alice sees a crowd of small flu viruses, which the *Bacillus* describes as unruly youth, Alice is pointed to another crowd of viruses: " 'If you want to meet some better mannered youngsters I would recommend the sons of Toby Mosaic. They are not very clever, but are very ambitious, and their only vice is in being too much addicted to tobacco.' "[50] The illustration here is a micrograph of TMV made at RCA from Stanley's specimen. Since the media spectacle of his crystallization of the virus, Stanley was practically synonymous with TMV. Marton would not be the first to suggest that Stanley was a "youngster" with more ambition than brains,[51] so it seems that the sons of Toby Mosaic represent Stanley and his Rockefeller Institute group. (There is double meaning at work in this mention of "ambitious" TMV, because Stanley and others considered the virus particles capable of self-assembly into higher units of protoplasm.) There is a third microbial character worthy of mention also. The *Bacillus* says: " 'Here comes a very distant relative, Frances Thioba. She is really rather pretentious. I really don't know why. Of course she is pretty, but very dumb.' "[52] Though I have been unable to identify the source of this specimen, it seems a fair guess that the *Thiobacillus* in question is Polevitzky, stereotypically stigmatized as stupid on the grounds that feminine attractiveness is incompatible with intelligence. (Here Marton might be able to claim that Frances Seymour, the *Thiobacillus*'s namesake soon to be introduced, was the intended target—if he ever needed to defend his satire. Naturally, the possibility of a second reading does not rule out the first.) One gets an overall impression that Marton was embittered when his former colleagues turned away from him, after Zworykin showed favor for Hillier and his model B.

Finally, "Alice" climaxes with a poem that speaks to Marton's Stanford situation, and can be read not only as an effort at popularization, but specifically as an appeal both to the public and to the Stanford administration for greater appreciation and continued support.

> The Walrus and the Carpenter were walking arm in arm,
> They wept like anything to see such antics on the Farm,
> "If this keeps up," the Walrus said, "the place will lose its charm."
>
> Around them as they strolled along were heavy tubes of brass,
> high vacuum pumps, high tension cones, and funny shapes in glass,
> And people working frantic'ly would hardly let them pass.
>
> "If this is what I think it is," the Walrus did declare,
> "Although there seems a wealth of stuff, there's really none to spare.
> You cannot build a microscope with only empty air."
>
> "A microscope," the Walrus said, "is what we chiefly need,
> Electron Optics, furthermore, is very good indeed,
> So let us build some microscopes with superhuman speed."
>
> "But wait a bit," the Walrus said, "before we have our chat,
> I quite forgot that O.P.M. has got a voice in that!"
> "A pity," said the Carpenter, "you're not a Democrat."
>
> "The time has come," the Walrus said, "to talk of applications,
> The uses of the microscope in our investigations,
> To take a peek at things that were our former speculations."[53]

Marton was eager to portray himself as "working frantic'ly," and to excuse his tardiness in completing the microscope by reference to insufficient resources and the politics of wartime industrial priorities (i.e., OPM, the Office of Production Management). In the final stanza he does seem to promise that the delays are over, though, and accept that the proof of his industry will be found in the pudding of fruitful applications on the Farm (Stanford's campus).

Not until mid-1943 did Marton have a microscope working and available to Stanford researchers (Plate 1.6).[54] At this time Marton began trying to make his microscope useful in war work.[55] Among his overtures to the government was an ingenious plan to use electron microscopes for submarine detection, as ship-mounted sensors of magnetic disturbances. But unlike Schmitt, Marton was unable to secure the patronage of government agencies like the OSRD even for this imaginative scheme (one arguably as plausible as Schmitt's artificial skin), at least partly because of his "enemy alien" status, which would have required that Marton work under an American citizen classified for military research, such as McBain— precisely the subordinate position Marton was trying to escape.[56] Marton was a political outsider in American science, in a formal as well as informal sense, because he lacked the business and government connections that allowed Schmitt to thrive in the wartime milieu (recall his war priority problems). He also showed a European failure to comprehend what American scientists liked to call "team play"—an interpersonal aspect of the cooperative ethos that was lacking in his early grant proposals—that made him seem egotistic to some at Stanford, just as he had to Dean Bunker at MIT.[57]

Marton did not accomplish a great deal, compared to original expectations, with the Stanford researchers who had planned to benefit by collaborating with him. In 1941 McBain had anticipated a microscope that could serve a "wide circle" throughout the university, especially considering that taking pictures should take "at most a matter of minutes," or so he thought.[58] Although one reason that few of the scientists listed in Stanford's proposal actually worked with Marton must be that the war changed their research priorities, another is that Marton did not involve himself enough with the research problems and practices of potential collaborators to facilitate applying the microscope. One project that McBain pushed was construction of a vacuum-proof specimen cell with walls thin enough that electron micrographs could be made of wet chemicals and living specimens. The cell was completed by January 1944, but no meaningful results seem to have come of it, presumably because the walls eliminated useful contrast. Marton later told Rockefeller official H. M. Miller that the failure of the cell idea soured McBain permanently on the microscope;[59] but if Marton had worked more closely with McBain to determine mathematically whether contrast would be insufficient (the sort of calculation that was his specialty), he might have nipped that disappointment in the bud. In an early 1944 progress report, Marton listed all the collaborative projects then underway: research on soap molecules and enclosed cells with McBain and colleagues in chemistry, microbe work with Schultz of bacteriology and his student Paul Thomassen, and polio work with Hubert Loring of biochemistry and his student Carlton Schwerdt. The last two resulted in one publication each (Plates 1.7, 1.8).[60] The polio micrographs allowed the first direct estimates of the virus particle size. This work, part of Schwerdt's doctoral project, employed McBain's ultracentrifuge to purify polio from infected monkey brains, a source that meant that virus would inevitably be mixed with other particles of cellular origin. Schwerdt recalls trudging across campus repeatedly from the preparatory lab to Marton's lab over a six-month period, every time he had a new batch of virus; Marton was unable to think of ways to improve the preparations.[61]

When the question of Marton's renewal came up at the end of the war, his two new biological publications, though respectable, appeared in context not so impressive, for reasons that will soon be apparent. His committee split essentially along disciplinary lines: the chemists and biologists preferred buying an RCA microscope to extending Marton's appointment beyond his five-year grant, while the physicists and engineers on the committee wanted to keep his electron-optical expertise at Stanford, though not enthusiastically enough to offer him one of their tenured slots. The physicists, who had immediately before the war been working hard on grant applications for a supervoltage X-ray source, presumably had a greater appreciation of Marton's finesse with electron beams; there may also have been a closer affinity with Marton on account of their shared cultural background in experimental physics. Marton left Stanford for the National Bureau of Standards when no arrangement with tenure was offered.[62] One of Marton's fundamental problems integrating into academia, despite his Rockefeller backing, was an ambiguity about who was in charge of the microscopy project. Paralleling this was an ambiguity about who should speak for the significance of the electron microscopic pictures: as the grant proposal indicated, Marton was to interpret micrographs physically, and biologists biologically, an arrangement fraught with both social and epistemological awkwardness. Marton lacked the background needed for facilitating application of his microscope to the research problems of potential collaborators, for instance by improvising specimen preparation techniques (like the Schmitt group's metal staining, which came from prior biological practice). He was unable to integrate his machine with existing biological practice.

Given his inability to participate as an equal within the academic subcultures of his biological collaborators, Marton sought to trade on his expertise with the electron microscope as an outsider, bartering access to this physicist's device for joint authorship with the life scientists and career stability.[63] Thus Marton drew a disciplinary boundary ("Electron Optics") around himself as part of an effort to make the microscope physicist, himself, an indispensable middleman, and—to modernize the economic metaphor—to raise the exchange value of his specific skills and of the general cultural capital belonging to him as a physicist.[64] The bell jar around the model A is perhaps emblematic of his strategy of valorizing microscope access. But biologists and chemists were not beating a path to Marton's door; he needed to meet them more than halfway to make his machine work for their activities. Moreover, to the biologists Marton's terms may have seemed especially unfavorable, given the bargain their East Coast colleagues had been getting from RCA. The reasons for Marton's defeat at Stanford are therefore complex, involving local cultures and economies of expertise (in which the combination of all his failings and liabilities told against him, even if none was decisive in itself), embedded in a global context shaped both by Rockefeller support and RCA competition.

"Selling the Electron": RCA-Sponsored Biological Electron Microscopy

Apart from the machines of Marton and Schmitt, the only electron microscopes given over to use mainly by biologists during the war were at RCA's New Jersey labs. When Hillier's prototype was operational in the summer of 1940, Zworykin rushed to Washington to arrange for RCA to finance a National Research Council fellowship for a postdoctoral researcher to work at Camden on methods for biology.[65] The fellowship was to be administered by the newly constituted NRC Committee on Biological Applications of the Electron Microscope, which initially consisted of Zworykin, Mudd, Stanley, and a handful of mainly local life scientists. Zworykin's strategy of allowing a select committee of biologists to oversee the introduction of the electron microscope circumvented the troubles Marton met in trying to retain a crucial role for his physics expertise while engaging in biology research: Zworykin did not require the biologists to take an alien (in terms of social connections, nationality, and academic subculture) into their midst as a coequal. As its postdoctoral fellow the RCA-NRC committee chose Thomas Anderson (Plate 1.9), a modest and diffident midwesterner with a Cal Tech doctorate in physical chemistry and broad postdoctoral experience (accumulated while looking in vain for a permanent academic post under Depression conditions), including a distinctly biophysical stint working on yeast radiobiology with Madison plant physiologist B. M. Duggar.[66] Anderson's multidisciplinary background and accommodating personality would make him an excellent mediator between the separate laboratory cultures in which the biologists and RCA's physicists worked. But Zworykin's chief reason for choosing him was that Anderson had a record of finishing and publishing work independently.[67] Publications—and other forms of publicity—were on Zworykin's mind from the start. As both sides understood from the beginning, RCA and the participating life scientists stood to profit from mutual assistance; Zworykin could win prestige for his electronics group, RCA might gain market share and government contracts on other products through favorable publicity even if few microscopes were sold, and the biologists might participate in major breakthroughs in their specialties.[68]

In addition to Zworykin, Stanley, and Mudd (the chairman), initial members of the NRC committee were: Milislav Demerec (geneticist at Cold Spring Harbor labs of the Carnegie Institution), Charles Metz (cytologist and geneticist, and Penn's zoology chairman), James Kempton (plant cytogeneticist with the U.S. Bureau of Plant Industry), and Caryl Haskins (radiation biophysicist at Union College in Schenectady). The committee was a set of academic men and institute researchers of established intellectual stature and considerable administrative power as well (particularly Mudd, Metz, Demerec, and increasingly, Haskins); indeed, Polevitzky was not a member of the committee until Zworykin insisted on adding her in the spring of 1941.[69] It was calculated to be a group with enough clout to win credibility among life scientists for experimentation using the electron microscope; Polevitzky's exclusion is just one indicator of the importance of status, because as a dental bacteriologist and little-known researcher she had less—a situation not unrelated to her gender. Such a group of eminences is typical of the committees responsible for fixing the standards of practice of academic disciplines through doctoral supervision and oversight of curricula, examinations, and funding.[70] That is, such committees are the stewards and gatekeepers of disciplinary culture, responsible for maintaining its continuity over time.

The basic plan was that Anderson would work in the labs of Mudd and Stanley, and possibly of other committee members, using RCA "chiefly for taking pictures"; Mudd and Stanley intended to supervise Anderson closely and ensure his proper training in bacteriology and virology, their specialties.[71] Demerec, Metz, and Kempton intended that the structure of chromosomes should become a high priority after Anderson mastered bacteria and viruses, and Metz offered to host such work to "throw light on the nature of the gene." This notion that the microscope would be useful to geneticists reflected the then-dominant assumption that genes had a complex enzymelike structure on which their function depended. Thus, just like Mudd and Stanley, the geneticists also planned to oversee closely the adaptation of the new microscope to their field.[72] Anderson set to his heavily supervised work immediately upon arriving in Camden in September 1940, for the first months with specimens of bacteria and viruses provided by Stanley, Mudd, and Polevitzky.[73]

At a meeting held at RCA in September to inaugurate the fellowship, visiting dignitaries, including Robert F. Griggs, chairman of the NRC biology and agriculture division, who came up from Washington, were given demonstrations of Hillier's microscope, of pictures of bacteria prepared mainly by Polevitzky, and also of the specimen supports used in lieu of glass slides (perhaps improved by Polevitzky, since she ran the demonstration). These supports—usually called "screens" or "grids" and still in use today—were prepared by spreading a thin film of dissolved collodion or other plastic over water, and picking up the film onto a disc punched out of metal screen after the solvent had evaporated. Zworykin announced RCA's intention of displaying micrographs at the next meeting of the American Association for the Advancement of Science. There was general agreement that "much was to be gained by getting good pictures into the scientific journals." Mudd called on the attending biologists to make the publicity drive a form of "missionary work" by showing their best micrographs at every scientific congress they might attend. (Already in October 1940, Mudd, Anderson, Polevitzky, and Hillier were giving papers at a meeting of the local chapter of the Society of American Bacteriologists.)[74] Griggs likened Anderson to a Robert Hooke embarking on an "Electron Micrographia" that would astonish the modern world. Griggs went on to exhort the committee members to let Anderson pursue particular research interests of his own because electron microscopy was not likely to become a science in its own right: soon "it will be necessary, for the Committee to . . . restrict its demands on [Anderson's] services in the interest of [his] consistent progress in one direction," urged Griggs.[75] This kind suggestion to let Anderson plan his own research would never be heeded by the committee, perhaps because the members' interests were not in Anderson's development and they each had too much to gain by employing him on their own projects. Or perhaps Griggs's notion of scientist as heroic individualist poorly fit the prevalent ethos of interdisciplinary teamwork and science-industry cooperation (which the war only served to strengthen). Or perhaps, more profoundly, startling and innovative application of the powerful new device was precisely what the committee was created to contain. All three explanations fit the evidence and all probably have some truth in them. In any case, this inauguration day ritually marked a collective effort to domesticate the electron microscope by powerful members of the scientific groups that would have to pass judgment on its acceptability.

Through the first half of 1941 Anderson studied more specimens from Mudd and Stanley, and undertook new projects with a handful of other biologists, despite chronic problems of microscope availability due to breakdowns, sales and publicity sessions,[76] and usurpation of any working microscope including Marton's for industrial and defense work,[77] such as rubber additive research and uranium separator evaluation for the Manhattan Project.[78] With Stanley, Anderson looked at several plant viruses and found not only that most particles had the uniform size and shape expected of homogeneous species of molecules (Plate 1.10), as Stanley had claimed viruses to be, but also had sizes consistent with what Stanley had estimated by kinetic methods (i.e., by diffusion and ultracentrifugation). The two also looked at TMV bound to antibodies, and at the precipitate obtained when secondary antibodies against the first were added. As expected, the virus particles appeared coated with a layer of antibody, and clumped together in the presence of secondary antibody. With Polevitzky, Mudd, and other Penn bacteriologists, Anderson looked at various microbes (Plate 1.11), finding not unexpectedly that the external envelope of these cells was much more solid than the interior protoplasm. With Polevitzky, Anderson obtained remarkably clear pictures of bacterial flagella consistent with the notion that these tiny protrusions were actually hollow tubes, which seemed briefly to suggest that hydraulic forces from the protoplasm might cause them to wave and thus propel the microbe.[79] This rather original theory was soon abandoned. There were few other surprises. Almost any picture might be publishable; on the other hand, samples handed over to him by his collaborators often presented Anderson with unforeseen stumbling blocks—for instance, he was given cross-reacting antibodies that vitiated an experiment to visualize the difference between binding by different specific antibodies. Anderson complained privately of "difficulties encountered first, in obtaining the [regular] use of a microscope, second, in obtaining suitably tested and prepared materials for study, and third, in the waste of time in petty matters of publicity, making prints of micrographs, and smoothing out political (personality) matters." He was engaged in a stressful juggling task of pleasing Zworykin by writing papers, printing micrographs, and showing off the microscope, while pleasing the biologists on his committee with progress on each of their favorite projects.[80]

Among the new users that Anderson helped in the later months of his first fellowship year were Mudd's and Metz's Penn colleague Les Chambers (with influenza virus) and Stanley's Rockefeller Institute senior Thomas Rivers (with smallpox virus). The contingent interested in gene structure also got its chance when Demerec, Metz, and Bernard Nebel, a plant cytogeneticist and Cold Spring Harbor regular, all took turns with chromosomes and sperm. A tally of the first nine months showed Mudd the heaviest user, followed by Stanley, Polevitzky, and Chambers.[81] At the committee meeting of June 12, 1941, Anderson succeeded in securing regular microscope time and a less chaotic work schedule by means of a calendar, to which visitors would be committed in particular time slots well in advance, generally one visitor per week. This improved Anderson's job satisfaction greatly because it protected him from being treated like a technician by the enthusiastic and imperious Mudd, who was frequently interrupting other projects to have his own pictures taken. In the summer of 1941, as he finished his first fellowship year, Anderson continued with Mudd, Polevitzky, Stanley, and Chambers, and began working with A. Glenn Richards, a Penn zoologist, on insect cuticle.[82] The insect and virus work came off especially well because such hard specimens are resistant to drying artifacts, whereas the work with chromosomes was frustrated by specimen thickness and drying problems.

The NRC fellowship generated favorable press for RCA, and Zworykin seldom missed a chance to pose with the microscope (Plate 1.9).[83] In news items, presentations at scientific conferences, and even to some extent in the experimental reports that were starting to flow from Anderson's work, the enthusiasm of Mudd and the other committee members showed through. By the time RCA renewed Anderson's fellowship in October 1941, the committee had been joined by Michael Heidelberger, an immunologist at Columbia Medical School (and former Rockefeller colleague of Stanley and Rivers), Gordon H. Scott, a physiologist from Washington University who was added for reasons having to do with politics within the NRC, and (officially at last) Polevitzky.[84] At the first committee meeting of Anderson's second year, Zworykin thanked the assembled biologists for the benefits their participation had bestowed on RCA's microscope project. Dividends of the collaboration, Zworykin said, included the sobering effect that Anderson's impressive but still primitive results had on the claims RCA made about the microscope's power, which would otherwise have been very exaggerated, and help in "stabilizing" the model B's design through use and user feedback. He was proud to announce that sixteen microscopes had been delivered as of that meeting, and that the government had granted model B production AA1 status, top defense priority.[85] Though Zworykin did not thank the biologists for these latter two circumstances, good publicity from the NRC affiliation could not have hurt (not to mention the presence on the committee of Haskins, who was becoming liaison officer at the OSRD and thus was close to such Washington decisions as production priorities).[86] Indeed, in the context of wartime propaganda the electron microscope was especially significant as an unclassified high-technology development with which to counter German claims of scientific superiority (see Chap. 6).[87] The mutually profitable relationship between RCA and the biologists was cemented by the wartime moral ideology of patriotic cooperation between industry and academia.

Through the 1941–42 academic year, Anderson continued as before, only more smoothly, working at RCA mainly with the small circle of committee members, who were essentially a tight clique centered on nearby Penn, the Rockefeller Institute, and Cold Spring Harbor (where many of them attended summer workshops). Occasionally Anderson would take on new projects with their friends and acquaintances. New biological users were Mudd's friend George Knaysi of Cornell, who imaged capsules around bacteria by fixing specimens with mordants, and Warren Lewis of the Wistar Institute, who looked at some embryonic tissues.[88] Bacteriophage specialist Salvatore Luria of Indiana was also granted microscope time, promoted by Demerec and Heidelberger over objections from Stanley that Luria's preparations were not adequately pure and that his methods were "controversial" (an objection stemming from the disagreement in estimates of certain virus sizes according to radiation target theory, versus diffusion and centrifugation as favored by Stanley).[89] Richards continued his insect morphology, Polevitzky and Mudd continued their bacterial work (now experimenting with antibody and metal-solution treatments of specimens), Stanley continued with his plant viruses, and Rivers brought in exotic animal pathogen specimens from several Rockefeller Institute colleagues. In January 1942 Scott and possibly also Heidelberger tried their hands at small projects as well.[90] Members of the RCA-NRC committee's circle frequently came back for seconds and thirds; for them, it was a bonanza of publishable pictures.[91] In June 1942, RCA moved Anderson, Zworykin's biologist daughter Nina, and a model B to Woods Hole, where they would be able to collaborate with and impress a wider crowd over the summer,[92] and the fellowship work could continue despite relocation of RCA's labs from Camden to Princeton.[93] In September, Anderson was rewarded with a research post in the Johnson Foundation biophysics unit of Penn's medical school, and his Penn salary was initially supplemented, somewhat irregularly, with NRC postdoctoral funds through the influence of one of his benefactors in Washington.[94] Even taking into account that the electron microscope's novelty made it easy to accumulate enough fresh data for publication, the diligence and productivity exhibited by Anderson are remarkable. Anderson's two RCA years gave rise to more than two dozen biological papers listing him as author, and still more on physical and chemical topics; but then, filling journals with results meeting certain standards was one of the expressed goals of Mudd and Zworykin.[95]

Picture Control and RCA's "Right Sort" of Microscopist

The original contract between RCA and the NRC stipulated that all micrographs made with Anderson's help be approved for publication, with respect to quality and interpretation, by his NRC oversight committee. This arrangement, in conjunction with the committee's control of access to microscope time, provided Mudd and Zworykin with the power to enforce the standards they strove to maintain. First, let us consider the role of access control through two examples. In early 1941, Kurt Stern of the Yale Medical School, who had visited RCA previously, and to whom both Mudd and Stanley owed favors for help with electrophoresis, contacted RCA about taking some more pictures of his chicken tumor virus for publication. Mudd wrote Stern that he hoped they might squeeze him in despite the "great difficulty" of microscope demand, and assured Stern that the review of micrographs and interpretations was only a "safeguard" rather than "editorial prerogative." Trusting the committee's judgment, Stern agreed without complaint, and in October his request for microscope time was put before the committee by Mudd and approved.[96] Academic status depends on just such a "circuit of continuous exchange" of professional favors among insiders: gift exchange cements bonds within a group and helps demarcate those who belong from those who stand outside.[97] And of course, many such circulated gifts are not only symbols marking the membership of scientific communities, but also practical resources that build solidarity by including members in a network sharing common experimental objects and tools.

Quite a different outcome befell Frances Seymour, a New York City physician and infertility researcher with a practice in artificial insemination. Like Stern, she had taken pictures at RCA before Mudd's committee began scheduling everyone, and when she requested more microscope time for her human sperm she was referred to Mudd. Explaining that she had learned of the micrograph approval process too late to stop a paper at the press, she pleaded for another opportunity and sent an offprint of her new paper.[98] It was not well received at RCA, where there seems to have been consternation not only at the quality of Seymour's micrographs, which could have been better focused, but also at the misuse of the extraordinary power of pictorial evidence to support controversial interpretations; in Anderson's words, "the artificial insemination gang at N.Y. . . . published some lousy pictures of sperm in which they saw suction mouths in the head, a body, and a tail."[99] The use of the word *gang* suggests unruly people of inferior class, hoodlums or racketeers outside the social order, and this was apt, at least in the sense that the committee's protocol had been broken. Mudd sat on Seymour's request for months, despite her repeated queries, contacting organizations like the Academy of Medicine in New York to find out whether she was "someone who is not the right sort, professionally."[100] In March 1942, Mudd had decided that Seymour was indeed not the "right sort," and rejected her request without formally putting it before the committee. Seymour's human insemination work might well have presented a distasteful image,[101] and she certainly was not the "sort" who summered at the Cold Spring Harbor or Woods Hole laboratories.[102] To pick up the thread of gift exchange among insiders, a number of ethnologists and sociologists have observed that shared tastes and coincidences of acquaintance are the tacit warp and woof binding together a field, and social circles within it. Outsiders and upstarts like Seymour lack the right attitudes and friends, so they tend not to be trusted to uphold standards of proper behavior.[103] It seems that proper manners and cooperative behavior were especially essential in micrograph presentation, if the electron microscope was to maintain a good image (and conversely, if the new device was not to become a vehicle to let disruptive outsiders into higher circles of life science).

Even after it was determined that an applicant was sufficiently reputable and trustworthy to merit access to the microscope and Anderson, the committee kept a close guard on how the results were presented in the final publication. Anderson took an active part in writing up the interpretations visiting researchers derived from his pictures, as those familiar with his well-known friction with Max Delbrück over their joint paper will recall. (Anderson had to soften Delbrück's interpretation of the spermlike appearance of bacteriophage, Plate 1.12, as strong support for the controversial theory that the viruses represented a sexual phase of bacterial cultures.)[104] Stanley himself was not exempt from requiring approval from Mudd and Zworykin for what he printed about his microscope findings—though in his case it was merely a formality.[105] The approval process, according to Mudd, was to ensure that printed micrographs would be those that looked best according to himself and Anderson, that no committee affiliate would "scoop" another affiliate's priority for any finding, and that no "naive" interpretations that might seem silly would be allowed to appear in print.[106] This filtration of all publications through Mudd and Anderson had the further effect that no affiliate's interpretation would conflict with that of any other committee affiliate. Such strict control over publication was necessary, Mudd and Zworykin agreed, in order to prevent interest from "burning itself out" and dying, which could happen when a spectacular new method breeds controversy rather than lasting "scientific accomplishment."[107] Like all of the strategies and precautions taken by Mudd's committee, access control and prepublication review reveal much about the tacitly understood conditions of instrument acceptance, which might under other circumstances remain unarticulated. The anxious behavior of the committee bespeaks the difficulties of and obstacles to successful domestication.

Anderson's Rules for Making and Interpreting Micrographs

The standardization of image production and image interpretation was another key aspect of the committee's agenda for integrating the electron microscope into experimental practice and establishing its reliability. Largely through the committee's exercise of power, Anderson's techniques, his aesthetic standards, and his style of interpretation all became models and benchmarks for the work of biological electron microscopists in the 1940s. Anderson's early preparative techniques were not complex: suspended or dispersed specimens were deposited on a grid, often rinsed with distilled water before or after deposition in order to remove salt crystals and other debris, and simply allowed to dry in air. As noted, with specimens originally dry and hard this treatment seemed unproblematic, but wet and soft specimens tended to suffer markedly; for example, Anderson underscored the drying artifacts and distortions in his own study of sperm, though he did not dignify Seymour's work by explicitly directing these observations at her.[108] For the next decade Anderson would make it his personal quest to develop a dehydration technique that might bypass the surface tension effects responsible for the patent havoc wrought by air-drying (Chap. 2).[109] Anderson also experimented with freeze-drying (not a satisfactory solution to the drying problem), with nontraditional embedding materials for finer sectioning, and as already mentioned, with reactive metal staining to enhance contrast.[110] None of his particular wartime protocols along these lines were spectacularly successful, and as later chapters will recount, they were not long in being superseded, but they pointed the way to further work on improved preparative techniques.

Anderson's rather basic aesthetic standards emphasized maximization of focus (so that micrographs would show the best resolution obtainable by his, and RCA's, microscope), minimization of background debris (so that casual or even expert viewers might not confuse dirt with the object under investigation), and selection of micrographs showing specimens suffering from the least obvious preparatory damage (so as to downplay visually the presence of artifacts, even when artifacts remain and are addressed explicitly in the text).[111] These aesthetic standards are still traditional fundamentals of practice in biological electron microscopy, and are still functional in making visual arguments with micrographs as convincing as possible.

Even more important, in his methods for interpreting micrographs, Anderson and his RCA-NRC overseers laid the epistemological foundations for subsequent practice in biological electron microscopy. The interpretive practices Anderson disseminated in the many committee-approved publications circumvented barriers to acceptance of the novel kind of microscope picture and paved the way for other biologists safely to take up the new machine. Thus Anderson's way of reading micrographs was, just like the limitation of access to elite practitioners, part and parcel of the social engineering carried out by the RCA-NRC committee in the interests of promoting electron microscopy; the favored interpretive strategies are similarly interesting for the light they throw on tacitly understood requirements of acceptance of a device or technique as an instrument. However rational it may be, epistemology also has a social history.[112]

Anderson's interpretive strategies played a part in the adaptation of the electron microscope to existing experimental practice and to established scientific knowledge, by finding a minimally disruptive place for the new device and the evidence it generated. First of all, Anderson carefully took precautions to ensure that the micrographs were in fact pictures of what the experimenter claimed. Appearances of a certain structure thought to be a virus had to be absent from uninfected, but otherwise identical, preparations from the animals or plants in which the viruses were cultured.[113] Similarly, the number of putative viral particles had to correlate with the number of infectious units (i.e., viruses defined operationally by a bioassay) observed in several similar preparations, if the pictures of particles were to sustain an interpretation as viruses.[114] This kind of strategy was already venerable among some biologists—at least as old as Koch's bacteriological postulate that a supposed disease agent had to be present in all diseased organisms—and quite generally regarded as prudent. By adhering to this tradition Anderson ensured that nobody would gainsay claims that micrographs in fact represented the entities they were supposed to depict. Thus embarrassing retractions would be prevented.

Anderson adopted a rather more original interpretive strategy— though one not without precedent in late nineteenth-century light microscopy—to deal with the large and unfamiliar set of artifacts introduced by electron microscopic technique: he would compare the same specimen prepared in as many different ways as possible.[115] For example, in one study of nerve cell protoplasm, Anderson compared smears (i.e., the wet specimen smeared directly on a grid) rinsed in various solutions, both freeze-dried and dried at room temperature, and sections of intact nerve fixed and embedded in a water-soluble wax (Plate 1.13). Although here it was admittedly "obvious that none of the pictures really represents the 'normal' state," because artifacts introduced by drying, freezing, solution treatments, fixation, embedding, and electron bombardment could not be evaluated against any image of unprepared material, yet the *similarities* among the various images must bear some "definite relation to the living state of the protoplasm." The conclusion that the nerve protoplasm was a colloidal gel composed of self-assembling rodlets, a view with considerable prior support, therefore was warranted by the presence of oblong particles of a certain size in all preparations.[116] These rodlets were just the lowest common denominator of all his pictures of nerve protoplasm. Anderson thus adopted and relied upon the untestable premise that agreement among many different techniques was more likely due to the specimen's native state than to coinciding artifacts. By deploying every possible preparative technique and reasoning in this way, Anderson was able to forestall possible controversy with other microscopists looking at the same specimens prepared differently. Unruly behavior of microscopists, and their micrographs, was thus prevented.

Another way Anderson exhibited epistemological caution indicates still more clearly that it was considered essential for the electron microscope, in order to gain acceptance, not to challenge established techniques, and therefore not to overturn the bodies of knowledge and the ways of life built around these techniques. Thus Anderson sought interpretations that maximized the agreement of his findings from electron microscopy with those from other methods, especially from other physics-based "molecular" methods of high status. For example, one of Anderson's RCA-NRC publications pointed out the correlation of structures he saw in nerve cytoplasm with those predicted by prior studies using polarized light birefringence on the same material (described above); another emphasized the close agreement in size estimates obtained by the electron microscope and analytical ultracentrifugation for the influenza A virus.[117] However, Anderson did not always refrain from using electron microscopic data to make a controversial point, provided the biophysical case was already strong on the side he was supporting. For example, with Luria and Delbrück he argued that the size of bacteriophage seen in the electron microscope agreed much better with the estimates given by their method of irradiating with X-rays than with the results obtained by a number of prominent virus workers by measuring diffusion rates and sedimentation in the ultracentrifuge.[118] (This is particularly interesting in light of the earlier efforts by Stanley, who relied on diffusion, to block Luria's access to the microscope because his radiation method was "controversial." Stanley's fears were borne out when the electron microscope vindicated target theory. One suspects that unconventional researchers wanting to work with Anderson but lacking friends like Demerec, even academics more reputable than Seymour, would be prevented from doing controversial work with the microscope by the NRC committee's efforts to find authoritative practices among the elite.) Great care was taken to minimize conflict with the kind of consequential researchers who wielded ultracentrifuges and other advanced technology: electron microscope findings were not allowed to run roughshod over their biophysical methods, even where disagreement was necessary. The same respect was not always shown toward prior results with the light microscope, a more plebeian instrument associated with medicine and old-fashioned life science.

Though background knowledge of the specimen's physical characteristics was—and still is—very seldom sufficient to allow mathematical analysis according to physical theory to play much of a role in the interpretation of electron micrographs, one of the best-known pieces of Anderson's RCA-NRC research was an exception. Working with Richards, Anderson found that the hard skins of iridescent beetles owed their colors to a simple diffraction grating structure, with spacing between lines in rough agreement to what optical theory predicted on the basis of color. This was anticipated on the basis of earlier work by entomologists. The iridescent scales of butterfly wings, however, harbored a surprisingly complex three-dimensional structure, the measurements of which were determined from stereo micrographs (made by changing the angle of incidence between specimen and beam, and then exposing a second plate). Each iridescent scale is covered with rows of vertical vanes, each containing regularly spaced thickenings in many layers, such that the entire assembly acts as a stack of diffraction gratings with similar line densities (Plate 1.14). Problematically, the spacing observed in micrographs was smaller than optical theory predicted on the basis of the blue iridescence of the scales. But this effect was attributable to shrinking caused by electron irradiation in the microscope: after observation with the model B, scales appeared black rather than iridescent, except for small blue areas where wires of the grid had shielded the specimen—an obvious indication of beam damage. Both the theory of iridescence by interference, and the value of the microscope for biology, were vindicated in one stroke when Anderson and Richards found that wetting irradiated scales with alcohol—which changes the color of unirradiated scales from blue to green (by increasing line spacing)—rendered the exposed parts of the specimen purple and the shielded spots green. This recovered iridescence not only suggested that the structures observed by electron microscopy were only shrunken and not destroyed by electron bombardment, and thus truthfully represented native morphology, but also allowed backcalculation to the spacing of the irradiated areas before wetting (from the observed purple of shrunken areas and from the percentage difference between calculated spacing in green and blue iridescent unirradiated areas). The calculated spacing in dry irradiated areas, 0.155 microns, was declared in "excellent agreement" with the spacing of "about 0.125 microns" observed in micrographs.[119] Thus the degree of specimen shrinkage due to electron imaging was determined, and more serious distortions ruled out.

That this fuzzy agreement was considered "excellent" bespeaks the uncertainty then surrounding magnification in electron micrographs (see Chap. 6). Nobody carped at the 20 percent error, and this experimental work was received as a triumph. Mudd considered the pictures of insect microstructure "fine art"; Zworykin expressed special enthusiasm for the three-dimensional reconstructions (no doubt partly because they highlighted the value of the model B's stereo feature); and Anderson—who could never be accused of boastfulness—described these findings as "spectacular" in his progress reports.[120] The feeling may well have been that such work, which integrated sophisticated instrumentation and quantitative theory to reveal the basis of biological phenomena in physical optics, exemplified the science of biophysics at its best. There was certainly an enthusiasm for these butterfly findings far beyond the importance of the iridescence phenomenon, presumably motivated by aesthetic approval of the quantitative approach utilized and of the satisfying explicability of life through physical laws. The committee made sure the butterfly work got plenty of notice; it was published in both biology and physics journals, and indeed it made the cover of *Journal of Applied Physics*.[121]

Thus, among Anderson's interpretive practices that are now traditional in biological electron microscopy may be counted reliance on multiple specimen preparation techniques to reduce the chance of unrecognized artifacts, correlation of electron microscope findings with those from other instruments, and—on the rare occasions when feasible—validation of observations by quantification based in physical theory. Some recent work in the philosophy of science has elevated these epistemological tropes to timeless strategies for gaining reliable knowledge about the world (see Introduction).[122] Correspondence realists may suppose that methods ideal for representing nature's one and only truth make themselves obvious; however, such a complacent attitude seriously underestimates the originality of Anderson's methodology. Relativists will likely conclude that the careful social maneuvers enacted through these epistemological tropes established the knowledge, and the methodology generating it, as sound. However, there can be little question that Anderson's work was seen as embodying and extending paradigms of sound experimental method, however inchoate and unclear their applicability to electron microscopy may initially have been, in a way that was creative and yet simultaneously constrained by material exigencies and by social structures. Perhaps we have here encountered a sign of the antinomic character of a science both fully social and fully natural mooted in the Introduction. Be that as it may, the interpretive methods that Anderson established—and subsequent practice has preserved—reflect the historically contingent circumstances of electron microscopy's hasty wartime birth among biophysicists anxious for their own status and that of their tools. The RCA-NRC committee was a key element in the contingent origin and promulgation of Anderson's methodology. As a set, Anderson's interpretive procedures, like the rest of the practices he disseminated by example in his many RCA-NRC publications, were designed to save electron microscope users from embarrassing disputes with one another and to prevent them from earning powerful enemies. As Mudd said, the idea was to prevent the microscope's value as a research tool from "burning out" by introducing its use without rocking the boat unduly.

Picture Control on a Continental Scale

At its January 1942 meeting the RCA-NRC committee added Schmitt and three physical scientists to its number, and moved to expand its authority in picture interpretation over all the sciences in North America.[123] Mudd and Zworykin contacted Ross Harrison, head of the NRC, and arranged to have him send a letter to the editors of 70 major biology, chemistry, and physics journals. The letter warns of the novel problems of electron micrograph interpretation and offers, under National Research Council auspices, the committee's services as reviewers of any submitted papers dealing with electron microscopy. It was approved and mailed to the editors, and also published in *Science*.[124] Not surprisingly, there were some objections to this monopolistic move. Stanford electrical engineer Karl Spangenberg aired his worry (presumably also his friend Marton's) to a journal editor that the industry-based committee members might steal ideas on microscope design. The director of the American Institute of Physics, acting as intermediary for an unnamed objector, inquired whether the committee's review activity was a camouflaged form of war censorship.[125] This latter is interesting in that it shows that Mudd's activities were the sort of violation of the ordinary academic intellectual freedoms that routinely found justification in the war effort. Anderson's response, which sought to mollify the objector by appealing to the cooperative ethos (he portrayed the committee's efforts as intended simply to help unsuspecting beginners avoid dangerous errors of judgment) was sufficiently vague that it did not necessarily correct the misconception that military security was involved.[126] But it is perhaps surprising how few objections were voiced; many readers of *Science*, and perhaps even some editors, may have assumed that the demands of military secrecy were behind the move and probed no further. Powerful elite committees in American science and industry were ubiquitous during wartime, and this context probably facilitated the efforts of Mudd and Zworykin to stabilize and—quite literally—to *discipline* microscope users so that a coherent and self-regulating body of practices would quickly became established among them.[127] And discipline needed to be imposed if self-discipline might not be expected to emerge spontaneously with the required speed. For as I argue more fully in Chapter 3, only when a standardized set of practices is established among a group of inquirers can it behave as a coherent scientific community (whether a discipline or a less formal assemblage), because only then can members build on one another's experiences and thus function in the manner of James's learning individual of the epigraph.

Though the RCA-NRC committee ceased its review function before the war ended, Mudd and Zworykin had other means to police the use of the electron microscope among experimenters (not to mention its image before the general public; Mudd monitored the press nationwide with a clipping service).[128] Most important was the Electron Microscope Society of America (EMSA), a users' group in which American electron microscopists shared ideas on how best to employ and maintain their machines. The first meeting of what became EMSA, held as a special section of a national chemical congress in Chicago at the end of 1942, was organized by University of Illinois physical chemist George L. Clark, with backstage instigation by Zworykin. Zworykin gave the keynote address, describing both technical improvements to the microscope and its usefulness to chemistry as well as biology. Organizers tried to "avoid any hint of commercialism," but RCA had a model B on display (which was supposed to be working but was not), and General Electric outmaneuvered RCA by setting up a working prototype electron microscope at the hotel next door. There were about 60 microscopists in attendance, representing almost all of RCA's customers and the handful of Americans using home-built machines.[129] The provisional head of the new group, and in 1944 officially elected EMSA president, was Zworykin's friend, new RCA-NRC committee member, and first model B purchaser Bowling Barnes.

Throughout the 1940s RCA engineers and scientists figured prominently among the group's officers, as, of course, did the earliest RCA customers.[130]

In two densely packed mornings, the foundational EMSA meeting of 1942 dealt with a variety of topics of general interest: trouble-shooting and modifying microscopes; strengths and weaknesses of electrostatic versus magnetic lenses (because the General Electric machine had electrostatic lenses, this discussion involved a tense showdown between GE and RCA engineers), and of increased accelerating voltages; preparation of films for specimen support; calibration of magnification and resolution; and ways of getting sufficiently low power pictures to compare with light micrographs.[131] We shall return to some of these discussions at Chicago in Chapter 6. The meeting also featured sessions on specimen preparation methods for metallurgical and biological applications. Anderson led the discussion of biology by describing his own newly devised methods of freeze-drying, and of cutting and handling thin sections embedded in a water-soluble synthetic wax, which he had worked out with Richards.[132] He also predicted that although viewing a wet biological specimen with an airtight chamber might be possible, the organism would be killed and contrast between the specimen and the surrounding water would be poor in any case. This prediction undercut the viability of McBain's plan to use such a chamber at Stanford, and when Marton challenged the assertion of specimen killing, Anderson exposed him as ignorant of biochemistry. Anderson then gave Marton some condescendingly simple advice on how to prepare biological specimens from suspension without leaving a background of debris and salt on the grid.[133] Nobody could contest Anderson's expertise. The attending microscopists were thoroughly introduced by EMSA to the RCA circle's techniques, interpretive practices, and standards of micrograph quality. They were taught how to make their machines behave like Anderson's, how to make pictures the same way he did, and how to read them in the same manner.

This first meeting was considered a resounding success by most attendees, EMSA was incorporated in 1943, and the second meeting was held in January 1944 in New York. Two of the eight biological presentations there were Anderson's, one was by Mudd and Polevitzky with Anderson, and another was by Schmitt and Hall. The third meeting, held in Chicago in November 1944, featured ten presentations dealing with new findings on biological specimens and specimen preparation, and again half of these were by members of the RCA-NRC committee. One was by Anderson (with Demerec and Delbrück), one was by Mudd's group, one was by Hillier, two were from Schmitt's MIT group, and two were by Marton. One of Marton's confirmed Anderson's findings on bacteriophage morphology and the other, with Edward Tatum, showed the effects of X-rays on bacteria (and seems never to have been published). Marton was falling behind, and the rest were marching in step with the RCA circle. The fourth meeting, held in Princeton in November 1945, included nine talks about biological specimens and their preparation.[134] Two were by Schmitt's group, and three others featured pictures of virus particles contrasted by the new metal-shadowing method worked out by Ralph Wyckoff and Robley Williams at the University of Michigan (discussed in Chap. 5); the shadow pictures by Anderson are remembered as exemplary.[135] By the end of the war, EMSA was a thriving group of RCA clients. Anderson was the acknowledged authority on microscopy of bacterial viruses, Mudd on bacteria, Schmitt on protein fibers, and Stanley was among the leaders on viruses of higher organisms. Clearly RCA had created a dominant concentration of expertise in biological electron microscopy with Anderson and his NRC oversight committee.

At the end of the war, then, RCA and its circle had emerged triumphant. Of the two main centers of independent practices in biological electron microscopy, Schmitt had been absorbed and Marton put out of business and moved to a service role in government (where he continued to contribute to electron optical theory and instrument design). The RCA circle's authority as founding fathers was recognized and sanctified in the nickname of the standard electron microscopy text of the decade following the war, "We the People." This book's first author was Zworykin and the biological section featured exemplary pictures by Anderson (from his work with Mudd, Stanley, and Luria) and Schmitt.[136] Zworykin's goal thus was realized when, through EMSA, the biological electron microscopists became a self-policing scientific community with a stable pattern of instrument usage; that is, when his microscope had become an accepted instrument for biomedicine. Mudd acknowledged as much by dissolving the RCA-NRC committee in April 1944 on the grounds that EMSA had taken its place.[137] RCA had cornered the market in electron microscopes, selling 58 model B units before replacing it with the streamlined EMU model in mid-1944 (Plate 1.15). The biological market for the EMU would grow substantially in the late 1940s when wartime government support for life science was carried into the postwar era and refocused on basic research, through a variety of funding programs mentioned in the Introduction.[138] Electron microscopes were soon within the reach of many life scientists. In 1950 there were already a conservatively estimated 220 electron microscopes in America, nearly all made by RCA, and perhaps half of them were being used for life science at least sometimes. The membership of EMSA then stood at around 350.[139] Only after 1950 would RCA begin facing significant competition, at first from Philips and Siemens, and by the later 1950s from Japanese firms also, though this meant little to biological practice with the instrument. The rest of this book explores the flowering of biological electron microscopy up to 1960.

However, this flowering could not have been foreseen before the war. Why then should Zworykin have gone to so much trouble establishing RCA's reputation in this particular sector of scientific instrumentation, given its apparently limited commercial prospects in 1938–39? The other American firms that built electron microscopes in the late 1930s did not push hard to commercialize the device. But RCA, with its massive and highly publicized investment in television and its failure to win government approval for full-scale commercialization of Zworykin's TV system before the national emergency stalled TV for the war's duration, was in a special situation.[140] The wartime electron microscope project enabled RCA to demonstrate technical proficiency in a newsworthy field of electronics closely related to video, during a time when commercial TV itself was off limits. Zworykin's presence in publicity photos of the microscope was a continual reminder of that relation to TV, and news copy often pointed it out. And in one of the more colorful episodes in business history, Sarnoff did ultimately win a long battle between 1945 and 1950 to secure market and government acceptance for RCA's technically questionable monochrome TV system. During early 1947 the FCC held hearings on archrival Columbia's incompatible color TV system, which if approved would make RCA's monochrome system obsolete, and RCA submitted its own jerry-built color system. The FCC chairman ruled that color TV should be shelved for further study, and a few months later was rewarded with a lucrative job as an RCA executive. Meanwhile RCA poured low-cost monochrome sets into the marketplace at cost, and even gave away the blueprints to all major manufacturers so as to disseminate RCA technology.[141] During the renewed FCC hearings on CBS's color system in 1949, Sarnoff staged a major publicity blitz highlighting RCA's advanced technologies. A full-page ad on page one of the December 1949 *Scientific American* depicts the electron microscope, radar, and RCA's television system as "leaves" on the same "tree of knowledge." The campaign's theme was "RCA—First in Television" and Sarnoff obviously considered the electron microscope pertinent to it. A few years later, *Newsweek* was correct in observing that Hillier's role (and equally, Anderson's) had been one of "selling the electron" for RCA.[142]

Conclusion: Traditions of Instrument Practice, the Channeling of Vision, and Paths Not Taken

I have pointed out how a number of features of the living tradition of biological electron microscopy can be traced to the contingencies of that tradition's wartime birth. The technical and epistemological standards laid down by Anderson's oversight committee, in the interest of continuity with and minimal disruption to established practice in life science, were promulgated not only by example but also through direct policing, followed by instruction of EMSA members. This disciplining work was certainly facilitated by the heightened wartime influence of elite advisory bodies in general—although standardization on some collectively approved practices would probably have come eventually. In addition to Anderson's now-traditional epistemological and aesthetic practices, the relative friendliness of the field to women (at least compared with other forms of early molecular biology) and the major role that EMSA has continued to play can also be cited as major holdovers of the microscope's wartime introduction and RCA's influence, respectively. But perhaps the most far-reaching consequence of RCA's wartime promotion of its commercial microscope is the institutional format in which the device found employment in biological research in the second half of the century, that is, as one instrument among several in the well-funded biologist's lab or department.

Arguing counterfactuals is a risky enterprise, but it is plausible to suppose that without RCA's wartime efforts, the instrument-centered facility planned by Marton along the lines of a big particle-accelerator lab might have taken root. The quality of Marton's microscopy at Stanford, if not the quantity, matched that of Schmitt and Anderson on similar objects (compare Plates 1.7 and 1.12), and he did have some satisfied clients. Supposing no model B success story, Marton would have had very little competition in biological and other academic electron microscopy research, so his work would have seemed more impressive, thus increasing his negotiating leverage toward the end of the war, and he would have attracted more collaborators on his own terms. The centralized electron microscopy institution he was trying to build might be common today. In such a laboratory, interpretive practices privileging the microscope over alternative instruments would likely prevail, and biologists using the microscope would have to accommodate their practices to the institution's requirements, epistemologically and otherwise. Of course, the shift in work pattern especially might have been jarring to a biologist in such a context; although not exactly a jack-of-all-trades, the life scientist has traditionally been capable of carrying out all steps of an experiment by him- or herself, or in a small group. In the event, RCA's early success at making the microscope an appliance allowed biologists to preserve that aspect of their way of life, and indeed ensured that electron microscopy would not become a discipline in its own right, along Marton's lines, but would grow *within* biological subdisciplines. It would be easy to overstate the degree of contingency involved in the adoption of new instruments: some conceivable ways of introducing the electron microscope to biological practice might never be acceptable to biologists under any circumstances (and perhaps what Marton required would have been so impracticable as to doom his project). But the particular ways in which the electron microscope was assimilated to biology, within the broad limits of the possible, might well have been otherwise. All these traces of the past (Anderson's practices, friendliness to women, EMSA's importance, horizontally integrated multitechnique biology laboratories including microscopists), which have persisted amid changes in theory and still inform present science through a cultural sort of founder effect, substantiate the saying that experiment has a life of its own.[143] Conserved features like these also raise the general issue of what conservative forces account for science's continuity with its past.

The present case represents an interesting specimen for exploring the requirements for innovation in experimental practice. I have shown how the methods for applying the electron microscope in biology were developed and authorized under the close supervision of Anderson's RCA-NRC committee, maximizing chances that the new instrument would be made useful in ways harmonious with preestablished practices and theories. Committee members guided Anderson's work in their own areas of specialization, granted microscope time only to researchers who would present the right image and set good examples, and controlled picture interpretation through prepublication review. Thus the committee had enough power to ensure that the new microscope would neither provoke too hostile a reaction in established fields nor disrupt their intellectual and social fabric. The rapid process in which the electron microscope was adapted for research in biomedical subdisciplines—domesticated, one might say—was accompanied by a partly deliberate channeling or restricting of microscopical practices along standard lines set by the RCA circle. Methods of specimen preparation giving results conflicting with established wisdom, or methods of picture interpretation challenging established authority relations, were countermanded from the start by the "calibration" entailed in the domestication process. Put another way, electron microscopy was only accepted when practices were standardized not just to an adequate extent, but in a manner that ensured compatibility and minimized disruptive impact for concerned biological fields (thus safely channeling the technique's application). The precautions of the RCA-NRC committee eloquently bespeak the presence of strongly conservative forces regulating whether, and on what terms, the electron microscope could be accepted as an instrument. And in the absence of reasons for regarding the electron microscope as peculiar among scientific instruments, except perhaps in the historical circumstances that forced the requirements of instrument domestication to the level of conscious work (such as the microscope's rather sudden introduction to biologists by industry, rather than slow evolution in the hands of scientists using it for their own experimentation),[144] one could generalize and conclude that this story casts doubt on the idea that new instruments ever can drive radical change in scientific culture or in scientific knowledge. Innovative devices and techniques must fit with established practice and established wisdom, or else never be taken up as instruments of science. And domesticating practices, like Anderson's ways of using the electron microscope, become incorporated with the traditional way of life as traditions themselves, instilled during training with the instruments, thus extending conserved patterns continuously into the future.

To conclude, I will attempt to capture the kind of conservatism entailed in the channeling of microscope usage, and the filtering of the visions produced by the new device, with a parallel. The sixteenth-century nun Theresa of Avila was a visionary mystic. This was a dangerous occupation in Counter-Reformation Spain, as attested by the fate of her fellow visionary and friend John of the Cross, tortured by the Inquisition and hounded to an early death for suspected heresy. But Theresa managed to do quite well for herself, despite a few scrapes, and died of old age as the revered and powerful head of an order. What was the secret of Theresa's success? She insisted on the best-educated Jesuit she could find as her confessor. All her visions went to him first, and only when the confessor judged them consistent with accepted doctrine did she claim them as true revelations; when the confessor judged her visions unorthodox, Theresa knew these to be Satanic deceptions, or as we might say, "artifacts."[145] Unquestionably life science, in its fundamental spirit, is more open to innovation than the Counter-Reformation Church of Rome; but Anderson's NRC committee did a job precisely congruent with that done by Theresa's Jesuit confessor nonetheless, always already assimilating the new visions of living things to established theory and practice before they even could count as knowledge claims. Schismatic or disruptive consequences for established authority are preempted and the visionary is simultaneously protected by the kind of channeling of vision enacted by both Jesuit confessor and NRC committee. Scientific instruments are, as Derek De Solla Price reminded us, "tools of artificial revelation" no less now than in the days of Galileo's telescope.[146] Some of the committee's channeling of the electron microscope's revelations, I hope I have shown, resulted from more or less deliberate efforts not to overthrow the biological status quo. But still more, I suspect, was unconscious, in that new experience is inevitably shaped by or (following James's horticultural metaphor) grafted onto the forms of knowledge already rooted in place.[147]

Here is perhaps the counterpart, in the realm of action, of the power of entrenched habits to impose familiar forms or *gestalten* on novel perceptions. I do not wish to push this point so far as to propose, like a Parmenides or Zeno of the history of science, that change is impossible. Even in the most "traditional" cultural groups, custom and tradition are constantly open to renegotiation.[148] I only wish to problematize the commonsensical notion that new experimental techniques can serve as engines of scientific advance simply by revealing revolutionary and unsuspected new truths. New techniques are often touted as agents of scientific change; however, the conservative mechanisms of calibration against existing methods and past knowledge—together with accommodation to existing cultural patterns and authority structures in a field—raise the question of how this ultimate change is worked. Conservative forces, however we may wish to theorize them, need to be reckoned with: Fleck's *Denkkollectiv* (the collective thinking of a scientific community), Kuhn's "paradigm" (the conceptual scheme informing observations and their interpretation in a realm of science), Wittgensteinian language games, and Bourdieu's *habitus* (the set of underlying dispositions giving rise to patterns or habits of work and social behavior specific to a cultural group) each seem to speak to some, but not all, of the linked phenomena of the conservation of knowledge-producing practices.[149]

Channeling of practice under the sway of conservative forces may ensure that very little of established knowledge or of scientific ways of life will need to be abandoned immediately on introduction of a novel instrument. Moreover, channeling may, by foreclosing palpably radical avenues, have long-term consequences in limiting what knowledge may ever come to light through an instrument. But even so, channeling does not on this account fully determine in advance the development of scientific knowledge. Change does come with new instruments, only not generally through the front door of instant revelations or revolutionary observations. Seen in this light, the interesting questions about technical innovation become: where are the other entrances, how much real change ever comes through, and by what mechanism do instruments open the passages to bring it in? The remainder of this book takes a detailed look at the growth through the 1950s of several research traditions emerging from the wartime RCA circle, in several branches of life science employing the electron microscope. In the conclusion I will return to this set of big questions about change and attempt to explain how this particular instrument worked its mutations on patterns of experimental work and thought, in a manner that might be extended to other instruments in other sciences. But for the next four chapters let us keep such questions in the backs of our minds while considering the work of the life scientists who took up the electron microscope and thereby joined, as Robley Williams jokingly put it, "the ranks of those who believe only what they see (or vice versa)."[150]

CHAPTER 2

Stuart Mudd and His School of Bacteriological Electron Microscopy

❖ Consider a colony of termites, whose motion, at first apparently Brownian, will much later appear to be directed toward the construction of a termitarium. . . .

Every termite, or almost every one, is a carrier, let's say, of a ball of clay. It does not bring it anywhere [in particular]; it positions it somewhere in the space under consideration. . . . The termites retreat and return to the task. The balls placed there are distributed sporadically, as if sown in a field.

Ancient cities are sown this way, in or by geography. . . .

It happens, or can happen, by some circumstance or another, that two termites deposit their bits of clay close by, maybe even on the same spot. That has an effect—we could call it a ball that's twice as high, twice as big. . . .

The termites leave and return burdened with new bits of clay. . . . They are going to put them down, preferentially, on the first ball, the highest, the biggest. Its effect is one of attraction.

The waves of termites do not stop. They go in search of balls and return to deposit them. . . . The piles grow even more when they are already voluminous—the effect of attraction. . . .

Thus it can happen that at a particular moment a giant ball attracts an ensemble of already-big balls and that as a whole this mine suddenly draws in all the workers; then the termitarium begins.

—Michel Serres, *Rome*

As Serres notes in his parable, when at the dawn of Western civilization cities sprang up around the hospitable Mediterranean, more cities were founded than could (by the laws of ecology, economy, war, etc.) ultimately survive.[1] This illustrates a general law of foundations. When the electron microscope became available as a new instrument for life science, opportunities for applying it suddenly opened and it began to be used in many ways, more ways—even leaving aside those ruled out from the start by the kind of adaptation to existing practice that Anderson and his RCA-NRC committee performed—than would later be recognized as successful foundations of a research tradition. Stuart Mudd (Plate 2.1), the would-be Romulus of one nascent research tradition in biological electron microscopy, was physically large and vigorous—a man of the sort who plays gridiron linebacker—and similarly imposing intellectually.[2] He also had an imposing personality in the sense that he was not one to keep his opinions and desires silent. Scion of a patrician family, liberal both politically and with his wealth—he supported research in his department with the excess interest income from his inheritance—Mudd was perhaps an ideal scientist for an institution like the University of Pennsylvania, which never fully adjusted to the postwar availability of large-scale government funding for scientific research. However well equipped Mudd was to carry on in an older, gentlemanly style of science, though, after the war nobody could outspend the NIH. The institutional obstacles Mudd faced in his efforts to keep up with the faster pace and more technologically lavish style of postwar biomedical research must be taken into account, together with the particular way in which he made and reasoned from electron micrographs, in explaining why Mudd's research program might appear something of a failure in retrospect. In the early days, after all, slight differences in terrain and natural resources can crucially determine how one city fares in its race against rivals to empire.

    In the 1930s (a period in which immunologists and microbiologists had not become fully distinct specialists within bacteriology departments), Mudd had become known as the advocate of an instructive theory of antibody formation, that is, a theory that the cells that produce antibodies, which protect the body from foreign microbes and macromolecules, do so on a template or scaffold consisting of the foreign, antigenic molecule itself.[3] But by 1940, work on antibodies had been taken up by sophisticated biochemists with whom a general bacteriologist like Mudd could not hope to compete, including Linus Pauling's powerful group at Cal Tech,[4] so the chance that Zworykin had given him in 1939 to move into the new field of electron microscopy came at an opportune moment for Mudd. The experimental way of life he brought with him was one steeped in the cytology and bacteriology of the early twentieth century, which depended on the selective staining reactions of exotic dyestuffs with particular components of cells to make visible tiny cellular structures barely at, and even below, the limit of resolution of light microscopes. In work of this tradition, inferences about the chemical constitution, and therefore the likely function, of the entities that took the stain were generally grounded in considerations of dye chemistry—and they generally stirred substantial debate, because considerations of dye chemistry could be made to support many interpretations.

In his research under the RCA–NRC aegis, Mudd had worked with Anderson, with Harry Morton, Mudd's colleague from Penn's Department of Bacteriology, and with some graduate students and juniors there, studying the morphology of a wide variety of medically important bacteria. Mudd's main interest had become the anatomical and physiological complexity of bacterial cells, and what this differentiation meant for the taxonomic place of bacteria with respect to the other forms of life. This was a lively topic in the 1940s, as a number of lines of investigation were starting to suggest that bacteria were not just the simple globules of protoplasm they had previously been taken to be. In addition to his RCA experiments, during the war years Mudd also worked on a number of government projects loosely directed at finding a preventive or curative treatment for dysentery, a disease with obvious military relevance, given conditions in foxholes, camps, and occupied territory. Mudd studied the sterilizing effects of ultraviolet light on *Eberthella [Salmonella] typhosa*, one bacterium responsible for dysentery.[5] With serologist Jules Freund he participated in trials on "mental defectives" and Penn medical students using Freund's new adjuvant (an additive that boosts immune responses to injected substances), evidently in a quest to develop a dysentery vaccine.[6] Mudd also worked with bacterial viruses in an effort to find a bacteriophage preparation that could prevent or cure dysentery when injected. Though the prospects for curing bacterial infections with bacterial viruses were quickly evaporating with the introduction of superior mass-produced antibiotics during the later war years, Mudd maintained his interest in phage. In the summer of 1945 he took Max Delbrück's famous course at Cold Spring Harbor.[7] Mudd tried to keep his hand in electron microscopy in connection with both the bacteriophage and ultraviolet projects, arranging to observe bacteria with an instrument at the Navy Medical Research Institute in Bethesda. Very little electron microscopy seems actually to have been accomplished in Mudd's war projects, however, and they ended abruptly in 1945 when his microscopist collaborator,

Lieutenant Frances Young, married and de-enlisted.[8] In the later war years, Mudd had also been able to arrange occasional sessions with Hillier and the microscopes at the RCA Princeton facility, such as when Mudd's student William Smith made some micrographs of *Bacteroides* there in the summer of 1944.[9]

Medical Microbiology at the University of Pennsylvania

During and immediately after the war, Mudd and Harry Morton had manned the Penn bacteriology department with instructors on short-term contracts. With the postwar surge in demand for science degrees, though, these junior staff left rapidly for tenure-track faculty posts, leaving Mudd feeling "bereft" and understaffed.[10] He managed to retain the industrious biochemist Manessah Sevag, who had successfully maintained himself through outside grants for many years, by finally finding him a tenured position. Mudd energetically courted Carl Robinow, a young British biologist who had pioneered staining methods that made bacterial nuclei appear more convincingly for light microscopy, and who had also gained experience in electron microscopy working with microscope physicist V. E. Coslett at Cambridge University. Hoping to continue a collaboration that would reinforce his own research into microanatomy of bacteria and bring electron microscopical expertise back under his command, Mudd offered Robinow a job in his department in 1948, but Robinow took one at Yale instead.[11] By 1950–51, the year the name of Mudd's department changed officially from bacteriology to microbiology, there were eleven instructors and postdoctoral researchers, while the permanent faculty consisted of Mudd, Morton, Sevag (whose work dealt largely with the biochemistry of antibiotic resistance), John Flick (an immunologist who worked on allergy), and Edward DeLamater (who worked on staining techniques for the bacterial nucleus), who held a split appointment in the Department of Dermatology.[12] In 1951–52, Joseph Gots (who had been working as a postdoc on bacterial metabolism in antibiotic action and bacteriophage infection) was added to the permanent staff.[13] Thus the departmental roster stood at half a dozen scientists of assistant professor rank or higher, and there it remained for the rest of the 1950s. Morton's research drifted away from morphological issues into classified, Air Force–funded research known as the Big Ben project (presumably concerned with germ warfare and Morton's specialty, the strange pleuropneumonia-like microbes).[14] That left only DeLamater and Mudd himself as researchers in bacterial cytology and physiology, or indeed in any sort of work that might conceivably at the time have been described as "biophysics" or "molecular biology" (see Introduction), even by stretching those already flexible labels. Despite his comparative dearth of resources, Mudd nonetheless managed to maintain an active program in bacterial microstructure in his department: in 1950 he counted ten graduate students working in this area, between himself and DeLamater.[15]

Mudd faced considerable resistance in keeping his department competitive in the postwar scientific world. From 1945 to the eve of his retirement as department chairman in 1959, he continually complained to the Penn administration of overcrowding, underfunding, and being hard pressed to maintain a serious scientific program when no real institutional support for research or graduate training was forthcoming.[16] Though the biophysics-dominated department run by Francis Schmitt at MIT was less than twice as big in terms of faculty, the scale of research seems almost an order of magnitude greater (see Chap. 4). For instance, whereas Mudd was feeling flush with a departmental budget of $69,000 for 1957–58,[17] Schmitt's department already had exceeded four times that level in 1949–50 and probably approached ten times Mudd's budget for 1957–58.[18] And whereas in 1957 Mudd was bitterly defending the 8,000-square-foot space allotment that would alleviate his department's overcrowding when it relocated to a new research building still under construction,[19] Schmitt had in 1952 moved his department into three times that space, entirely dedicated to research, in a posh new building.

Mudd thus lacked support from a medical school that was not highly committed to graduate training and research, in a university that was not prepared to take full advantage of the new postwar government support for science. Nonetheless he did what he could to bring outside grants into his department. Sevag, whose biochemical work on drugs to treat bacteria and viruses was supported by the Macy Foundation from 1941 to 1947, was able to switch to NIH funding from 1947–48 onwards, and secured ONR funding from 1952–53 onwards to expand his metabolic research with isotope techniques.[20] Gots's work with bacteriophage was at first supported by the National Foundation for Infantile Paralysis, but he was able to switch to NIH funding in 1955,[21] just when the NFIP was cutting its support for basic virology research because of the new Salk vaccine (see Chap. 5). Morton handled his military funding for Big Ben on his own, and DeLamater brought his own grants supporting his light microscopic cytology.

Encouraged in 1948 by an admiral friend to apply for some of the millions earmarked for biophysics in the Atomic Energy Commission (AEC) budget, Mudd looked into matters and in 1950 decided that he and DeLamater might be able to get their studies on bacterial nuclei funded by the agency.[22] Mudd approached his acquaintance (and family friend, apparently), AEC's biology and medicine director, Shields Warren, reassuring Warren that though he "did make Senator Joe McCarthy's smear list" as a promoter of U.S.-Soviet cultural and scientific exchange, he was no communist.[23] With a bit of coaching from Warren's staff, Mudd managed to represent his bacterial cytology research as radiation biophysics ("The Internal Organization of Normal and Phage-Infected Bacterial Cells, with Especial Reference to Activation of Latent Phage Infection by UV Radiation"), and the AEC funded it starting in March 1953, which allowed Mudd to buy his department an ultracentrifuge.[24] Despite grumblings that Mudd's research was straying far from the AEC mandate in radiobiology, which Mudd countered with cosmetic name changes (e.g., "Internal Organization of Normal and Phage Infected Cells as Influenced by Radiation") to dispel that impression, the AEC continued funding Mudd into 1959, the year he went into semiretirement.[25] But even the largest grants in Mudd's department barely exceeded $10,000 annually, a tiny amount compared to what the departments with the best facilities and equipment, such as Schmitt's, were bringing in by the mid-1950s. And although some of Mudd's students were able to start successful careers in bacteriophage research—the centerpiece of the budding field of molecular genetics—his smallish group did not fare very well in a field so fast-paced and competitive that one always had to be looking over one's shoulder, as we shall see.[26]

Mudd's difficulties in obtaining resources for research in the Penn Medical School definitely had an impact on the way his research program in electron microscopy developed. A clash between Mudd's own forceful personality and that of the quiet Anderson evidently having alienated the latter (who, as noted, had found it necessary as the RCA-NRC fellow to protect his time from Mudd's demands for personal favors), Mudd was unable to induce Anderson to make any pictures for him, or to give him and his students access to the Johnson Foundation model B, which was Penn's only electron microscope for ten years—despite appeals to Anderson's superiors. Thus, with the end of his war projects, Anderson's "uncooperativeness" left Mudd without regular access to any microscope.[27] In the summer of 1946, Mudd and Cornell bacteriologist Georges Knaysi spent some time with Hillier at the RCA Princeton facility, largely for renewed testing of the utility of higher accelerating voltages for bacteriology.[28] Through 1947 and 1948, Mudd continued to finagle occasional sessions at RCA for himself and students, Hillier supplying not only the microscope but most of the operating expertise. A break seemed to open for Mudd in November 1948, when he and Zworykin privately and informally discussed the possibility of establishing an electron microscopy center at Penn, with the purpose of exploring and demonstrating the new instrument's utility in medicine. Zworykin was considering what Mudd described as "a principal effort . . . to bring the possibilities of the electron microscope to the attention of physicians,"[29] and Mudd hoped to profit by helping Zworykin, just as he had by chairing the RCA-NRC committee in 1940. Mudd drew up a joint proposal with Penn's anatomy department in collaboration with RCA, officials at the NIH were sounded out informally as possible patrons, and the plan was officially started through channels of the Penn Medical School administration in December 1948.[30] The proposal itself and its exact fate have so far proved elusive, but it is most likely that the Penn administration balked at the operating expenses that would be entailed in case adequate grants were not forthcoming in future years.[31]

By June 1949, Mudd had arranged to use the electron microscope at Philadelphia's Franklin Institute, a laboratory lacking biological facilities, on a fee-for-service basis, and he continued to use it as necessary for the next several years despite concerns about the technical competence of the Franklin staff.[32] But this was unsatisfactory, largely because Mudd lacked high technical competence himself, rather than because of the expense. "The electron pictures that I get are principally the results of carefully planned campaigns to intrigue the interest of Dr. Hillier" at RCA, Mudd was still complaining in 1952.[33] After 1954 he was able to borrow occasional time on the second electron microscope acquired by Penn, a new Philips 100 in the pathology department.[34] Mudd was finally able to purchase a microscope for his own department at the end of 1957, a Philips 75-B, chosen because it was deemed "simple to operate and to keep in good condition."[35] Mudd never had gained the proficiency of the first generation of electron microscopists, mostly younger, that came from personally mastering the RCA model B. The conservatism of his style of electron microscopy throughout the period covered here may owe as much to his spotty instrument access and lack of deep familiarity with microscope operation as it does to intellectual considerations binding Mudd to older ways of picturing microbes.

Mudd's Research Programs in Bacterial Morphology and Physiology

Through the second half of the 1940s, while Mudd and his students were deprived of all but sporadic access to an instrument, the field of preparation techniques for biological electron microscopy had been briskly moving. In 1944, Robley Williams and Ralph Wyckoff at the University of Michigan first showed how biological material could be coated with a thin film of metal so as to render surface detail in sharp relief, and the technique quickly became popular, particularly for the study of viruses and other fine, dispersed specimens (see Chap. 5).[36] Albert Claude and Keith Porter at the Rockefeller Institute had shown remarkable detail from the interior of animal cells grown in tissue culture on microscope grids, then fixed and contrasted by osmium vapor treatment.[37] Claude's group and a number of others were experimenting with novel microtome designs and specimen embedding techniques in an effort to cut sections (i.e., slices) of tissue thin enough for penetration by the electron beam (see Chap. 3).[38] The thin-sectioning methods would soon be extended to bacteria, right on Mudd's doorstep at RCA. And surface replica methods developed for metallurgy were being applied to biological specimens too (see Chap. 6).[39] The pictures from this burgeoning variety of techniques were quite different from the kind with which Mudd had begun, those simple shadow images of air-dried microbes that Mudd liked to compare with medical X-rays of the specimens.[40] Mudd stayed with the original image type, however, modifying cytological and cytochemical staining methods in an effort to locate biochemical activities in the shadowy interiors of the whole cells he was viewing, mimicking and expanding on what was possible with contemporary light microscope technique.

Mudd's most striking wartime work, done in collaboration with Knaysi and using an experimental high-voltage microscope (up to 300 kv, but in practice 150–200 kv) that Hillier developed at RCA, was the finding of small bodies in the bacterial protoplasm (Plate 2.2) that the investigators interpreted as primitive nuclei.[41] Though a controversial subject, the bacterial nucleus had already won a number of believers among bacteriologists, including Knaysi himself, on the basis of light microscopical cytology,[42] and Mudd was an easy convert. In two large, high-profile articles published in the *Journal of the American Medical Association* (JAMA) in 1944, Mudd summed up the state of microbiological electron microscopy—almost all of it Anderson's work that Mudd had overseen through the RCA-NRC committee. There Mudd argued that the evidence from many species showed that bacteria are complex creatures not terribly unlike the cells of higher organisms, with differentiated cell envelopes, simple "nucleoprotein" nuclei within the protoplasm and also more numerous smaller granules of unknown function, and fine hairs or flagella on motile forms. Although this view of high bacterial complexity was controversial, it was not radical, and the parallel theme in the article was rather orthodox: that bacteria, Rickettsial microbes, and viruses stand in a continuum, both morphological and biochemical, of decreasing size and complexity.[43] This compilation of the latest microbiological electron microscopy for the medical audience received terrific attention: 20,000 offprints were ordered and distributed with the help of the Macy Foundation (with whom Mudd had connections as recipient of a modest grant until a few years after the war, on chemotherapy for viral infections).[44] The JAMA pieces established Mudd as an authority for many years to come.

Mudd's high-voltage finding with Knaysi of what appeared to be bacterial nuclei was followed up by Knaysi in the summer of 1946, working in Mudd's lab in collaboration with Richard Baker of RCA (a postdoctoral fellow there under Hillier from 1945 to 1947). Knaysi and Baker, growing large *Bacillus* cells in a special low-nitrogen starvation medium that strikingly decreased the electron density of the protoplasm, were now able to see two distinct types of dense bodies, large and small (Plate 2.3). The numerous small bodies appeared to be associated with the cell envelope and have a beaded or stringy composition, but their function remained mysterious; the larger bodies, only one or two per cell, were interpreted as nuclei because they appeared in electron micrographs to behave during cell division much like nuclei of higher cells, and because bodies of comparable size and number appeared by light microscopy in bacteria grown in the same starvation medium and then treated with standard nuclear stains.[45] That same summer, Mudd, with student Andrew Smith and Hillier, experimented with new preparative methods and microscope hardware in an effort to resolve what seemed to be a fine mesh or trabecular lattice structure in the bacterial protoplasm (Plate 2.4).[46] In July 1947, Mudd gave a presentation at an electron microscopy meeting in Stockholm in which he elaborated on the new electron microscopic evidence supporting the claims he had made in his 1944 JAMA papers: that bacteria have some sort of nucleus and smaller granular bodies of unknown function associated with the cell envelope, and also, in some species, elaborate specialized structures such as spores and flagella. He supposed that further advances in electron microscopy might reveal still greater complexities in these organs, as well as entirely new fine structure such as his protoplasmic latticework.[47]

In the late 1940s and the 1950s, the period immediately before the institution of the prokaryote/eukaryote distinction that now sharply divides bacteria from the cells of higher organisms, many microbiologists were advocating the idea that bacteria were more complex than traditionally supposed, and shared all major biochemical mechanisms with higher organisms.[48] This implied that bacteria were suitable experimental organisms for work that could be extrapolated to higher organisms. In this context, Mudd had become another champion of the sophistication of bacteria and their organs, which he believed to be as complex as, and in many case homologous to, those found in the cells of higher organisms. He simultaneously positioned himself as the leading proponent of the electron microscope as the preferred instrument for revealing this sophistication; indeed, Mudd's Swedish talk was soon published in *Nature*, strengthening his image as one of American electron microscopy's official spokesmen. Before Mudd's 1947 summer in Europe was over, he became acquainted with certain new developments in electron microscopy beyond the RCA-dominated North American scene. Mudd had learned about recent experiments by Carl Robinow (who, as mentioned, had been doing electron microscopy with V. E. Coslett in Cambridge) that showed impressively well defined nuclear bodies in bacteria stained for the light microscope and in electron micrographs as well, where they appeared as lighter, less dense areas in the bacterial protoplasm.[49] Mudd invited both Robinow and Woutera van Iterson, a young Dutch bacteriologist who had been working with both A. J. Kluyver and electron microscopist J. B. le Poole in Delft, to visit his lab the next summer and to work with Hillier and Knaysi, whom he was hosting again, at RCA.[50] He also began to grow alarmed at the influence of the argument recently advanced by the respected South African bacteriologist Adrianus Pijper, to the effect that flagella observed on motile bacteria were not the whiplike organs of propulsion they had traditionally been taken to be, but excreted slime spun into threadlike form as a by-product of mobility.[51] As

James Strick has shown, over the next several years Mudd went to extraordinary lengths in his energetic campaign to combat Pijper's theory, not so much from a belief in a particular rival theory of flagellar structure or function, but from a commitment to the epistemological priority of electron microscopy over the light microscopy with living specimens on which Pijper built his case.[52] That it was Pijper's epistemology that Mudd saw as problematic is supported by Mudd's willingness to entertain the unorthodox theory that flagella are organs of nutrient absorption rather than motility, when suggested by Hillier on the basis of electron microscopy,[53] though in this case Mudd's commitment to the notion that flagella were complex organs of some sort was not challenged in the same way as it was by Pijper's excretion theory. Mudd's epistemological commitments will be further treated below.

Mudd and the Bacterial Nucleus

Mudd soon became embroiled in a controversy on the reality and the structure of the bacterial nucleus in which, as in the flagella story to some extent, the complexity of bacteria and their status with respect to higher organisms was at stake. In 1948, in writing up the work Mudd and Andrew Smith had begun with Hillier in the summer of 1946, the authors added a good deal of discussion about the nucleus issue to their discussion of the latticework of the protoplasm, and included some light micrographs of bacteria stained for nuclei by Robinow during his 1948 RCA visits.[54] Their electron micrographs had been made from bacteria prepared, like those in Knaysi and Baker's demonstration of nuclei from the summer of 1946, by the new RCA method of growing bacteria on a film of collodion overlaying nutrient agar, followed by simply lifting the film onto bare wire mesh from which grids are made.[55] Thus the bacteria were air-dried with less disruption from surface-tension effects than if they had been suspended in distilled water, and in this case no fixative treatment was used; both aspects of this preparation protocol meant less manipulation and thus less likelihood of artifact. There were one or a few distinct light zones in almost every bacterial cell (see Plate 2.4). The authors argued at length that these should be interpreted as the same putative nuclei appearing as the dark, electron-dense bodies in the Knaysi and Baker electron micrographs of starved cells, and seen as the distinctively colored spots in Robinow's light micrographs of stained bacteria and in the various stained preparations made by other proponents of bacterial nuclei. Thus, Mudd and company contended, the bacterial protoplasm must be denser than the nuclear region, and must contain ribonucleic acid (RNA), which takes nuclear stain and hides chromosome-containing nuclei; staining procedures like Robinow's must work by selectively removing RNA from the cell and thus showing chromatin better; and starvation may deplete RNA and other dense material from the protoplasm so that the nuclei in such cells appear as more rather than less electron-dense than their milieu. Thus electron microscopy, certain findings from ultraviolet microscopy, and the balance of evidence from a variety of work in light microscopic cytology all agreed in indicating that the nuclei of bacteria are real. This interpretation won general acceptance before long. Mudd, Hillier, and Smith did not yet take a stand on the controversial issue of the particular form in which bacterial nuclei existed—whether they were vesicles bounded by membranes or otherwise.[56]

Mudd and his graduate students took up the nucleus issue with enthusiasm in 1949. Mudd had Robinow come to Penn that March to give a seminar and to teach his group new light microscopy staining techniques.[57] Andrew Smith modified the similar techniques that Robinow and DeLamater used to stain bacterial nuclei: his simplified combination of fixing with osmium tetroxide vapor, soaking in hot hydrochloric acid, and treating with formaldehyde as a mordant prior to staining with basic fuchsin—all standard elements of the traditional cytologist's technical repertoire—showed what looked by light microscopy like distinct nuclear bodies in several species of microbe.[58] By carrying out the staining on cells grown on thin film that could be transferred to grids, Mudd and Smith were able to follow what was occurring at each step in this staining procedure by electron microscopy (now using the Franklin instrument). The idea was to show light and electron microscopic cytology are "most rewarding" when applied in a complementary way, and specifically to use electron microscopy as a check on staining procedures for light microscopy.

Mudd and Smith determined that osmium fixation increased the contrast between dense protoplasm and lighter nuclear zones that was barely visible in unfixed cells, but did not alter the shape or size of nuclei; that acid treatment, which reversed the contrast situation by making the nuclei appear as dark zones against a lighter background, did so by condensing the nuclear bodies and by stripping dense material from the cytoplasm (as was already believed); and that formaldehyde treatment and fuchsin staining do not alter the postacid density picture very much (Plate 2.5). That the light zones in unfixed and osmium-fixed bacteria were the same as the dark zones in acid-treated bacteria, and that both were nuclei, was supported by the staining of the latter with fuchsin and other stains for nucleic acids; the only other support was that there were usually one or two of these light, dense, or fuchsin-colored spots (depending on staining and imaging procedure) per cell, which is what one would expect in dividing cells, given the appearances of nuclei in cells of higher organisms. All of this amounted to a validation by electron microscopy of certain light microscopic staining methods for bacterial nuclei, and a confirmation that the dark bodies, light zones, and colored spots in various preparations of bacteria were the same entities—bacterial nuclei (indeed if they were not all the same, real entity, this would raise the question of which staining methods introduced artifactual appearances). Additionally, Mudd and Smith argued that their nuclei were vesicles, on the grounds that the distinct margins and low density to electrons implied a sac limiting diffusion of the denser cytoplasm into the nuclear region.[59] As significant as the specific results, though, was the logic embodied in the experiment: Mudd was putting the electron microscopist in a position to arbitrate disputes about cytological methods for light microscopy. This, and the overall message vindicating DeLamater and Robinow, brought Mudd into conflict with some bacteriologists subscribing to different cytological methods for light microscopy.

Mudd was invited to head a section at the Fifth International Congress of Microbiology, held in Rio de Janeiro in August 1950. One of two papers he presented there, representing the latest results of his coauthor DeLamater, claimed to show that the whole elaborate machinery and activity of cell division in higher organisms was observed in a large *Bacillus* species. DeLamater had used a well-known light microscopic nucleic acid stain for nuclei (though the true degree of its specificity was open to dispute, as with all other stains), after the not unusual fixation in osmium vapors and the increasingly common acid treatment to reduce obscuring RNA in the cytoplasm. The most novel step was subsequent freezing in superchilled alcohol or similar solvent to remove water prior to mounting. The summary of Mudd and DeLamater's presentation reads:

[By these light microscopy] techniques it has been possible to define the nucleus as a vesicle within which the chromosomes are clearly delineated. The vegetative nuclear activities have been found to consist of the following: (1) The nuclei undergo a true mitosis; (2) A centriole is produced at the beginning of prophase which divides to form the two polar bodies; (3) The chromosomes, which are minute beaded threads in the interphase nucleus, contract, condense, or coil up to form dense granules on the metaphase plate; (4) It has been demonstrated that there are three to four discrete chromosomes which divide at metaphase and proceed to the poles; (5) Spindles have been demonstrated and photographed as areas of increased density, lying between the chromosomes and the polar bodies; (6) The divisional stages appear to consist of interphase, prophase, metaphase, anaphase and telephase [sic]. . . . In summary, it is felt that the first clear-cut evidence for mitosis in bacteria is herewith presented.[60]

Mudd and DeLamater knew the talk and the pictures accompanying it (see Plate 2.6) might hit the congress like a bombshell, and by keeping DeLamater's priority claims modest, took pains not to antagonize any more than necessary some of the eminent attendees, such as Emmy Kleinberger-Nobel, who had been staining bacterial nuclei for years without finding mitotic figures.[61] Mudd's other paper, giving the latest electron microscopic evidence that bacterial nuclei are vesicles, may even have come as a bit of an anticlimax. The assertion that bacterial nuclei exist and divide was not the biggest shock—several researchers claimed to have already deciphered the division behavior of bacterial nuclei—but rather the notion that DeLamater's preparative techniques were so superior that his light micrographs actually represented the individual chromosomes and mitotic figures.

For the next year following Rio, DeLamater did manage quite effectively to capture the attention of the biological world, through publications showing the mitotic figures from a number of bacterial species, and through talks including a presentation of his results at the annual American Association for the Advancement of Science (AAAS) meeting in April 1951.[62] In August 1951, he and Mudd gave a paper together at a symposium on the cell nucleus at the Brookhaven National Laboratory, where the two apparently found their distinguished audience fairly receptive.[63] The main event of the summer, however, seems to have been DeLamater's June confrontation at Cold Spring Harbor with bacterial cytologist Kenneth Bisset, whose view of bacterial cell division conflicted with DeLamater's.[64] Bisset spoke first, presenting a summary of the way the nuclei of various bacteria appeared in his own stained preparations, and interpreting them all in terms of a few major variations on an alternating-generation life cycle that is common in fungi. Mudd and one of his recent Penn graduates gave Bisset a hard time in the question period, suggesting that Bisset's preparations were not optimal, and that his life cycles were only "intellectual constructions" produced by arranging snapshots of dead bacteria into arbitrary sequences.[65] (DeLamater was a little less vulnerable to this charge, as he was at the time claiming to see his mitotic sequences in individual living bacteria, by phase microscopy, corroborating what he saw in his unique stained preparations.) Bisset later described his experience in this Cold Spring Harbor session as a "well-concealed ambush" that demanded revenge.[66]

DeLamater followed Bisset at Cold Spring Harbor with pictures said to be of mitosis in four bacterial taxa, and of conjugation in *Bacillus* (which seemed to mesh with the exciting results in bacterial genetics presented by Joshua Lederberg at the same meeting). Bisset countered with his own criticisms of DeLamater's staining techniques, and especially attacked the pictures supposedly representing mitotic spindles, suggesting that DeLamater had mistaken nuclei for centrosomes and cell plates for chromosomes because he had incorrectly assumed his bacteria to be unicellular. Bisset was a great proponent of the multicellularity of bacteria, a notion that fit with his homologies of bacterial life cycles to those of fungi.[67] Mudd tried to assist DeLamater, but there was not much Mudd's electron microscopy could offer the debate, which turned on largely undecidable questions about what various light microscopic stains did in the cell; moreover, DeLamater himself had cast aspersions on the usefulness of electron microscopy to study bacteria except with peripheral structures, because cell thickness made contrast in the interior problematic. Bisset had conceded that bacteria may have vesicular nuclei, avoiding conflict with Mudd's evidence. Bisset even conceded that "the mechanism of chromosome movement probably accords quite closely with what is found in other types of cell." The point was just that DeLamater's light microscopic observations were a travesty—"Dr. DeLamater's figures are not spindles and . . . his 'chromosomes' are not even composed of nuclear material."[68] In February 1952, *Nature* carried the rancorous exchange for all the scientific world to behold.[69] In the Sixth International Congress of Microbiology, in Rome during August 1953, Bisset again took it upon himself to "slug it out" with DeLamater, and thoroughly antagonized Mudd, who had been hoping that the Cold Spring Harbor squabble would not be repeated for a European audience.[70] The debate about staining dragged on, and as Mudd observed, did nothing to counter the tendency among those "in more objective and quantitative areas of science . . . to regard cytologists as a lot of subjective primadonnas who cannot agree among themselves about anything." For the moment, though, the bare existence of the bacterial nucleus seemed firmly established.[71] When the dispute over bacterial chromosome behavior finally quieted years later, showing DeLamater's claims to have been mistaken, it was largely through genetic rather than cytological evidence.

Although he kept defending DeLamater both in public and in a lengthy and increasingly hostile correspondence with Bisset because, as Mudd put it, "I believe to the depths of my conviction that he is correct scientifically," and although he had worked hard to keep DeLamater at Penn (for instance, by paying part of DeLamater's salary out of his own departmental budget and by convincing the medical school dean that DeLamater's demonstration of mitosis in bacteria made him too valuable to lose),[72] Mudd's relations with DeLamater had cooled drastically by the end of 1952. DeLamater seems to have begun to feel that Mudd was receiving, whether by failing to attribute work to DeLamater when speaking for the two of them or just passively by the Matthew Effect, too much of the credit in their collaborative projects.[73] According to Mudd in a February 1953 letter, DeLamater had a "tremendous complex against authority and a powerful ego drive." On the other hand, DeLamater would not have been the first to object that Mudd had a tendency to take liberties with the experimental data of others,[74] though there is no reason to doubt Mudd's word that credit did not concern him and his motive was only dissemination of what he took to be the truth (his same attitude in the campaign against Pijper). By May 1954, DeLamater was going so far as to misrepresent his departmental affiliation in order to stress his independence from Mudd.[75] Mudd had already pulled back from active research on the bacterial nucleus, conceding that turf to DeLamater's territoriality, and concerned himself more with defending his own work, as described below.

Mudd had used his second paper at the 1950 Rio conference as a platform to indicate the general conclusions he wanted drawn from his own and other recent findings. It begins:

Energy-yielding and biosynthetic mechanisms, biochemical requirements and deficiencies, intracellular organization, means of genetic constancy and of genetic change which are characteristic of the cells of higher plants and animals are in process of discovery and investigation in bacteria. Similarities are more impressive than differences in these fundamental manifestations of the organization and functioning of protoplasm in bacteria and the cells of higher beings. Activities and mechanisms of living cells in many cases are manifested in bacteria under circumstances favorable for study and elucidation. Understanding of these general facts can not only vitalize bacteriology with clearer insights, but also can lead to more fruitful utilization of bacteria for the enrichment of our understanding of life in its more complex manifestations.[76]

Mudd went on to say that the hypothesis that bacteria were fungi that had devolved into simpler forms was being substantiated by the latest findings[77] (unlike Bisset, however, for Mudd bacterial complexity did not imply bacterial multicellularity). His by no means lonely vision of the continuity of bacteria with higher forms, and the corollary importance of bacteriology for agriculture, medicine, and other life sciences,[78] informed Mudd's keen pursuit throughout the 1950s of another of his electron microscopic discoveries, one he could truly call his own: the bacterial mitochondrion. For Mudd, the continuity of bacteria with higher forms entailed that bacteria should have microanatomy similar to that of higher cells.

Mitochondria in Bacteria

As will be discussed in greater detail in Chapter 3, the mitochondrion is a granular body found in cells of higher organisms, in which are located—it came generally to be believed by 1950—most of the enzymes associated with oxidative phosphorylation and other energy transactions of cellular respiration. Classical cytology identified mitochondria by certain staining reactions, notably the traditional Janus green, and also by more recently developed indicators like tetrazol dyes and the Nadi reagent, although it took the isolation of mitochondria by means of the ultracentrifuge in the interwar period to eliminate all doubt about the entity's real existence. Before Mudd, bacteria were not thought to have mitochondria. The story of the bacterial mitochondrion and the controversy surrounding it—although it did involve concepts about the complexity and status of bacterial cells—hinged less on theory than on proper electron microscopic epistemology. Thus, as in the story of the mitochondria of higher cells (Chap. 3) and the story of Pijper and the flagellum, the ways in which claims about the nature of bacteria should be settled was at least as significant an issue as the content of the claims themselves.

Mudd and his group, in the course of their work on the nucleus, seem to have come in the first part of 1951 to the conclusion that bacteria do have mitochondria. They noted the presence of dark granules in electron micrographs of several different bacteria, distinct from nuclei in that these are lighter than surrounding cytoplasm and larger than the granules (Plate 2.7). They then decided to try some standard mitochondrial stains, presumably reasoning that if bacteria have nuclei and other complex features of higher organisms, then they might have mitochondria too. In species of *Mycobacterium*, including tuberculosis bacilli, Mudd's team found that their mitochondrial stains revealed granules by light microscopy, especially at the ends of the rod-shaped cells, which was the prevalent position of the dark granules in the electron micrographs. Thus they concluded that the specifically stained bodies in the light micrographs were mitochondria, and were the same as the dense granules seen in electron micrographs at cell poles.[79] As with bacterial nuclei, there was a major problem deciding what, if any, features seen in electron micrographs corresponded to things seen with the light microscope; here distribution was the prime indicator that the electron-dense bodies were the same as the granules taking mitochondrial stains in light preparations. They repeated their light microscopy procedures in other species of bacteria, with similar results, showing the staining bodies to be ubiquitous, as they should be if they really were mitochondria. Furthermore, the dense bodies cum mitochondria appeared to have a bounding membrane, as they should if they were like mitochondria of higher organisms, judging by the ring left over when the bodies themselves had been boiled away by excess electron bombardment in the microscope.[80] Mudd and DeLamater publicized these new experiments at the aforementioned Brookhaven nucleus meeting in August 1951 (ostensibly because the positive identification of one granule type distinct from nuclei implied that DeLamater could not be confusing another granule type in his putative nuclear preparations), and at the EMSA meeting in November 1951, held in Philadelphia.[81]

In 1952–53, as Mudd extended his mitochondrion experiments, he and his mitochondria got caught in one of Bisset's published attacks against DeLamater's nuclei. Bisset reiterated his charge that most of what DeLamater interpreted as nuclear material was part of the cell envelope, and added that what Mudd thought were mitochondria were clumps of protoplasm shrunken away from the cell wall. He addressed primarily the staining procedures, and the electron microscopy weakly if at all; indeed, Bisset's only argument seems to be that he saw different structures in his unfixed and wet light microscopy preparations than those seen by his opponents, who all used dried or fixed cells, and drying and fixation are known to cause shrinkage. Hence, Bisset claimed, his observations were less likely to contain artifacts, simply because he did not use methods that cause shrinkage.[82] (To the obvious counterargument that with a light microscope one can distinguish objects better when they are fixed and stained, Bisset's best defense would have to be that whatever he could distinguish—even if poorly—in his minimally prepared bacteria must be real.) The implication that any method involving dehydration crucially distorts specimens totally undercut the validity of electron microscopy for biology, because biological material is always wet and has to be dehydrated before being placed in an electron microscope. Mudd reacted violently, much as he was still doing against Pijper's similar threat to the electron microscope's preeminence. Previously, Mudd had been cajoling the younger Bisset, in a constructive if patronizing tone, to abandon his mad "course . . . to the point of no return" whither Pijper had gone with his slime-twirl interpretation of flagella "which few people even bother to refute any longer."[83] But now that Bisset was challenging his mitochondria and the dried electron microscopy preparations on which they were based, it was "Monsieur, en garde!"[84]

Irritated from being accused of "egregious confusion" (in attacks experienced as "vicious," in the words of DeLamater to another bacteriologist), Mudd even refused to send Bisset a strain of large *Micrococcus* that he and DeLamater had been using to show stained organelles more clearly by light microscopy, until such time as Bisset joined the ranks of "objective scientists who are more concerned with what is right than who is right."[85] The fear was that Bisset would deliberately produce and publish unfavorable preparations of Mudd's organism, supposedly according to the DeLamater and Mudd methods. Mudd's published riposte to Bisset was comparatively calm and measured, and based entirely on electron micrographs of granules that he thought could speak for themselves "concerning their reality and the validity of Bisset's assertions." The main argument was that because one could see the cell envelope clearly, and could discern where small amounts of shrinkage had occurred, massive shrinkage and misidentified cell envelope could not account for the granules in the electron micrographs. However, to the question of the granules in his light microscope preparations, and of their identity with those in the electron pictures, Mudd added nothing.[86] In correspondence with Bisset, though, on this point Mudd insisted that because his light microscopic stains could pick out the cell envelope, there could be no confusing it with his granules. It emerged that no necessary conflict was present here: Mudd was content that his mitochondrial granules were real, and was prepared to concede that they were concentrated at growing points in the cell envelope, whereas Bisset was content that the staining granules were found at new septa (i.e., growing points) as he had insisted, and was prepared to concede that they might be mitochondria or equivalents.[87] The heat of this controversy in the absence of essential disagreement on substance points to the centrality of methodological commitments, although egos and Bisset's greater theoretical investment in bacterial multicellularity were both involved as well.

Mitochondria were very much on the collective minds of biologists in 1952–53, as electron micrographs of their internal fine structure were beginning to come through new methods for ultrathin sectioning, and a fierce controversy over that structure erupted between two of the leading electron microscopy groups (see Chap. 3). Thanks largely to Mudd himself, then serving as president of the Histochemical Society,[88] that organization sponsored a symposium on mitochondria, featuring many of the field's leaders, in Chicago in April 1953. Mudd described the symposium as "very hot," and was excited to be giving the paper on bacterial mitochondria.[89] In Chicago, George Palade, who had been working with Keith Porter at the Rockefeller Institute, showed electron micrographs of mitochondria from animal cells, embedded in plastic and sectioned an order of magnitude thinner than had been possible only a couple of years earlier (see Chap. 3). Mudd considered them spectacular and they seem to have made a large splash at the meeting.[90] Having failed to convince Hillier to make him thin sections of *Mycobacterium*, which he hoped would show granules with the same distinctive internal structure that Palade and others were seeing in animal tissue,[91] Mudd could only rework his existing evidence: techniques of light microscopic cytology showed granules in bacteria that stained like mitochondria, and the positions of these stained granules (hard to make out because they were at the limit of resolution of light microscopes) supported their identification with the dense granules seen in electron micrographs of whole bacteria.[92] This was still the fundamental weak point in Mudd's case for bacterial mitochondria: he could not map the granules in his stained preparations—near or past the limit of resolution of the light microscope as the granules were— onto the dense granules in his electron micrographs with any degree of certainty.

Even while Mudd was doing battle against cytologists who, like Bisset, downplayed the value of electron microscopy, some of the electron microscopists closest to Mudd were beginning to part ways with him and his rather traditional cytological style of research. Undercutting the value of Mudd's pictures of air-dried bacteria, though not actually the whole enterprise of studying bacteria fixed and dehydrated and mounted whole on grids, was Mudd's erstwhile helper Anderson. Anderson was for a decade a lonely skeptic about the appearance of bacteriophage that he had first brought to light during his RCA fellowship, with Luria and Delbrück (Chap. 1). When longtime virus researcher Jacob Bronfenbrenner first saw that the bacteriophages he had been studying looked like a ball on a string (Plate 1.12), he reportedly slapped his forehead and exclaimed, "Mein Gött, they've got tails!" In writing up these electron microscopical findings, Delbrück had wanted to endorse the notion that bacteriophage is to host bacterium as sperm is to egg—that is, that phage are a part of the bacterial life cycle instead of parasitic invaders—even though this theory was by that time quite unpopular among the crowd with which he moved, occasioning friction over interpretation with Anderson.[93] That bacteriophage swim with their tails, as sperm can be seen to do with a light microscope, overnight became a widely known "fact." One 1948 textbook showing an RCA-circulated picture of phage with bacteria (Plate 2.8) carried the caption: "This photograph shows the phage particles swimming toward the bacteria."[94] In 1950 Anderson finally showed, by his new method of drying specimens at pressure and temperature above the suspending medium's critical point so as to bypass surface-tension effects, that the observed orientation toward bacteria was a drying artifact, presumably caused by the pull of receding water in a drying droplet containing the phage. Moreover, bacteriophage actually attach to host cells tail- rather than headfirst (Plate 2.9).[95] Here Anderson was able to draw on formal physical theory to justify his concern about surface tension, and to point him in the direction of a new preparation method to resolve the issue, although inferences based on formal theory could not decide the state of microscopic affairs a priori. In any case, Anderson's finding implied that drying artifacts needed to be taken into account more seriously in interpreting at least the finer details of micrographs from air-dried specimens like Mudd's.

More directly relevant to Mudd's concerns in bacterial cytology, at RCA Hillier had been supervising George Chapman, a Princeton graduate student, in a project to build an ultramicrotome (i.e., cutting machine) and to use the new plastic embedding material, methacrylate, in order to make thin sections, a fraction of the thickness of a cell (see Chap. 3), of bacteria for electron microscopy. Chapman began in the second half of 1950, and was largely occupied through 1952 with making the ultramicrotome cut reliably in the desired thickness range of 0.1 micron, more than an order of magnitude thinner than achieved routinely in traditional cytology. By the beginning of 1953, finally, he and Hillier had the micro-tome working and, through some advice from Porter, had hit upon a way of preserving the bacteria that prevented their explosion during the poly-merization of the embedment.[96] This was fixation in osmium tetroxide, at the high concentration of 2 percent, buffered for osmolarity with sodium chloride and for pH as well, for the incredibly long period of 16–20 hours. The electron micrographs of thin-sectioned bacteria Chapman and Hillier made were stunning for their excellent contrast and the level of detail they revealed in the bacterial interior (Plate 2.10).[97] On the question of the nucleus, these pictures showed no membrane boundary (or "vesicle") and no classical mitotic apparatus, as DeLamater's work would have suggested, but only an indistinct mass of material with a fi-brous texture that occupied roughly the same position as the low-density zones in earlier electron micrographs of whole bacteria. No plasma mem-brane between the cell wall and cytoplasm was visible in the pictures either, contrary to expectations. (Years later, the membrane did appear in bacteria prepared with the next generation of embedding plastics.) Things in the bacterial interior looked rather different from higher cells, where membranes surround most inclusions.

There was also a new entity in Chapman's pictures, a small body found especially near the ingrowing points of the cell wall septa during division, four to eight per cell. These new entities, which Chapman and Hillier called "peripheral bodies," usually appeared empty except perhaps for a small granule or two; the authors proposed that they were involved in secreting new cell wall material. They also suggested that the presence of the peripheral bodies at the growing septum had been mistaken by some light microscopists (including Bisset) for a membrane preceding the com-pletion of the new cell wall, because they were of a size barely resolvable by a light microscope, and when lying in a line, might look membrane-like. This reasoning undercut the arguments of light microscopists, like Bisset, that whatever could be seen with a light microscope in living bacteria could not be artifactual. Chapman and Hillier noted that they saw no mitochondria with internal structure resembling those found in higher organisms, but pointed out that bacteria may differ widely in their mitochondria, so that the large ones that Mudd had seen in other species might be otherwise.[98] When Mudd first saw the peripheral bodies in these pictures, he felt vindicated and wanted to know why Chapman and Hil-lier would not call them mitochondria, and he was very disappointed when the authors insisted on remaining agnostic about that issue.[99]

Mudd was not worried that the peripheral bodies might not be mito-chondria, because their empty appearance was admitted even by Chap-man and Hillier to imply that the original interior structure had been destroyed in the heroic preparative procedure for thin sectioning (espe-cially through the extraction of lipid-rich structures like membranes dur-ing the lengthy soaking of the specimen in organic solvents prior to resin infiltration). The unexpected and implausible lack of a plasma membrane, coupled with Mudd's own evidence for a nuclear membrane, suggested that in general, membranous structures were poorly represented by the thin-section micrographs—in other words, that the new technique was unreliable in this respect. As Mudd wrote to a former student: "You know . . . that protoplast, nuclei and mitochondria do not exist naked without a differentiated layer about them. Protoplasm doesn't operate this way. Jim Hillier is a great physicist and has an amazing understanding of biology for a physicist, but . . . this time he has made a serious mistake."[100]

However, the failure to find the internal membrane structure charac-teristic of mitochondria in Chapman's thin sections put pressure on Mudd to close the gap in his argument, that is, to find some other way of showing that the dense granules in his own electron micrographs were the same as the granules showing mitochondrionlike staining properties in his light preparations. At first he accumulated more circumstantial evidence. Having defined bacterial mitochondria as "cytoplasmic granules, dense by electron microscopy, possessing limiting membranes, and possessing coordinated systems of oxidative-reductive enzymes," he pointed to evi-dence from light microscopy that there existed granules staining for mito-chondrial reactions, which could be distinguished from cell envelopes, which contained lipids (as membrane-bound bodies must), and which in some cases also contained high-energy phosphate compounds (which he took to indicate that phosphate, associated with the cell's energy currency, was accumulated in some mitochondria). To link these data to electron microscopy, he again pointed to the membranelike remnants of his dense cytoplasmic granules when volatilized in the electron beam, and to similar appearances obtained when phosphate and lipid were chemically pre-pared as droplets and volatilized in the same way.[101] This reasoning did not establish, however, solid evidence of the right kind of enzymatic activity *in the same picture* as the electron-dense granules; it was still possible that the tiny bodies that stained so suggestively in light preparations were entities altogether different from the dense granules appearing in electron micrographs.

In April 1954, Mudd took a trip to Japan, during which he seems to have encountered audiences sympathetic to his bacterial mitochondria, a trip that would soon prove fruitful in bringing scientists from that country to work in his lab.[102] At the start of the next academic year, Mudd tried to close his interpretive gap between the electron micrographs and the stained light micrographs through an approach involving collaboration with Chapman, who had taken a job with Hillier at RCA after his Princeton doctorate. In September 1953, Chapman was left to his own devices just two months after he took up his new postdoctoral position—Hillier having left RCA to direct research at another firm—and found himself with a free hand and sole responsibility for drumming up further interest in the electron microscope, particularly as an instrument for routine medical diagnosis (where Zworykin now hoped to find a large market). A further duty of Chapman's was to work with Katherine Polevitzky—since 1951 Katherine Zworykin—on the new television microscope that RCA labs had created for electronic magnification and projection viewing of light microscopy preparations.[103] Impressed by the clarity with which Chapman's projected TV images of bacteria showed granules (and no doubt tempted by access to Chapman's electron microscope and expert services), Mudd proposed a way out of their mutual "blind alley" of interpreting the granules, by comparing living unstained bacteria and those stained for mitochondrial reactions with the TV microscope. At the same time, Chapman would work on making thin-section electron micrographs of the *Mycobacterium* species that, in Mudd's electron pictures of whole mounted cells, showed the clearest dense granules. The joint project seems to have begun in 1954–55, but nothing ultimately came of it.[104] Even if it had come to fruition, it still would only have provided yet more circumstantial evidence.

Another more promising approach that might lead out of the "blind alley" to a positive identification of the electron-dense granules as mitochondria seems to have occurred to Mudd the next academic year, possibly through his new Japanese visitor and Rockefeller fellow, Kenji Takeya. Mudd and his collaborators used potassium tellurite as a staining agent for oxidation-reduction reactions, a chemical that, like the tetrazoles, leads to the appearance of colored granules when living bacteria (or higher cells) are incubated in its presence sufficiently long to reduce the substance metabolically. The difference was that this stain is a metal salt, and the reduction products that produce colored granules in light micrographs are metallic tellurium, which because of its high density is distinctly visible in electron micrographs. Comparing mycobacteria stained with tellurite and with his other cytochemical indicators of mitochondrial activities by light microscopy, Mudd saw good agreement, thus validating the tellurite stain and confirming positive reacting granules in the microbes. With the electron microscope, needles of reduced tellurium adorned the electron-dense granules he had been calling mitochondria, just as Mudd expected. There were, however, a few needles in the cytoplasm away from the granules, too (Plate 2.11).[105] The tellurite acted just like a classical cytochemical stain (in fact, it was one) that could be applied to electron microscopy and thus make the needed connection to his light cytology. The localized tellurium needles were as close as Mudd would get to proving that electron-dense bodies in bacteria existed and harbored biochemical activities found in the mitochondria of higher organisms. Indeed, by Mudd's ever-looser definition of bacterial mitochondria, these pictures effectively proved that they exist. However, the microbiological community—even including Mudd's own students—was growing dissatisfied with the very loose definition of bacterial mitochondria that had been forced upon Mudd, both by the failure to find bodies that looked just like animal and plant mitochondria in thin-section electron micrographs, and by the variable results biochemists had been getting when looking at the enzymatic activities of particles purified from bacteria.[106] If mitochondria existed in bacteria, the biochemists thought that it should be possible to isolate them by methods similar to those that worked on higher cells, and that they should have similar properties as well.

From the early 1950s, biochemists had begun applying to bacteria the new ultracentrifuge-based experimental procedures, largely developed at the Rockefeller Institute (see Chap. 3), that were becoming popular for the study of cells from higher organisms. These procedures involved rupturing cells so as to release the cytoplasm and other contents, and from this homogenate or soup separating components according to density by passages in an ultracentrifuge. After separation, each fraction of the homogenate was then analyzed according to chemical composition and catalytic activities, by dye reactions and other enzymological tests. Some experimenters found that the activities that characterize mitochondria in higher cells appeared in a particulate fraction of bacteria, as would be expected if bacteria have mitochondria, whereas others found that the mitochondrial activities were found in the fraction containing the empty cell envelopes or "ghosts." Results varied greatly from species to species, and with fractionation procedure—the gentler cell-disruption methods tending to locate mitochondrial activities with the ghosts.[107] This confusing mass of contradictory data was, of course, hard to interpret, but already by 1954 the suggestion had appeared that, rather than supposing that different species of bacteria had mitochondrial particles of widely different properties, it would be more satisfactory to suppose that bacteria have no distinct mitochondria and that the cognate biochemical activities are found on the interior of the cell envelope (it was still not clear that bacteria had true plasma membranes).[108] For subscribers to this view, the problem became finding protocols that would demonstrate mitochondrial activities in cell ghosts, in many groups of bacteria, and particularly in those bacteria that had given results showing activities in a particulate fraction by other protocols. By the late 1950s, such efforts had made great headway, eroding biochemical evidence for Mudd's particulate mitochondria in bacteria.[109] Moreover, the reduction of tetrazoles to granules of colored formazan— Mudd's favorite light-microscopic staining technique for mitochondria— had even been demonstrated in ghosts after the release of cell contents.[110] This showed that whatever Mudd was looking at was attached to the cell envelope.

On the other hand, many electron microscopists (both within and beyond Mudd's circle) who were tinkering with preparation techniques for thin sectioning so as to improve on Chapman and Hillier's pictures of bacteria, were finding distinctive membranous bodies in a number of bacterial groups, especially in the mycobacteria in which Mudd had first claimed to have discovered mitochondria.[111] Mapping between the electron micrographs and the nonpictorial representations of mitochondria that biochemists made with their ultracentrifuges was uncertain, and left as much room for divergent interpretation as the mapping between electron micrographs and cytochemical preparations for light microscopy. Mudd's response in defense of his bacterial mitochondrion, under siege by biochemists who had progressively forced him to gut his definition of the entity (mainly by finding differences between the relevant biochemical properties of bacteria and higher cells in the ultracentrifuge), was to point to the new electron micrographs and concede that these membranous mitochondrial equivalents were in many species attached to the cell envelope.[112] This concession that Mudd's own views should be "integrated" in such a way with those of his antimitochondrion opponents, submitted in 1960 at the time of Mudd's retirement, effectively marks the end of the bacterial mitochondrion. By admitting that the entities could be attached to the cell envelope rather than floating free in the cytoplasm as they do in

higher cells, Mudd had yet again allowed his definition of the mitochondrion to become looser, and this may have been the last straw for some of
his supporters. But more important, the membranous bodies in electron
micrographs of thin-sectioned bacteria had taken on a different and more
prominent appearance by 1960 through improvements in specimen preparation technique, and although some called them "chondrioids" and
followed Mudd's interpretation of them as mitochondrial equivalents,
others preferred to attribute alternative functions to the bodies, and different names. The name that stuck was "mesosome."[113] The bacterial
mesosome has had a long and checkered career that extended into the
1980s, and thus well beyond the time frame of this account. I have
sketched a later part of its story elsewhere.[114]

Conclusion: Mudd's Pictorial and Epistemological Practices

The electron microscope tremendously encouraged and enlarged the
scope of investigations into the differentiated organs of the bacterial cell,
which for the first time could be imaged with adequate resolution. Indeed, in making possible a far greater appreciation of the morphological
differences among classes of microbes, it contributed to the clarification
of distinctions between true bacteria and pseudobacterial groups such as
Rickettsia, yeast, and slime molds—and ultimately to a new conception of
microbiology based on the new taxonomic category of "prokaryote"
(defined, as it is, largely on the basis of internal morphology). In the
context of burgeoning studies in microbial genetics, light and electron
microscopic research on the nuclear apparatus achieved great currency
and at first pointed to many similarities between the reproductive mechanism of bacteria and cells of higher organisms—although, as noted, much
of this work eventually came into conflict with accumulating genetic
evidence and later had to be discarded or reinterpreted in light of the
genetics.

Mudd's own approach did not aim so much to address the equivalence
of the genetic and biochemical properties of bacteria and higher organisms, but rather the more general but related problem of bacterial complexity. Adapting the new microscope to deal with the newly fashionable
issues of bacterial differentiation and reproduction, Mudd's research program sought the *structural properties* of the organelles of bacteria by electron
microscopy, in tandem with light microscopic techniques to determine

the *biochemical properties* of those cell components through their staining properties. The approach drew heavily on preexistent and familiar intellectual and technical tools, not only in Mudd's own use of light microscopy, but also in his electron microscopy. Technically, Mudd's way of making images with the electron microscope preserved many of the features of the way bacteria were treated and portrayed by the traditional light microscopy of medical bacteriologists and histologists (or cytologists), with their practices of chemically treating bacteria fixed to a microscope slide in order to "read" biochemistry from the stained microscope image.[115] And as with standard light microscopy techniques of the bacteriology lab, Mudd did electron microscopy on whole, dried (and/or fixed) microbes, often chemically pretreated. His electron micrographs recorded the shadows of all the components throughout the entire depth of the bacterial body, as an X-ray records the depth of a patient's body, just like bacteriological light micrographs. Indeed, Mudd's electron micrographs of bacteria look much like highly enlarged monochrome light micrographs, but with finer detail (compare, e.g., Plate 2.5 bottom left and bottom right). Intellectually, Mudd's program was rather traditional too, in that his interpretive practices aimed for the same "reading" of biochemical information from the micrographs—first from light micrographs of the same bacteria he was viewing in parallel by electron microscopy, and eventually, directly from electron micrographs as well.

By the end of the 1950s, however, bacteriologists were turning with increasing preference to evidence from ultracentrifuge-fractionated bacterial homogenates, rather than staining in situ, to characterize cell components biochemically. Mudd never adapted his program to this bundle of techniques, despite the presence in his department of an ultracentrifuge and a biochemist (Sevag) he might have enlisted in the project. He stuck to his tandem electron and light microscopic cytology. In the controversy Mudd encountered over the mitochondrion, in which biochemists wanted to map the enzymological findings from their ultracentrifuge experiments onto bacteria in a contradictory way, Mudd was challenged by the uncertainty in his own mapping between the light and electron pictures. The way Mudd responded was by finding a cytochemical stain that visibly indicated biochemical activity in both light and electron micrographs, and thus bridged the gap between the two pictorial spaces. This was a perfect solution from within Mudd's program. However, most electron microscopists had by then stopped making images of whole-mounted cells. In the thin-section pictorial medium that came to prevail as standard among bacteriologists, which Mudd never adopted despite some half-hearted attempts to learn,[116] the pictures were very different and basically incommensurable with those that Mudd was still making. There was not much room for the cytochemical stain in thin-section images (ferritin in the later 1960s and immuno-gold technique in the 1980s introduced more biochemically specific "stains" visible in sections). Mudd's research program was superseded through changes in technique that made his pictorial data hard to compare, by the end of the 1950s, with that obtained by others using the thin-section methodology that had rapidly become standard for bacteriology. Slow progress in finding cytochemical "stains" visible by electron microscopy, compared with the high productivity of ultracentrifuge-based studies, may have doomed Mudd's program.

As with Anderson's preliminary work, technical innovation went together with innovations in epistemology. From the start, the overall goal of Mudd's practices of picture-making was to adapt cytological methods to the electron microscope, and thus to make images that could be read and interpreted similarly to light microscopic observations. This envisioned adaptation of experimental practice (including pictorial style) involved some changes from traditional bacteriological cytology in the novel way electron micrographs were to be interpreted as compared with the way light images had been, and also in the way light microscopic evidence should be interpreted so as to accommodate electron microscopic observations of the same object. For instance, ordinary light microscopy records the different colors or opacities of objects to light, creating the expectation that any significant structure will be darker (or contrastingly colored by stain), whereas electron microscopy records the densities of specimens under the electron beam and offers no a priori reason to expect that significant cellular structures will be especially dense. Although this may seem obvious, it came as something of a revelation when nuclear bodies of bacteria, which in stained light microscope preparations appear as darker, colored zones, appeared in electron micrographs as bodies lighter than their surroundings (because of the higher water content, thus lower density, of the nuclear material). Robinow insisted, probably correctly, on credit for the insight that predicted this.[117] Mudd's 1950 move to evaluate cytological procedure by examining specimens at each preparatory step with the electron microscope—his response to the perpetual uncertainty, insoluble by cytologists within their staining idiom, about which of many preparation procedures gives the truest picture—can be taken as an example of how he wanted to alter the way conclusions were drawn from light microscopy.[118] Previously the only arbiters of disputes over staining protocol were chemical trials on questionable, simplified model specimens, but now Mudd wanted electron microscopy to serve as umpire. A very plain example of the way Mudd worked to weaken the overall epistemological authority of light microscopy, in order to give a leading place to electron microscope data, can be found in a 1949 paper on the nucleus that he coauthored with Hillier. After reporting that light cytology gave consistent evidence in favor of bacterial nuclei, both through numerous staining experiments and even through ultraviolet microscopy, the authors went on to make the bold statement: "[Agreement of] the investigations cited above should not imply that the details of the structures are demonstrated accurately by those instruments. In fact the main contribution of the present work with the electron microscope is that the actual structural differentiations are more complex and on a much smaller scale than would be suspected from light microscopy."[119] The greater resolving power of the electron microscope gave it supreme authority in judging structural questions, no matter how many different light microscopic methods might agree otherwise (and so much the worse for any methods giving conflicting results!). The priority of electron over light microscopy also comes through clearly in the way Mudd argued against Pijper's unconventional view of flagella, and in the way Chapman and Hillier dismissed prior light microscopic observations of a plasma membrane preceding the nascent septum on cell division (accounting for the latter observations as inadequately resolved "peripheral bodies").[120] Mudd shared this valorization of the electron image based on resolution, of course, with other biological electron microscopists. In fact the others, who unlike Mudd were not seeking to apply light and electron micros- copy in tandem, can probably be said to have emphasized the supremacy of electron microscopy more, and to have counted light microscopic evidence less.

Thus Mudd's school of electron microscopy differed significantly from traditional bacteriological and cytological light microscopy, and also from the thin-section electron microscopy that overtook it (about which more will soon be said), both in the look of its pictures and in the kind of reasoning by which their meaning was determined. The research program of Mudd's school aimed to locate biochemical function in the cell compo- nents visible in electron micrographs by traditional cytochemical meth- ods, and ideally by cytochemical stains actually visible in the electron pictures. One can portray Mudd fairly, then, as a traditional cytologist eager to adapt existing preparative, pictorial, and epistemological practices originating in staining-based light microscopy to the electron microscope in a conservative way. So, what is one to make of the ultimate decline of Mudd's X-ray style of electron microscopy and the repudiation of bacterial mitochondria that went with it? There are several factors to consider. There were technical limitations to Mudd's program, in that not many cytochemical stains visible in electron micrographs existed at the time. The research program of cell biology (see Chap. 3), and the bacteriology that emulated it, relied for chemical identification of cell components on biochemistry performed on cells homogenized into soup and fractionated with an ultracentrifuge, not on stains applied to whole (and sometimes living) cells. Structural information came from thin-section electron micrographs in which, because the limits on resolution imposed by specimen thickness and superimposed structures were lifted, far more detail could be seen. But one could not read the physiological meaning of all the wonderful detail portrayed—all the granules, filaments, dense layers, and so forth—directly from the thin-section image in the way Mudd could, at his best, read it from his pictures. Thus the pictorial and experimental style that overgrew and outlived Mudd's school was not simply better able technically to accomplish the same thing, but accomplished a related but different thing in a completely different way, with cell fractionation.

Mere material resources and effort are not always adequate to overcome technical limitations, but these factors certainly deserve a place in any account of the Mudd program's downfall. Perhaps if Mudd had been younger, more innovative, or less engaged in other research and administrative activities, he might have reached and surpassed his tellurium success much sooner. If he had had the kind of support from Penn that would have facilitated full and rapid realization of the electron microscopy center that Zworykin proposed in 1948–49, his chances would have increased greatly. Certainly, Zworykin would seem to have chosen wisely in choosing Mudd, if the goal was indeed to develop methods for electron microscopy that would move easily into the medical context and thus sell plenty of electron microscopes to hospitals. Mudd's methods deviated relatively little from established bacteriological practice, both pictorially and in terms of preparative procedure, and Mudd preserved a greater place for traditional light microscopy in tandem with the electron microscope. Thus Mudd's conservative way would have been relatively friendly to hospital practitioners. In the event of thin-sectioning's preeminence, fewer microscopes were actually sold to hospitals than might have been, and one reason must be the difficulty and novelty of the preparation technique involved. Thin sectioning even today takes a year or more of hard apprenticeship to learn competently; and in those days of inferior materials, primitive microtomes, and improvised protocols, craftsmen who had mastered the technique sufficiently to produce good pictures quickly and consistently (both essential for the hospital context) were very few and far between. To sum up, then, Mudd's program was a failure only in the context of competing programs growing up around him. If Mudd's school had had the same kind of strong start enjoyed by the other early electron microscopy schools, to which the remainder of the book is largely devoted, there is no telling whether it might have gone much farther, and whether it might even have descendants today. In an elaboration of James's horticultural metaphor, one could say that many new shoots may sprout from a newly grafted bud, but not many will grow to become limbs. Small and unpredictable advantages at crucial, early stages can translate into large cumulative effects—as Serres discusses, too, shifting the metaphor now from organic development to human history. Analysts of history ought not to succumb too easily to the temptation to reason from actuality to inevitability, as we tend to when surveying the foundations of Rome's ruined competitors and looking for a profound difference in Roman character or Roman actions determining the great city's triumphs. The smallest differences may suffice.

CHAPTER 3

The Rockefeller School and the Rise of Cell Biology

✤ If we give attention to the fact that any ability of control whatever depends upon ability to unite these disparate appearances [of things] into a serial history, and then give due attention to the fact that connection into a consecutive history can be effected only by means of a scheme of constant relationships (a condition met by the mathematical-logical-mechanical objects of physics) . . . we recognize that it is only with respect to the function of instituting connection that the objects of physics can be said to be more "real." In the total situation in which they function, they are the means to weaving together otherwise disconnected beginnings and endings into a consecutive history. . . .

[I]n the practice of science, knowledge is an affair of *making* sure, not of grasping antecedently given sureties.
—John Dewey, *Experience and Nature*

This epigraph from Dewey sums up much of what, in his pragmatist view, characterizes knowledge in natural science (for which physics serves as exemplar). Science is built in such a way that over time new knowledge extends the old, connecting ever greater numbers of elements into a coherent "consecutive history." In so doing, science seems to explain ever more, and it also allows ever increasing powers of prediction and manipulation of the world. All this can only take place when phenomena—or any raw experiences of the world—are *abstracted* from the infinite complexity with which they present themselves to our naive senses, and abstracted *systematically*. Before reason can begin to deal with the empirical, experience must be broken into chunks of a size and simplicity conducive to easy handling. The actions through which this abstraction from experience is carried out—the selective attention and other operations, both mental and technical—must be performed by the scientific observer or experimenter according to a consistent method, to insure that the "histories" built from experience will cohere.[1] In practice, because more than one researcher is engaged in a given scientific enterprise at any time, these same considerations mean that systematic methods of abstraction must not only be public but to a great extent *standardized*. Because all the practitioners in a scientific discipline observe and report on the world in a standard manner, they produce facts that are interchangeable in the sense that all facts are of the same type and can easily be fitted together into coherent "histories" or representations of the world, no matter which practitioner happened to make which observation. Otherwise, scientists would be hobbled, like cartographers trying to make an atlas from maps made according to differing, incompatible conventions. The same may be said of science's power to manipulate the world: because practitioners in a given experimental field all use sufficiently similar methods, materials, and instruments that they can replicate the essential features of one another's trials, each can, like a navigator equipped with readable maps made by predecessors, learn from another's mistakes and benefit from another's successes, so that as a community they improve their power to isolate, generate, and measure the phenomena that concern them. Standards render their routes of action in the world secure.

Thus, maintaining science's "scheme of constant relationships" (that constancy that allows connections, combinations, and substitutions to be made among the objects of scientific knowledge) all depends on the abstraction that makes objects of knowledge from inchoate experience being carried out systematically, in a fixed and standard manner. Physics, as Dewey points out, abstracts mathematically, so as to work upon its mental objects logically and its physical objects through mathematical reconstruction. But every science has its own proper, standard manner of systematic abstraction—or else the realm of phenomena treated by the science must lack stability and the science's grasp on that realm will not increase, and for these reasons we might even categorize the situation as prescientific. (Indeed, this is one explanation of Kuhn's "preparadigm" situation, and of the need for what sociologists call normative consensus.)[2] So it was for the first biologists confronting the astounding complexity that the electron microscope revealed in the cells of higher organisms: what they saw in the cell kept shifting and slipping away until standard methods both of making and of reading electron micrographs were chosen. In effect, vision had to be standardized, and—insofar as a unified discipline was at stake this time—to a far greater degree than already accomplished by Anderson's guidelines. In this chapter we consider the work of the cytology laboratory of the Rockefeller Institute, and its role in establishing the standards around which the discipline now called "cell biology" had formed by the end of the 1950s.

The captain of the Rockefeller cytology laboratory in this formative period was Keith Porter, a self-reliant, cool-headed, and perhaps taciturn Nova Scotian who was given, by a conjunction of circumstances, a unique opportunity for leadership when rather young. Many other life scientists in the later 1940s and early 1950s were eagerly trying to adapt the new electron microscope to the established (and overlapping) scientific endeavors of cytology and histology, the sciences of cell and tissue structure, but none were more influential than the Rockefeller group of Porter and his partner through the 1950s, George Palade. I will show that the particular methods they developed (and especially the methods of micrograph interpretation), which the community of electron microscopists studying cells chose as standard, grew out of the history and organization of this laboratory's research program, and very much bore its stamp and that of its Rockefeller context. By describing an alternative road not taken by the majority, I illustrate how methods that become standard are not necessarily the only rational and coherent ones available, and how in this case the choice of standard method was neither arbitrary nor strictly predetermined by suitability to the subject matter, but linked to a certain notion of the proper aims and limits of the discipline carrying out electron microscopical investigation of the cell—"suitability" and indeed the conception of the subject matter itself emerging as consequences of such choices. I suggest further that the standard methods may have achieved general acceptance partly because they made for a disciplinary boundary and modus vivendi for cell biologists that were functionally well adapted to the biomedical research institutions in which many of these scientists found a place; for with the postwar growth in government funding for biomedical research, many more institutions could afford the rich material infrastructure of the Rockefeller Institute and to some extent the disciplinary ecology that came with it. Methods, then, are shaped by both intellectual and political goals in the "total situation in which they function," that is, in theoretical, technological, institutional, and national context. Thus we are reminded that what Dewey aptly calls science's work of "making sure" (in both senses, "fixed" and "certain") is constitutively social. For how best to proceed is specified by no pregiven rule, and must always be decided—especially in a discipline's formative stages, but even in the ordinary course of research, as new techniques and theories potentially affecting the direction of inquiry become available—through potentially divisive choices on the part of the community of practitioners.[3]

The Cytology Laboratory in the Rockefeller Institute

Since its founding in 1901 through the magnanimity of John D. Rockefeller, the Rockefeller Institute in New York City had been America's leading biomedical research center. The vision informing its organizational structure was of an ideal community of scholars dedicated to discovering the causes and cures of illness, unhampered by teaching responsibilities or limited budgets. Naturally, with no student population and ample resources, the Rockefeller laboratories were mostly staffed by paid technicians and salaried scientists (though always leavened with visiting postdoctoral researchers), which led to a more hierarchical atmosphere than was common where graduate training was the stock-in-trade. There was an ornate dining hall, where only the scientific staff were permitted to lunch, with good china and silver, for example, and two separate cafeterias where the technicians—many young ladies of New York's better families—ate lunch safe from the company of the manual laborers.[4] As Sinclair Lewis's account of the "McGurk Institute" illustrates (which Lewis placed, symbolically, on the top floors of a lofty skyscraper, whereas the Rockefeller Institute is a set of relatively low buildings, by Manhattan standards), the Rockefeller Institute had something of a reputation as an ivory tower remote from the common human lot, divided into fiefdoms by its eccentric prima donna researchers—each more dedicated to his own theories and sophisticated experimental apparatus than to the alleviation of human suffering.[5] Elitist or not, the Rockefeller Institute was powerful; under the cautious directorship of Herbert Gasser, it survived the lean Depression and war years without losing its leadership role, and in the immediate postwar period it could boast a distinguished staff that included Oswald Avery, discoverer that DNA carried genetic information in bacteria, Peyton Rous, discoverer of a cancer-causing virus in fowl that still bears his name, and a firmament of other biomedical eminences.

What became the cytology laboratory began as a part of James Murphy's cancer research group. Throughout the 1930s, Albert Claude had been working under Murphy on a project that started as an effort to purify the Rous sarcoma virus by use of a high-speed centrifuge. Separating all the particles from a soup of homogenized tissue according to size, and seeking to identify which particles retrieved from fractions of the homogenate were normal constituents of the cell and which were the virus, Claude was led into a project of mapping the locations of enzymes and other biochemicals to the various components of normal cells (i.e., finding where particular biochemical properties were located within the cell), an enterprise that involved him in collaboration with several biochemists at the Institute.[6] In 1939, the Murphy lab was joined by Keith Porter, a 27-year-old trained at Harvard in experimental embryology who, like Anderson, was not finding a regular academic position under Depression conditions. At first Porter continued the experiments with amphibian eggs that had been his doctoral work, transplanting nuclei into the eggs from related species in an effort to discover cytoplasmic factors responsible for development and perhaps inheritance.[7] Interest in self-reproducing cytoplasmic entities (or "plasmagenes," as they were sometimes called at the time) capable of influencing development was a unifying theme in the Murphy lab research, and was the connection to cancer, for it seemed that this deadly form of deviant development might be caused by rogue plasmagenes.[8]

After 1941, work in the Murphy lab slowed as male staff went to war and not enough skilled women could be hired, but Porter, unenlistable because of a persistent tuberculosis infection, was able to continue his research hampered only by his sanatorium visits.[9] Apart from his frog project, with Murphy's encouragement Porter taught himself the then still-primitive art of growing animal cells in culture, so that Claude might be able to test whether any of his particle fractions altered cell development (the way plasmagenes should) by just adding them to cultures,[10] and when this did not work, so as to observe cellular changes that accompany the spontaneous transformation to malignancy that often occurs during tissue culture. In 1944, Porter and Claude, whose biochemist collaborator, George Hogeboom, had discontinued their enzymological analysis of centrifuge-fractionated cell components because of a war project, availed themselves of the opportunity to try electron microscopy at the Interchemical Corporation in New York. (According to one story, Interchemical's research director, Albert Gessler, who was interested in involving his lab in cancer research and had read one of Claude's publications, invited them over to look at their specimens with the company's RCA model B, one of the first batch made.)[11] The paraffin sections prepared by

standard cytological techniques were several microns thick and thus completely opaque to the 60 kv beam of the model B; however, Claude was
able to obtain fair pictures of the particles he was isolating from cell
homogenates.[12] Better still, Porter was able to provide very flat tissue
culture cells grown directly on specimen grid films, and he fixed them by
a traditional cytological procedure over osmium tetroxide (OsO_4) vapors
before drying them for viewing. Near their margins the cells were thin
enough to be penetrated by the beam, so some interior detail was visible.
By making a montage of electron micrographs representing a whole cell
as seen at low power, and by placing it side by side with a light micrograph
of the same kind of cell prepared by traditional staining (Plate 3.1), Porter
and Claude made a convincing visual case that their electron microscopy
procedures represented the same entity and introduced no drastic artifacts
(or at any rate, none in the range of light microscope resolving power),
which supported, or at least did not rule out, the veracity of novel structures visible only at high power (Plate 3.2).[13]

Electron Microscopy and Cancer Research

In 1945–46 Claude was working with Interchemical's Ernest Fullam,
trying to devise a microtome and new preparative techniques so that slices
of cells far thinner than ever before could be obtained. Their early pictures
of tissue sections, showing impressive detail even though too thick and
riddled with holes (Plate 3.3), attest to what a long way preparative procedures for tissue still had to go.[14] Claude's interest in the electron microscope was, as he put it in his 1948 Harvey Lecture, to use the instrument as
a "guide or check" on the results from fractionation and biochemistry, to
make sure that centrifuge fractions represented definite parts of the cell
and not just arbitrary collections of particles.[15] When Hogeboom finished
his war work toward the end of 1946 and returned to the centrifuge
fractionation project with Claude, he was assigned as a helper (and initially an unsalaried volunteer) George Palade, a young medical doctor
from Romania who had just arrived in New York.[16] Meanwhile, Porter
refined his methods of growing tissue cultures on thin plastic films so that
"almost any type of cell" could be fixed in osmium vapor and imaged by
electron microscopy. With Claude, he prepared some cells infected with
the Rous virus in order to search the electron micrographs for those
expected "small bodies of uniform size" not present in micrographs of
uninfected cells, that is, the profiles of the elusive cancer virus particle.

Also, with Hogeboom he sought (as Mudd would; see Chap. 2) cyto-chemical stains that would leave signatures in electron images of cells through local deposition of a metal, indicating where specific biochemical activities were located.[17] Reading the laboratory's progress reports for these first few postwar years, one gets the sense that Porter was groping for a way to make the electron microscope an element of a productive experi-mental system, that is, a bundle of methods and objects of inquiry suited to addressing a certain type of problem, even as he struggled to make sense of the bewildering, intricate, and shifting elements of cell structure that the instrument was showing him.[18] (How he made sense of what he saw is discussed below.)

The Murphy group would not obtain a microscope of its own until 1948, but that did not stop Porter. Borrowing time on the RCA EMU at the nearby laboratories of the International Health Division of the Rockefeller Foundation through the good graces of Edward Pickels, his senior collaborator there, between 1946 to 1949 Porter managed to turn his whole-mounted tissue culture cells into a productive experimental system for cancer research.[19] The experimental logic was a straightfor-ward adaptation of Koch's method for bacteriology: look for a micro-scopic entity present in all diseased organisms and absent in healthy ones. In this case, the organisms in question could only be cells that could be cultured on electron microscope grids. With Claude, he succeeded in finding the Rous virus particles in cultured chicken tumor cells, and with Helen Thompson, a postdoctoral fellow with Murphy, he also got pic-tures of telltale particles in cultured cells from a mouse mammary tumor, one of the few other cancers for which there was evidence of transmis-sion by a virus.[20] Turning their attention to another more typical kind of tumor, not associated with an infectious form of cancer, Porter and Thompson found no subcellular structure distinctive of the tumor cells. They found only a difference in the quantity of a certain dense granular body of widely ranging size, suspicious in that malignancy seemed associ-ated with a greater number and variety of the granules. But after finding the bodies abundant also in rapidly growing normal cells such as those from embryos, they cleared the particles of responsibility for cancer and dubbed them "growth granules," theorizing that they might be primordia of the formed structural elements in the cytoplasm (again, just as a plas-magene might be).[21] With the advent of new types of electron micro-scopical image from thin sectioning, these entities were hard to identify surely and would soon, like Mudd's tellurite-reducing granules, be left behind among the curios made obsolete, although never fully explained, when different representational techniques became the standard.[22]

By 1952 Porter had virtually abandoned the increasingly trendy but mostly fruitless "particle hunt," popular with cancer researchers buying new electron microscopes and keen to discover the viruses causing the dreaded disease, that he had done much to launch.[23] Part of the reason for discontinuing this project may be a nasty misunderstanding Porter had that year with the powerful and imperious Dr. Rous himself, who proposed a "collaboration" that would, Porter felt, have reduced him to a microscope and tissue-culture technician in the search for viruses in Rous's cancer cell lines.[24] But one can find other reasons, in that techniques that would allow thin sectioning were opening new kinds of research opportunities to explore the architecture of any normal cell, not just cell types that could be cultured on grids (see below). Uninhibited by Porter's withdrawal, the cancer particle safari continued through the 1950s, fueled by NIH largesse. RCA did nothing to discourage this trend, no doubt a major source of microscope demand: one 1957 full-page RCA advertisement in a scientific journal carries the headline "Cancer Research Marches Ahead" and, below a photograph of a well-known biologist operating the latest RCA model, bears an inset of a "small area of a tumor cell" full of particles, which seems to be a Porter micrograph from ten years earlier.[25]

Transitional Years, Research on Technique

The laboratory went into a period of upheaval with Claude's departure in 1949 and Murphy's retirement in 1950 (and death shortly thereafter). The young cytologists were orphaned. The lab's biochemists, Hogeboom and Walter Schneider, left for posts at the National Cancer Institute (of the NIH), but Gasser stepped in to "salvage" Porter and Palade, giving them relatively secure if provisional positions in a freestanding cytology laboratory with only Porter and no senior member of the Institute in charge.[26] In addition to their new EMU, they were given a lower-resolution Philips console microscope and a service role in using it to provide Institute researchers with micrographs.[27] The lab staff was reduced to four technicians and assistants, a secretary, and a dishwasher.[28] As this was happening, in late 1949, Palade—who had been immersed in Claude's tissue fractionation experiments with Hogeboom and other biochemists—now began to work assiduously with Porter on electron microscopy technique, with the intent of developing new tissue sectioning methods. That year a group at the National Bureau of Standards had reported that embedding specimens in the plastic methacrylate instead of wax made them tough enough to cut thinly, and Palade and Porter decided to dust off Claude's microtome.[29] In these transitional years of 1949–51, Palade continued doing his fractionation experiments as Porter tinkered with the microtome, Palade with embedding procedures, and—especially—both with fixative preparations that would be suitable with the plastic embedment. Their guide in this last project was mainly Porter's tissue culture cells, which could be observed during the course of fixation by darkfield microscopy (in which even particles below light's resolution limit show as bright motes) and by ordinary light microscopy, and then examined with the electron microscope. Preparations that by light microscope techniques showed the least apparent changes in the cell during its passage from living thing to preserved specimen, especially the least clumping and coagulation of the cytoplasm "ground substance" (conceived at that time as a colloidal gel homogeneous in the size range even of their electron microscope's resolution), were judged best. Porter did not put all of his eggs in the thin sectioning basket: through 1952 he continued to look at dispersed muscle, fibrin, and collagen fibrils (the last both as secreted in tissue culture and as deposited from solution), a project that began in 1947 and had put him in a direct and generally losing competition with Schmitt's MIT group (see Chap. 4); and into 1952 he continued to look by his old technique in whole tissue culture cells at the "endoplasmic granules" that might be responsible for malignancy (and at the innocuous "growth granules," their later incarnation).[30] The laboratory was juggling, then, three different experimental systems involving cells and the electron microscope, both the tissue sectioning and the fibril self-assembly systems derived from the original whole-mount cultured cell system. In this scientifically and administratively unstable period they fended off what seemed, at least in retrospect, a takeover attempt by Rous.[31] Porter and Palade were searching for a way to make electron microscopy of cells an independent discipline, and not simply a technique subservient to physiology, microbiology, or any other discipline.

    In 1952 the cytology laboratory's first post-Claude successes reached publication, and quickly brought national recognition. One of the crucial achievements was a formula, hit upon by Palade, with which small pieces of animal tissue could be reliably fixed, without obvious change in the tissue structure, prior to the elaborate preparative procedures for thin sectioning: the soaking in a series of alcohol solutions to remove water, then soaking in methacrylate resin to replace the alcohol in the tissue, then hardening with a catalyst that turned specimen and resin embedment into rock-hard plastic. In their many preliminary trials, Porter, Palade, and visiting fellows helping them had found that strong, pH-buffered OsO_4 solutions seemed generally superior to other reactive metal concoctions and to formaldehyde, and gave the same sorts of image in tissue cultured cells that Porter had earlier achieved with osmium vapors.[32] With pH indicators Palade confirmed that unbuffered osmium causes a potentially destructive acidity when it first contacts tissue during fixation. Trying OsO_4 solutions buffered to various pH levels, he found that the appearance of cells was very sensitive to pH and that a certain recipe buffered with the chemical veronal gave the most homogeneous appearance of the cytoplasm ground substance (Plate 3.4), which had been the criterion for good preservation of whole tissue culture cells (see above). Also, in those earlier whole cell images the bodies called mitochondria appeared narrow, whereas in the sections from veronal-buffered OsO_4 fixation the mitochondria also appeared narrow. In tissue homogenates like those used to isolate cell fractions, mitochondria showed the least swelling at this pH also (mitochondria are dealt with in greater detail below).[33]

This recipe, then, gave appearances closest to those Porter obtained with cultured cell specimens fixed in osmium vapor. To sum up, the Palade fixative formula was established as reliable on the theory that the cytoplasmic ground should appear homogeneous, on the empirical grounds that living tissue culture cells observed by visible light microscopy methods showed minimal change during fixation at this coarser resolution, and on the assumption that anything larger than the narrowest mitochondria represents swelling—in accordance with their appearance in Porter's tissue culture micrographs. Thus Palade's methods were in large part a calibration against the standard of the Rockefeller lab's prior pictures. Though critics might have questioned the sufficiency of the theoretical and empirical grounds for this calibration, the methods were received with much more gratitude than criticism. Soon microscopists everywhere were preserving their specimens with what came to be known as the "Palade Pickle" recipe and, not surprisingly, they too were seeing similar things.

Soon also they were buying the commercial version of a microtome Porter had built in 1952–53 with Joseph Blum, of the Rockefeller Institute machine shop, to cut their tissue.[34] Many were experimenting with microtome designs in the immediate postwar period, in an effort to build a device capable of cutting slices less than 0.1 μ (or 100 mμ) as required for adequate beam penetration—which was more than an order of magnitude thinner than could be reliably obtained with machines developed for light microscopical histology. Working with Blum himself, Claude had developed a microtome just after the war, and Palade and Porter had refitted it with a motor and modified it to take the glass knife introduced at MIT (see Chap. 4). Claude's associates at Interchemical, Fullam and Gessler, guided by rigorous theoretical calculations showing that high speeds would localize strain to the knife edge and give cleaner slices, during the war had developed a microtome that operated at 57,000 rpm and delivered a cutting speed of 1,100 feet per second. This ultramicrotome was the first offered commercially, with an advertisement suggesting that purchasers were making a safe investment because the device could easily be converted into an ultracentrifuge.[35] The section quality was not remarkably better than that delivered by slower microtomes, and the clouds of flying downlike sections it produced had to be collected with a fine butterfly net.[36] Porter's friend and collaborator in Seattle, Stanley Bennett, had worked out, in another exercise of mathematically rigorous physical theory, that vibration and knife chatter were inversely proportional to microtome mass. The design seems never to have been published, but Porter recollects that Bennett had a "monstrous" machine weighing "about a ton" but "never did anything with it," which suggests no great improvement in results.[37] As noted, Chapman and Hillier had developed their own microtome at RCA (see Chap. 2). But the Porter-Blum design (eventually marketed by Ivan Sorvall, Inc.) stood out for elegant simplicity and reliability in a wide "array of devices . . . some crude, others almost comically complex" in use in 1954.[38] It soon became a standard tool on the biological electron microscopist's benchtop.

The new thin-sectioning approach that sprang from these new tools and techniques gave images at once less confusingly complex, because the layers of superimposed structures in whole cells were minimized and more richly textured (see, e.g., Plate 3.4). Both of these qualities of the new kind of picture were conducive to recognizing and describing novel structures, and indeed, exciting reports of novel cell structures came pouring from the Rockefeller lab's thin-sectioning circle. The rise of the laboratory's reputation brought with it a solidification of its institutional status. Porter was promoted to associate member in 1950 and Palade in 1953 (see Plate 3.5). By the time Porter and Palade's protector, Gasser, had retired as director in 1953, to be succeeded by Detlev Bronk in 1954, the Laboratory of Cytology in the basement of Smith Hall had become one of seven official units in the Department of Laboratories. Bronk initiated the process of turning the Rockefeller Institute into Rockefeller University, and the first class of graduate students was admitted in late 1955. During that year, Porter and Palade were able to acquire a new Siemens microscope (for Porter himself) and another EMU to accommodate the students that began crowding what had become the Mecca of cytological electron microscopy, and to separate themselves from the service microscope, which was moved out of their labs and across the corridor.[39] We now turn to the paradigmatic ways Porter and Palade used their new tools, and to some examples of what they found as they learned how to make sense of what they saw in the cell.

Keith Porter and His Endoplasmic Reticulum

One of the most striking novelties that appeared in the first electron micrographs made of whole tissue culture cells was something quite apart from larger structures expected on the basis of light microscopy (e.g., nuclei, mitochondria), something permeating what had generally been thought of as a completely structureless cytoplasmic ground substance, especially the cytoplasm's interior region or "endoplasm" (Plate 3.1). In their wartime whole-cell publication, Porter and Claude described it thus:

Another formation appearing in the ground substance of cells fixed with osmium is a lace-like reticulum. This can readily be seen at relatively low magnification . . . in the lower right corner of the cell [Plate 3.1, left]. It looks as if the reticulum had extended into the thin margin and even into the fine processes of the cell from the more dense center where possibly its three dimensional dispersal and overlapping act to confuse the details of its organization. . . . At higher magnifications [Plate 3.2], vesiclelike bodies, i.e., elements presenting a center of less density, and ranging in size from 100 to 150 mµ, can be seen along the strands of the reticulum just mentioned.[40]

Because structures like this network also appeared in the condensed, ill-preserved cytoplasm of cells fixed in chromic acid, and also (though this is not explicitly stated) because the membranes implied in "vesicular" structures are not plausibly explicable by the spontaneous aggregation of fine particles, the entity had a chance at least of being something real. What to make of it was another question.

After a year of looking at the endoplasm in many cancerous and nonmalignant cells in culture, Porter had ideas about its origin. A 1946 letter to Pickels sums up his interpretation at the time: "The endoplasm of the cell may appear in a variety of forms, but most frequently as a complex reticulum of strands. . . . [T]he endoplasm frequently contains small dense bodies (100 mμ) which appear to be self duplicating and are probably the source of the endoplasmic material. . . . [T]umor cells are characterized by the presence of abnormal quantities of endoplasmic material which in these cells appears in a more finely divided form."[41] The reticular structure now seemed a product, connected with a certain state of cell differentiation, of the dividing "endoplasmic granules," which were loosely identifiable both with particulate "microsomes" isolated in the centrifuge by Claude and with "plasmagenes" supposed in theory to influence or determine cell fate (if any existed).[42] A year later, Porter was convinced that there were both different kinds of "endoplasmic granules" appearing in tumor cells, and different arrangements, too: "there are fewer long strands of dense granules in the malignant unit; instead they appear in twos or short strands. When strands are present, they are generally 'wiggly' in contour."[43] But since the reticulum was present in both malignant and nonmalignant cells, its dispersion or "wiggliness" was probably just a secondary sign of the activities causing malignancy. So much for the issue of reticulum origin; there was now an idea about its function independent of the granules and of whatever their role in determining the cell's differentiated state might be: "Endoplasm . . . is the most active part of the living unit. It is present in one form or another in every cell so far examined. It makes up a fibrillar apparatus concentrated around the central body and sends fine extensions of its substance throughout the cell. The vesicular structure of the small units suggests a secretory activity."[44] This basic topological insight that a system of membranes might be a system of containers and conduits would outlast what looks in retrospect to be an artifact of poor specimen preparation combined with high theory: that seeming association of the reticulum with growing and dividing "endoplasmic granules" (or plasmagenes) thought to control cell differentiation.

The very presence of the reticulum in all cells was one reason to believe simultaneously in its reality and its importance, because if a part of a living thing is essential to life, it must be ubiquitous. Also, the fact that despite variations in preparative procedures it always can be found implies, by Anderson's rules, that it is probably not an artifact—although Porter's thinking along these lines suggests that he had already become so accustomed to osmium fixation that this preparative procedure had almost faded into the background as a given, since only in the osmium preparations of one kind or another that he had tried did the reticulum have its characteristic form. (And reliance on osmium would cut against the reasoning from ubiquity, because this could be explained as a ubiquitous reaction of protoplasm with fixative.) At any rate, by early 1948 fresh grounds for belief in the entity's existence in living cells had been obtained from time borrowed on another laboratory's new instrument, a prototype phase microscope, which operates at visible light wavelengths but produces contrast based on even small differences in refractive index:

The endoplasmic material and structure . . . continues to show in all [electron microscope] preparations and there has been no important change . . . in our interpretation of it. It characteristically appears in the form of small vesicles and strands which may be strung together to form a reticulum. . . . During the past year yet another observation supporting its reality has been made. Though many of the endoplasmic vesicles are large enough to be seen in the light microscope, they remain invisible probably because they differ only slightly in refractive index from that of the surrounding ectoplasm. For segregating objects thus similar, the new phase contrast microscope is especially useful. . . . As was hoped, it was found that the larger units of the endoplasm could be made out. In addition, it was seen that the vesicles are involved in a streaming motion which might carry them from one part of the cell to another.[45]

Of course, the identification of the dark strands and vesicles in fixed cells, visualized clearly by the electron microscope, with the shadows and bright streaks near the phase microscope's limit of resolution was uncertain; nevertheless, it remains true that something should have been visible in the cytoplasm by phase microscopy if the reticulum were real, and something was.

The next change in Porter's interpretation of the reticulum came in 1950 or early 1951 from the first plastic-embedded tissue sections he and Palade were making (Plate 3.4). Whereas before, the reticulum in whole cells had seemed a network of fibers dotted with vesicles, attempts in their own and other labs to identify the profiles of the same entity in the new thin section images failed until the fibers or filaments were reconceived as deceptive appearances of membranous structures: "It then developed that the 'filaments' were in fact fine tubules, or strands of vesicles identical in size and form to those constituting the endoplasmic reticulum of the cultured cells."[46] So, under the influence of the new kind of image, the endoplasmic reticulum was by early 1951 repictured as a "complex canalicular system" permeating the cytoplasm with a "lacelike reticulum of channels and vesicles."[47] And as if to answer unspoken doubts about its reality (perhaps especially that noted above, that ubiquity might simply indicate a general reaction between protoplasm and osmium), doubts perhaps enhanced by the change in the character of the entity required for mapping it between whole-mount and thin-section electron micrographs, Porter now argued that in cells cultured in the same dish and fixed together simultaneously the endoplasmic reticulum looked very different in different cells.[48] Because this nonuniformity can be more plausibly explained by presuming differences in physiological state among the cells (e.g., stage of cell division) and supposing a role for the entity related to those differences, than by supposing an artifact induced in all protoplasm yet at the same time terribly sensitive to even the subtlest physiological differences, this militated for the reticulum's reality on an intuitive level (the detailed theory of OsO_4 reaction with cell constituents being woefully insufficient for any rigorous assessment of likelihood of artifact formation here).[49] To sum up then, the grounds for belief in the endoplasmic reticulum in 1951 were of three types: the evidence from darkfield and phase microscopy that changes gross enough to see are not induced during osmium fixation of cultured cells; the reticulum's presence in both thin-section and whole-mount electron micrographs (modulated negatively by the fact that osmium fixation was necessary for its clear demonstrability in both, and perhaps also by the difference in its appearance in these two sorts of picture, but positively by the argument from different appearances in similar cells under identical fixation conditions just rehearsed); and the appearance in living cells under darkfield and phase microscopy of cytoplasmic structures at least consistent with what is seen in electron micrographs. If these arguments now seem compelling, it is with benefit of hindsight, but they did have some force.

The evidence in the next few years that made the endoplasmic reticulum's existence convincing to the general biological community involved two additional kinds of reasoning: identification of the viewed entity with a fraction isolated by centrifugation and characterized chemically, and correlations with function according to the traditional logic of anatomy and physiology. Already by their early 1952 annual report, Porter and Palade had done substantial exploratory work along the latter lines, looking for the endoplasmic reticulum in the many animal tissue types that had suddenly become accessible to the microscope through the new thin sectioning preparation procedures.

Observations on this cytoplasmic system have been continued and elements of the system have been found in all of about 40 cell types thus far examined with the single exception of the mammalian erythrocytes. The volume of the system varies considerably from cell type to cell type as does its relative disposition. . . . In cells showing a greater quantity, e.g., kidney and intestinal epithelia, and especially liver cells, canalicular elements are frequently encountered in addition to vesicular ones. Where the system occurs in greatest volume canalicular elements predominate. Here belong the glandular epithelia, particularly those producing a protein-rich secretion.[50]

Here, in a germinal idea that would later grow into a mass of publications by the Rockefeller group and others on the endoplasmic reticulum in one or another of the countless thousands of tissues composing the animal kingdom, is anatomical logic that pointed to the endoplasmic reticulum having a function in protein production and secretion. The more a cell is devoted to manufacturing protein for export the more endoplasmic reticulum it has, so it must need reticulum for protein manufacture. This kind of morphological reasoning, which makes sense of an organic structure by finding correlations between variations in its appearance and known variations in physiological function, not only explains what a structural entity is in terms of that correlated function, but also (as I have pointed out elsewhere) silently reinforces belief that the entity really exists. For how could an artifact show variations in structure so closely attuned to physiology?[51]

From this point at the start of 1952, the research programs of the Rockefeller laboratory show a purposefulness seemingly absent previously, in the period after Claude's departure of groping for a new electron microscopical experimental system. The morphological case for the endoplasmic reticulum was pushed forward vigorously on many fronts at once (though the results were publicly presented from late 1952 through 1955). It may well be that the endoplasmic reticulum was deliberately made, as the first completely novel structure revealed in cells by the electron microscope, the test case for the instrument's reliability and the methodological paradigm for using it properly; certainly, in public appearances for general audiences such as his 1956 Harvey lecture, Porter used it as his prime example.[52] The strategy for interpreting an appearance reliably visible in cultured cells by several microscopical methods as a functionally crucial organ of all the cells that make up the higher organisms in the world, depended on a double mapping operation or translation. First, the reticulum, which was, after all, well characterized only in its appearance in whole cultured cells, had to be recognized in thin sections by electron microscopy, so that all the arguments supporting its reality in the cultured cells would also apply to the endoplasmic reticulum as it appeared in the thin sections, which were the sole access to the many kinds of cells that exist only in tissue. Arguments that the endoplasmic reticulum only exists in cultured cells (perhaps as an artifact of the peculiar conditions under which a cell taken from the body is made to grow in a dish), or that the structures seen in tissue thin section were unreal or deceptive (perhaps generated by the many transformations tissue undergoes before it can be visualized in the electron microscope), could therefore be simultaneously blocked. Thus the reticulum in cultured cells had to be unequivocally identified in the thin-section tissue micrographs, which was not trivial—and which, as noted, already involved a reconception of their appearance in whole mounts. Second, the reticulum seen in electron images had to be linked with the immense accumulation of information from a century of light microscopy on stained and sectioned tissue by classical histologists and anatomists and accumulated in great atlases, by locating it in these traditional light microscopy preparations. The science of cells and tissues would not have to be rebuilt, then, from new electron microscopical foundations (indeed, it seems doubtful that wholesale destruction and replacement of old knowledge would have been accepted by the general biomedical community). Once one knew which elements in existing descriptions of cells corresponded to the electron microscopic reticulum, one could translate from existing light microscopical evidence to the reticulum, for instance in deducing the new entity's function, and back; hence the need to identify the reticulum with some part of the cell apparent in classical light microscopy preparations. So, to sum up, the reticulum in electron micrographs of whole-mounted cells had to be systematically related or mapped to appearances of cells in two other media (or "representational spaces," if one prefers),[53] the medium of classical histology and the medium of thin-section electron micrographs. Only by these two intermediate linkages could one pass from thin-section pictures to the atlases of classical histology.

The first link was to find in cultured cells, by comparing electron micrographs with light micrographs of cells of the same variety stained by traditional techniques, that the cell regions rich in endoplasmic reticulum corresponded roughly to regions that took traditional basic dyes such as toluidine blue (Plate 3.6).[54] Though not in itself conclusive, this provided Porter with a warrant for supposing that the regions of cells in tissue sections that took these same basic dyes might also be endoplasmic reticulum; if this could be established, it would be quite significant, since there was a large body of literature relating the "basophilic" zones in histological sections to physiological changes. But to test this, it was necessary to compare tissues prepared with basic stains for light microscopy with the same tissues in electron micrographs, so one first had to know what the reticulum looked like in sections (and, of course, that it actually existed in tissue). A given cell type grown in culture and visualized as a whole mount could be compared with the same cell type, cut from tissue and thin sectioned for electron microscopy; but the appearance of the endoplasmic reticulum differed so much that so far this was insufficient to build correspondences between the two representational spaces (Plate 3.7). (Many explanations could be offered for the discrepancies, such as the flattening of cultured cells or changes in physiological state attending tissue culture.)

Thus, the second link after tentatively identifying electron microscopic reticulum with light microscopic basophilic matter was, instead, to determine what endoplasmic reticulum looked like in thin sections of the *cultured cells* that could be shown to contain it in whole mounts by electron microscopy and other techniques. What had appeared as a network of vesicles in the projection of the whole volume of a cell, now appeared as a set of isolated circular and oblong hollow profiles where the straight and narrow section plane connected with short segments of the reticulum's curvy channels (Plate 3.8).[55] This, at least, was the interpretation that reconciled the two appearances; though it stretched the visual imagination, it was plausible, and Porter and Palade did some rough serial sections to convince themselves. With heavily educated seeing, it now seemed "evident that these profiles represent sections through vesicular elements, the sizes of which coincide with those displayed by the endoplasmic reticulum [in whole mounts]. This similarity plus others and the fact that these profiles are the only structures in the section remotely like those of the reticulum, identify the two as different views of the same cytoplasmic component."[56]

Once the expected appearance of the endoplasmic reticulum in thin sections was established from cultured cells—that is, once one knew roughly what to look for—electron microscopical thin sections of a given *tissue* could be compared with traditional histological sections of the same tissue stained with basic dye, thus completing the translation. Porter was now enabled to "demonstrate a correlation between the distribution of these profiles representing the reticulum in [electron microscopical thin] section and the basophilic component," that is, to show that in tissue cells with much basophilic cytoplasm the quantity of endoplasmic reticulum was also great, that in cells with a certain distinctive distribution of basophilic material the pattern of reticulum in thin section was similar, and that when a distinctive distribution of basophilic material was altered by treatment of the animal with drugs the appearance of reticulum in thin sections was altered in similar ways.[57]

By this elaborate labor of finding equivalences that forged links for mapping the endoplasmic reticulum between whole-mount electron microscopy of cultured cells, cultured cells observed by light microscopy both living and stained traditionally, light microscopy of stained tissue sections, thin-section electron microscopy of tissue cells, and even centrifuge fractionation and biochemistry of cell parts, as discussed below (see Plate 3.9), Porter and Palade developed the bridge principles needed to map from one representational space to another. Incongruities between the spaces and imperfect correspondences between "the same" entity in different media remained, but it nonetheless became possible to move from a tissue feature known from classical histology, such as the Nissl body in neurons, to thin-section electron micrographs, and to explain the feature in terms of endoplasmic reticulum and other electron microscopic entities (Plates 3.10, 3.11).[58] Porter had extended both the established logic and the accumulated findings of anatomical sciences into the uncharted territory opened by the electron imaging of the cell. Many would follow him along this route by superadding electron microscopy technique to a grounding in classical anatomy and histology, fusing them into a single morphological approach.

Another aspect of the Rockefeller cytology group's approach to studying cell structure was less conservative, indeed almost unprecedented: the linkage, mentioned above, that Palade especially is responsible for forging between the visual representational spaces of microscopy and the chemical representational space of cell fractionation by means of the ultracentrifuge. It all began with Claude's efforts to purify the Rous tumor virus by centrifugation, which had evolved, by the time Palade was working with him in 1946–49, into an experimental system for dividing up the contents of the cell by size, and assigning various biochemical properties to each fraction.[59] Claude and his biochemist collaborators would subject the raw, homogenized cell soup—on which biochemists had traditionally performed their experiments without further refinement—to multiple steps of centrifugation that divided it into four fractions: the nuclei and other debris that quickly formed a pellet at the bottom of the centrifuge tube, the "large granules" visible by light microscopy in stained cells (such as fat storage granules and the filamentous mitochondria, to be discussed below), the small granules or "microsomes," and the supernatant solution. They would then assay for a certain enzymatic activity or chemical constituent in all four fractions, expressing the results in each as a percentage of the activity or quantity in the raw homogenate. By such methods they had found, before Claude's departure from the Rockefeller, that the microsome fraction contained almost all of the cytoplasm's "nucleoprotein" (as it should if plasmagenes were among the microsomes), which was of the ribonucleic acid (RNA) variety, and that this RNA-rich fraction stained strongly with basic dyes.

Even in his 1948 Harvey Society lecture, Claude had suggested that the endoplasmic reticulum in cultured cells might correspond to the microsomes,[60] and after Porter's identification of the reticulum with the classical "basophilic component" this became an obvious, even irresistibly tempting hypothesis. Electron microscopical examination of the microsome fraction from the ultracentrifuge (see Plate 3.9) disclosed irregular bodies that could easily be vesicles broken from the reticulum; "with a few reservations," Palade felt by early 1952, "this finding justifies the transfer of the biochemical data obtained on the microsome fraction to the endoplasmic reticulum."[61] But other biochemists, fractionating their homogenates with the ultracentrifuges they were recently able to buy, were finding in some tissues that much of the cytoplasmic RNA was sedimenting in a different fraction consisting of particles smaller than the microsomes of Claude's group.[62] This implied that Claude had been mistaken in locating most of the cytoplasmic RNA in the microsome fraction, that Porter and Palade were mistaken in identifying the biochemical properties of the microsome fraction with the endoplasmic reticulum, or both.

Porter and Palade reconciled the localization of the RNA in the microsome fraction with the conflicting biochemical results in a way that preserved the identity of the microsome fraction with their endoplasmic reticulum, thus allowing them to continue referring to ultracentrifuge biochemistry as the source of information on the chemical properties of the things they visualized by electron microscopy. They noticed, with the higher resolutions their new thin sectioning techniques were making possible, that in some of the tissues where the basophilic component was most concentrated, the dense masses of endoplasmic reticulum supposedly identical with the basophilic areas had reticulum membranes encrusted with small dense particles on the order of 10–20 mμ (100–200 Å). (See, e.g., Plates 3.10, 3.11.) In other tissue, where the cells stain diffusely with basic dyes, the endoplasmic reticulum is not the main location of the small dense particles, which are instead scattered throughout the cytoplasm. Based on this correlation between the distribution of the staining properties in histological sections and the distribution of dense particles seen by electron microscopy in thin sections, taken together with the RNA-richness of the small particles separable from microsomes by ultracentrifugation, Porter and Palade attributed stainability with basic dyes not to the reticulum per se but to small RNA-rich particles that are often bound to the reticulum.[63]

By the beginning of 1955, the distinction between endoplasmic reticulum encrusted with the "ribosomes" or RNA-rich particles ("rough ER"), and reticulum not encrusted ("smooth ER") was clear in their minds. Palade and Philip Siekevitz, whom Palade hired in 1954 as his collaborator to bring the activity of a "real biochemist"—whose vocabulary referred strictly to his fractions and subfractions rather than to cell structures—back into the cytology laboratory, were hard at work on the biochemical experiments with the microsome fraction.[64] Their work through the late 1950s would do much to elucidate the role of the ribosome-enriched microsome fraction, and thus the role of the endoplasmic reticulum, in the chemical reactions wherein protein is synthesized from its amino acid building blocks.[65] By the beginning of 1957 they had found, with microsomes isolated from pancreas cells, that freshly synthesized enzymes destined for secretion were concentrated in the endoplasmic reticulum, evidence for Porter's early notions on the purpose of the vesicles.[66] Palade was mapping biochemical results from the test tube onto the structures in the microscope image, and at the same time using the electron microscope to reinforce the fractionation research program he had helped develop with Claude, by trying to show that the centrifuge was isolating real cell components and not just an arbitrary and heterogeneous collection of fragments.

By 1956, Porter and Palade, now recognized authorities on electron microscopic method, had achieved a remarkable level of eminence compared to their uncertainty and vulnerability of only five years earlier. That year both were made full tenured members of the Rockefeller Institute. In January, a massive conference of almost all the North American electron microscopists (and a number from abroad) working on cells took place in upstate New York, organized by Porter with NIH funds expressly for the purpose of setting "high standards of quality for the newcomers in the field," which was now burgeoning.[67] In 1955 they had begun producing the *Journal of Biophysical and Biochemical Cytology* (renamed the *Journal of Cell Biology* in 1962), which Porter had conceived from the start as not only an ideal medium for electron microscopists because of its more numerous and higher quality glossy plates, and shorter publication lag in the fast-paced and increasingly competitive field, but also as a vehicle for promoting the integrated approach between electron microscopy ("biophysical cytology") and fractionation biochemistry ("biochemical cytology") they favored. "Cell biology" did not occur to anyone as a title, and indeed the discipline now known by this name had not yet coalesced.[68] (Porter later recollected that the original unwieldy title was foisted on him by Schmitt, that enthusiast for all things "biophysical"; see Chap. 4.)[69] They were setting and disseminating standards of practice in their field, as conference organizers and editors and as doctoral and postdoctoral supervisors, building a community of electron microscopists by widely sharing their way of studying cells.

As noted, when Porter gave his own Harvey Society lecture in 1956, just eight years after Claude's, he used the endoplasmic reticulum as the exemplary story of how one maps an entity between the representational spaces of electron microscopy, classical histology, and fractionation biochemistry. He was able to joke at the start of his talk, speaking for biological electron microscopists: "Investigators still appear on the fringes of our society who question the authenticity of what the rest of us accept as factual, but such terrorists are neither numerous nor long-lived. They soon join us in extolling the virtues of our OsO_4 and our new gadgets."[70] Porter's endoplasmic reticulum and the Rockefeller style of reasoning that had established it were then widely accepted, indeed so much so that others were taking up study of the entity and already it was "drifting out of [his] clutches," though into friendly hands.[71] However, not everyone was impressed with the reticulum or with Rockefeller methods. One 1956 review article saw it thus:

At the low resolution used in Palade and Porter's first work it was impossible to differentiate membranous or vesicular structures of different kinds and therefore everything occurring in the cytoplasm that appeared as tubules or vesicles was described as the "endoplasmic reticulum." . . .

This mixing of various structural elements under one and the same term "endoplasmic reticulum" would be justified [only] if it had been demonstrated that we are dealing with homologous components from the point of view of function or cytogenesis, or if we were not able to differentiate between them structurally.[72]

Who dared to accuse the Rockefeller cytologists of poor technique and sloppy thinking—some wild-eyed, backward "terrorist" without proper appreciation of new technology and osmium fixation? No, it was Europe's most respected electron microscopist of cells and tissues (and the "cancer researcher" in the RCA ad mentioned above), Fritiof Sjöstrand (Plate 3.12).

Palade, Sjöstrand, and the Mitochondrion

So the Rockefeller cytology group developed techniques of making electron micrographs that were already becoming standard for the study of cells and tissues by the mid-1950s. "Everyone puts his tissue into Palade's mixture and describes what comes out," Porter wrote in mid-1955, expressing private misgivings that their standard fixative might not be absolutely perfect.[73] In parallel, Porter and Palade also had developed an epistemology—a system for interpreting micrographs to make knowledge from those pictures—that was similarly becoming standard. As I have described, Rockefeller interpretive modes involved an empirical comparison or "calibration" of preparative techniques for tissue specimens against specimens of cultured cells observed living or during preparation; a maximum reconciliation of electron microscopy with the established findings of light microscopical histology and an adoption of the traditional reasoning from variations in morphology and variations in physiology to general interpretations of function; and a correlation between the morphology seen by electron microscopy with the biochemistry of cell components fractionated in the ultracentrifuge, such that test tube biochemistry provided the information used to "color in" the structures outlined topographically by microscopy. The Rockefeller way of interpreting micrographs meshed well with the group's way of producing the pictures and its involvement in fractionation, and together these ways represented a productive basis for an experimental system, complete except for an interesting and convenient cell type of the researcher's own choosing. Described thus, as an adaptable, open-ended, and easily transportable bundle of practices, it seems no wonder that the Rockefeller ways became popular with biologists wishing to investigate cells and tissues with electron microscopes.[74] There is no question that it "worked" in generating experimental findings and continues to work today among the practitioners of what is now called "cell biology." But was the Rockefeller way a standardization of the only properly scientific—or even (admitting multiple possibilities) the epistemologically optimal—bundle of practices, and the one that had inevitably to triumph in the end?

To raise questions about this issue (i.e., whether there is any absolute sense, independent of context and history, according to which one of several workable experimental modes can be said to be superior), which is, after all, relevant to all experimental sciences, and to examine more closely the process of standardization that accompanies the foundation of research traditions, I will describe the different manner of making knowledge from pictures developed by Sjöstrand. The difference between the Rockefeller way and Sjöstrand's was almost exclusively one of interpretive practice, since the two schools produced electron micrographs in almost the same fashion. Indeed, from the Rockefeller point of view, Sjöstrand's techniques were equivalent, though from Sjöstrand's point of view there were crucial minor differences that distinguished his as the superior and more careful practices. Such a case in itself raises doubts about the degree of technological determinism that can be attributed to experimental systems, because the two camps had practices that were so similar technically, yet they used the results so differently that they came to conflicting conclusions about the natural world. Such a case also inherently foregrounds the roles of individual purpose and social context; for as we shall see there were consequences for the proper boundaries and goals of the nascent discipline of cell biology, and these stakes may even have determined whose camp—and which epistemology—emerged as dominant. To illustrate the differences between the competing ways in which the Rockefeller group and Sjöstrand proposed to standardize interpretive practices, nothing could serve better than the dispute between Palade and Sjöstrand over the structure of another of the cell components drawing early attention, the mitochondrion.

The mitochondrion's discovery is usually attributed to Richard Altmann of Leipzig, a cytologist who in 1890 named a certain protoplasmic granule the "bioblast." He believed these particles to be the fundamental units of life, and Altmann's ensuing insanity and demise has been attributed to his devotion to this theory and the difficulty of defending the reality of the hard-to-see granules. Around 1900 the Janus green method for staining the granules even in living cells, and thus for reliably demonstrating their existence by light microscopy, won a number of converts over to belief in the entity.[75] In the early part of this century quite a few histological studies attended to these bodies, called "mitochondria" or "chondriosomes" or many other names, and a variety of ideas about their physiological function emerged. A move beyond purely microscopical study of mitochondria was made by Robert Bensley and Normand Hoerr of the University of Chicago in 1934, when they showed that the bodies, somewhat the worse for wear and lacking some staining properties that distinguished them in the whole cell, could be separated from the other cell components in an organ homogenate by centrifugation.[76] This fractionation method was taken up by Claude and developed in his efforts to determine the contents of tumor virus–infected and normal cells, as mentioned above. One of Palade's first jobs under Claude was to work with Schneider and Hogeboom on improving fractionation technique, and in 1947 they published an improved method of isolating mitochondria by centrifuging homogenized tissue in a heavy sucrose solution, which preserved their Janus green staining properties.[77] Electron microscopy and biochemical results from his "large granule" fraction, in which most of the enzyme activity associated with aerobic respiration could now be found, had convinced Claude by the time of his 1948 Harvey lecture that mitochondria were, as Bensley's school had proposed, the "power plants" of the cell, and that they were bounded by membranes that kept water–soluble enzymes from leaking out.[78]

Immediately after the Second World War, biochemists showed an intensifying interest in mitochondria, stimulated not only by the inherent interest of the bodies but also by the new availability of funds with which to buy ultracentrifuges, now available commercially for the first time from brands like Sorvall and Spinco (Specialized Instruments), with which to isolate them. Speaking no doubt for many biochemists, two early Spinco purchasers recall "they will never forget their sense of wonder and delight, almost of reverence, when they first unpacked the slick new instrument that completely changed their lives."[79] Much as cytologists were adopting Porter and Palade's bundle of tissue preparation techniques for use with their new electron microscopes, these biochemists took up the fractionation methods that Claude's group had developed, particularly the Hogeboom-Schneider-Palade sucrose technique, the Claude four-fraction scheme, and the strict accounting procedure by which activities are expressed as a percentage of the total for raw homogenate. Thus the Rockefeller group established standards for the practice of "biochemical cytology" just as they did for "biophysical cytology." We shall eventually return to the relation between electron microscopists and biochemists studying mitochondria.

Fritiof Sjöstrand could in many respects be considered the European counterpart of Keith Porter. During the war, as a medical and then doctoral student at the Karolinska Institute in Stockholm, he began working with an electron microscope built by physicist Manne Siegbahn.[80] He experimented with ways to prepare very thin tissue sections, and in 1943 won recognition for micrographs of muscle that were published in *Nature*.[81] In 1947–48 Sjöstrand visited as a postdoctoral fellow in Schmitt's MIT group, which was actively studying macromolecule assemblies by measuring features directly from micrographs and checking these findings by X-ray crystallography and other biophysical methods (see Chaps. 1, 4).[82] From Schmitt, Sjöstrand may well have acquired both that style of direct molecular measurement and the term "ultrastructural" for such research.[83] On his return to the Karolinska Institute as a faculty member, Sjöstrand devoted renewed attention to specimen preparation methodology. He had connections with the circle of RCA clients through Schmitt, and must have made more North American acquaintances when he gave a paper at the 1948 EMSA meeting in Toronto.[84] In late 1949, when Sjöstrand returned briefly to attend the EMSA meeting in Washington, he visited Hillier at RCA and also the Rockefeller lab, where Porter shared with him his techniques for preparation of cultured cell specimens and the two traded ideas on microtome design. In early 1950 Sjöstrand could report that his new microtome was cutting well and his new RCA EMU was on its way,[85] purchased for his lab with difficulty, since Dutch and German machines were becoming available by then and there was pressure on him to buy one of Siegbahn's commercial instruments as well.[86] Soon Sjöstrand's microtome design was produced commercially, one of the first featuring thermal advance (from expansion of a heated metal bar) rather than mechanical, and it "for a time . . . dominated the European scene," just as the Porter-Blum model ruled America.[87]

Palade and Sjöstrand appear to have begun intensive work on the mitochondrion independently, despite intermittent communication between the two while it was underway.[88] Sjöstrand felt himself in a race to achieve genuinely high resolution pictures of tissue thin sections, and considered that he had won when at the October 1952 EMSA meeting in Cleveland he showed pictures of cellular structures, including mitochondria (Plate 3.13), that were so well received that, he recalls, he was called upon to give eight spontaneously invited seminars in American labs over the seven remaining days of his visit.[89] Palade had the first major study on the topic in print, in the November 1952 edition of the American histological journal *Anatomical Record* (submitted, of course, before Sjöstrand's EMSA presentation).[90] There Palade reported using his new buffered fixative on a wide variety of rat tissues, embedding them in methacrylate, and cutting sections on Porter's microtome with a glass knife. The images of mitochondria he obtained present a wide variety of outlines, from spherical, to oval, to very elongate (Plate 3.14). Palade concluded that the true shape is sausagelike; however, he pointed out that if a set of such sausages is randomly oriented, then the majority of outlines in a section will be oval because the section plane is oblique to their long axis. In certain tissue such as the proximal kidney nephron, however, the mitochondria all lie in one direction, so the investigator can intentionally cut them the long way (i.e., longitudinally) or in any other desired fashion, by keeping track of the tissue fragment's original orientation after removal from the animal. Palade made no such special effort to present longitudinal views.

Palade endeavored to establish two main points in this first study. First, the mitochondrion is totally enclosed in a membrane, as suspected from electron microscopy on whole cells and on fractions, and from some biochemical findings. Second, all mitochondria have internal membrane leaves or lamellae, aligned more or less orthogonally to the long axis, and these lamellae appear to be protrusions of the external membrane. This point fit well with the opinion of some biochemists working on the oxidative reactions that take place in isolated mitochondria, that the many enzymes involved in these pathways must be supported in some "definite spatial relationship," in the words of Palade's colleagues Hogeboom and Schneider.[91] Palade thought the membranes inside mitochondria might bear these enzymes "in the proper order in linear series or chains" like assembly lines to break down substrates. He clearly was considering the significance of his findings to the biochemists and trying to make microscopy useful to them. Subsidiary to Palade's second point, that these membrane protrusions into the mitochondrion interior exist, was the issue of whether they are true septa, dividing the length of the body into closed chambers, or simply baffles that do not go all the way across. In some mitochondria (e.g., in Plate 3.14, m_1), it looks as though most of these lamellae do go all the way across. But in others (e.g., in Plate 3.14, m_2 and m_3) where the outline is narrow, the lamellae go across, but where it is wide, they do not. Palade based his argument that these lamellae are just baffles or ridges—he dubbed them "cristae"—mainly on oblique cuts like m_2 in Plate 3.14. His interpretation was that the cristae extend inward from the mitochondrion walls part of the way, so where the section plane skims close along a wall they seem to go across, but where the cut goes near the center of the body, a free channel down the middle of the mitochondrion is visible. Finally, just at the limit of resolution in Plate 3.14, the cristae themselves appear to be multilayered. Palade estimated that each (possibly layered) lamella was 1.5 times thicker than the mitochondrion's limiting membrane.

Sjöstrand's first publication on the internal structure of mitochondria appeared in *Nature* early in 1953.[92] Using techniques similar to Palade's except that he employed his own microtome, still fitted with a steel razor rather than glass, Sjöstrand showed images of longitudinally sectioned mitochondria, cropped and carefully selected for the limited space given by the journal. These were among the pictures he had shown at the 1952 EMSA meeting (see Plate 3.13). The resolution, claimed as 30 Angstroms, was obviously better than Palade's. Similarly, noting that there were stacks of internal double-layered membranes going crosswise through the mitochondrion interior, Sjöstrand here differed from Palade's first report mainly in not supporting the observation of a free channel, and also in giving a larger estimate for the outer, limiting membrane's thickness. This membrane is a double laminate, he claimed, just like the lamellae, and is of equal thickness. Sjöstrand argued that the thickness and appearance correspond to what would be anticipated, based on the then-current "sandwich" theory of membrane biochemistry, from two layers of protein coating a lipid bilayer. Sjöstrand showed an interest in the relevance of his results to biochemists but for him, unlike Palade, this involved testing molecular structures predicted by biochemical theory by examining micrographs directly.

In Chicago in April 1953, the fourth annual meeting of the Histochemical Society featured the symposium on structure and chemistry of mitochondria that Mudd organized and eagerly chaired (see Chap. 2). Palade presented an expanded version of his structural study using new micrographs of higher resolution, which Mudd considered as good as Sjöstrand's, though Hillier still judged them inferior.[93] Palade's new pictures of mitochondria changed his interpretation of the structure little, however, except that he now agreed with Sjöstrand's report that the outer membrane appears double.[94] Palade reiterated the argument, given earlier, that if the lamellae were true septa, they would go all the way across in

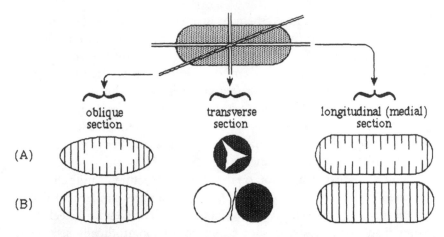

Fig. 1. Views expected of mitochondria sectioned at various angles, (A), if interior membranes are baffles not compartmentalizing the body's interior; and (B), if interior membranes are true septa. Drawing by author, after Palade, "An Electron Microscope Study of Mitochondrial Structure."

all longitudinal sections and even in oblique ones, this time illustrating it to facilitate understanding of the micrographs (Fig. 1). Note the predicted appearances, on the two models, of a transverse section. He then showed micrographs depicting the predicted cross-sections for bodies filled with baffles or ridges and a model to communicate all three dimensions at once (Plate 3.15).[95] According to Palade's interpretation of the mitochondrion's structure, the cristae were ridges extending only part way inward, leaving a channel down the center free for diffusion. The ridges appeared to be infoldings of the inner of the twin bounding membranes, a suggestion with biochemical implications, discussed below. Palade thought Sjöstrand's description of the lamellae as a system of "double membranes" to be similar to his own notion of cristae, but unlike Sjöstrand, Palade only described topography and did not address questions of membrane composition using electron microscopy.

Because his view of the mitochondrial interior derived from sections was consistent with findings from ultracentrifuge studies, Palade thought it "permissible to attempt a correlation of the biochemical data yielded by studies on mitochondrial fractions with the structural pattern found . . . *in situ*" by electron microscopy. This was his goal, to forge a link between his interpretation of the structure and biochemical evidence, and he accomplished it by arguing from the latter to the former:

It is known that the mitochondria contain a number of oxidative enzymes which, being insoluble or structure bound, have to be part of the solid framework of the organelles. If the present electron microscope observations are correlated with already known findings such as those yielded by the fractionation of disintegrated mitochondria . . . the insoluble enzymes have to be located either in the external, limiting membrane, or in the internal membrane that forms the cristae by its folding. The good coordination shown by the many enzymes in the oxidative and phosphorylating processes carried out by isolated mitochondria indicate that the enzymatic units . . . are disposed in the proper order in enzyme chains, an arrangement comparable in design and efficiency to an industrial assembly line. . . .

[T]he cristae represent the most probable location for the enzyme chains, for the following reasons:

a) A number of experiments indicate that a substrate has to penetrate inside the mitochondria before being acted upon.

b) Particles of the general dimensions of the cristae and containing most of the succinoxidase system of the mitochondria have been isolated from suspensions of disintegrated mitochondria.

c) In the case of cristae the enzyme systems would be less exposed to disruption than if a part of the limiting membrane.[96]

Palade went on to urge the plausibility of the central free channel in terms of biochemical theory: it made sense in terms of efficiency for the series of enzymes supposedly bound to the cristae to be bathed in a common pool of soluble enzymes and substrates. The cristae increase the interior surface area of this "diffusion chamber" and therefore enzyme exposure to the common pool in that central chamber. And since the cristae are folds of the inner membrane, another peripheral chamber might exist between the inner and outer membranes of the mitochondrion where rate-limiting reaction products might be pumped from the interior.

Palade's reasoning in this effort to "correlate" electron microscopy with biochemistry takes for granted both biochemical theory and results ("known findings") from fractionated mitochondria, and employs the visual evidence from electron microscopy simply to provide a three-dimensional space in which to locate the biochemistry. By refraining from any effort to visualize the enzyme systems or membrane architectures, that is, by not treating the same scale of structure as the biochemists, Palade removed virtually all potential for conflict between the evidence from electron microscopy and that from biochemistry. The two experimental approaches spoke to separate domains: the former to questions of arrangement, the latter to questions of molecular structure and function.

When a number of the traditional cytologists at the meeting expressed doubts that mitochondria, with all their bewildering and mutable appearances through the light microscope in various tissue contexts, represent an entity with a fixed and uniform identity, Palade defended the biochemists' project of purification and analysis by arguing that constancy of mitochondrial internal structure implied constancy of function.[97] Thus Palade offered the biochemists a visual means of defining the mitochondrion as a stable entity, and a way of mapping the enzymological data derived from fractionation into three dimensions, but not a competing method for investigating molecular structure.

In 1953 Sjöstrand, with his student Johannes Rhodin, published a detailed study of mitochondrial structure in a report on the mouse proximal kidney nephron, one of those few tissues in which mitochondria have a particular orientation.[98] Submitted to the journal in August 1952 and thus before Palade's first publication appeared, it is a portion of Rhodin's doctoral thesis;[99] if Sjöstrand and Rhodin were aware of Palade's efforts, they did not acknowledge it. In any case, they obtained some more pictures with excellent resolution (see Plate 3.16), in which all the mitochondria were longitudinally sectioned. Their interpretation of mitochondrial structure was based almost exclusively on such longitudinal views, because only with section planes perfectly normal to the membrane surface can membrane components be precisely measured from micrographs. Sjöstrand did not, however, set out the reasons for his selectivity plainly in any one publication, which may have contributed to misunderstandings of his views. Those reasons are part and parcel of Sjöstrand's strident insistence on precision thin sectioning and fine tuning microscopes for maximum-resolution pictures. His reliance solely on medial longitudinal sections of mitochondria can be explained as follows: the electron micrographs in question are shadow images or profiles of structures made by an electron beam passing orthogonally through the section. In general, membranes whose planes deviate from vertical, as defined by the beam path, will leave a wider shadow the greater the deviation, and will at a given deviation angle leave a wider profile in direct proportion to section thickness, exacerbating the problem. Thus, if one of these membrane structures is the mitochondrial wall lying nonvertically in the section plane, its profile will lead one to conclude that its cross-section is wider than is actually the case, and moreover its shadow may overlap that of a nearby object (such as the edge of another, internal membrane) and thus give the false appearance of connection. Therefore images obtained from the thinnest possible sections and from perfectly medial sections (in which the curved mitochondrial outer wall deviates minimally from the beam path) give more accurate measures of membrane thickness and may reveal closely appressed inner and outer membranes to be noncontiguous. Thicker sections from the same specimens, or sectioning planes in which the mitochondrial wall deviates from the beam path at a greater angle, may yield false interpretations of contiguity because of superposition. Low resolution microscopy may lead to the same false results even in perfectly sectioned specimens, through blurring.

The interpretation Sjöstrand and Rhodin derived from their medial longitudinal sections was this: the mitochondrion is completely enclosed in one two-layered membrane. Inside there is a "system of internal double membranes" stacked transversely to the long axis. The main difference from Palade's model was that the lamellae go all the way across in most cases and are not part of the bounding membrane of the body. Instead, "at the edge of the . . . inner double membranes the two constituent membranes are connected," that is, the lamellae are actually a stack of independent membranous plates. In a diagram, the authors gave a hypothesized molecular architecture of the membranes based on measurements from their micrographs (see Fig. 2). Sjöstrand and Rhodin did not attempt to muster biochemical evidence for their interpretation, but gave considerable attention to prior light microscopy evidence and histochemical studies, linking the force of their argument to these latter sources, although the matter did have implications for biochemists. In addition to their hypothetical membrane architecture, they, like Palade, suggested that the lamellar organization they described might plausibly support an array of mitochondrial enzymes (though it was not consistent with the Palade two-chamber picture). For Sjöstrand much more than for Palade, the microscopist was in a position to speak to problems in the biochemical domain, such as the size, shape, and distribution of macromolecules in the cell components.

In a subsequent publication with another student, submitted in April 1954, Sjöstrand responded directly to the differences between Palade and himself concerning the mitochondrion. The quality of the micrographs was again high (see Plate 3.17), but only two mitochondria were shown in the publication, both medial longitudinal sections. Here as elsewhere, Sjöstrand stressed the superior resolution of his group's micrographs, evincing the attitude that only the best thin section pictures, no matter how few, had any weight at all as evidence. The authors acknowledged

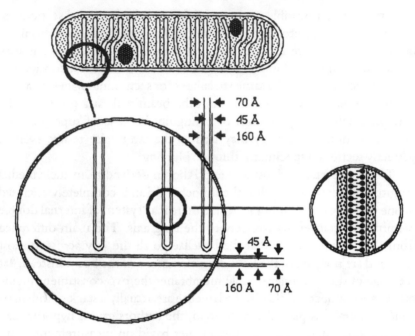

Fig. 2. *Top*, view of longitudinal section of mitochondrion in situ; *below left*, large inset, close view of junction zone between bounding mitochondrial membrane and edge of internal membrane, with membrane dimensions; *below right*, small inset, hypothesized molecular architecture of mitochondrial membranes as lipid core surrounded by protein. Drawing by author, after Sjöstrand and Rhodin.

that in this different tissue (pancreas) the lamellae do not go all the way across, but still denied that the internal membranes are infoldings of the inner bounding membrane.

Densely packed inner membranes or plates are seen mainly oriented perpendicularly to the long axis of the mitochondrion. . . . One end of the membrane is in contact with the outer surface membrane and the other end is in most cases free from this membrane. . . . There is with few exceptions no continuity observed between . . . the inner and the outer membranes.

No support has been found for the opinion of Palade regarding the internal structure of mitochondria. There are no indications that the inner membranes represent folds. . . . The inner membranes are individual structures with only topographical relations to the outer membrane.[100]

The lamellae, then, are independent plates not physically continuous with the limiting double membrane of the mitochondrion, just as Sjöstrand had concluded earlier. Thus the Swedes confirmed in pancreas the structure they had previously described in kidney mitochondria, and they maintained that the molecular architecture proposed earlier for the membrane might still be "tentatively assumed" to be true, even though they now admitted that some septa may have gaps. They argued that Palade had probably seen a free channel in his kidney mitochondria because of poor tissue preservation (which might have caused collapse or displacement of the septa). Sjöstrand this time offered no arguments from biochemical or histochemical studies for his view of the mitochondrion, reasoning entirely from electron micrographs and specimen preparation methods. This kind of evidence, for Sjöstrand, was sufficient even to address molecular structure on a molecular scale.

In July 1954, at the Third International Conference on Electron Microscopy in London, the conflicting positions of Sjöstrand and Palade crystallized and hardened. Sjöstrand, one of three plenary speakers, gave a summary of recent progress in biological electron microscopy, and used the occasion to critique Palade in a way that could only have been taken as challenging. After a few paragraphs relating to improvements in ultrathin-sectioning technique, Sjöstrand entered into a discussion of proper interpretation of the new sort of image thereby obtainable.[101] He chose the mitochondrion for a good example of interpretation mistakes—Palade's interpretation mistakes, that is. Citing the independent work of Palade and himself, he showed a drawing of his stack-of-plates model, either with irregular or perfectly fitting membrane septa, as "the very characteristic organization" of the body that had "been revealed" (see Fig. 3). Here Sjöstrand once again proposed a molecular configuration of the membrane derived from his micrographs. Sjöstrand then displayed some of his excellent electron micrographs, apparently showing that the internal membranes are not extensions of the inner mitochondrial envelope. For Palade and Porter in those days, Sjöstrand was their prime adversary, and when he brandished a "fist full of slides" like these, they were very jealous.[102] Nothing could demonstrate more plainly that both sides in the controversy basically agreed about how best to make an electron micrograph than the Rockefeller group's acknowledgment that Sjöstrand's pictures were better. But Sjöstrand had few such superior micrographs, and the Americans reportedly began to suspect he was carefully cropping images to get several pictures from one good negative, and even printing from upside down negatives to increase his apparent supply of good mitochondrial images.[103]

In any case, while showing his excellent pictures and describing his

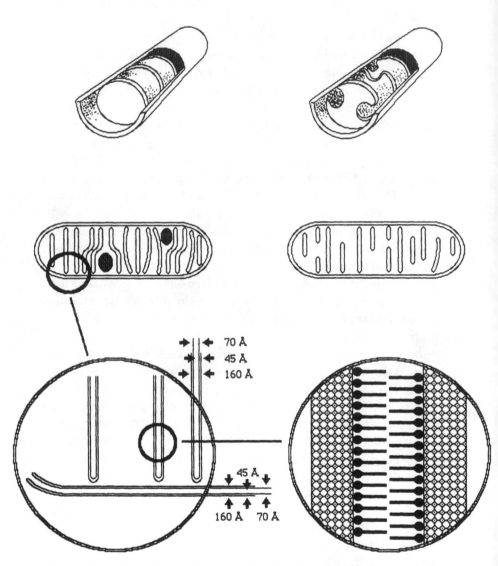

Fig. 3. Illustration of structure of mitochondria found in different tissues, *left top and left center*, with regular internal membrane plates (i.e., septa); and *right top and right center*, with irregular internal membrane plates; *bottom left*, inset, close view of junction zone between bounding mitochondrial membrane and edge of internal membrane, with membrane dimensions; *bottom right*, inset, hypothesized molecular architecture of mitochondrial membranes as lipid core surrounded by protein. Drawing by author, after Sjöstrand, "Recent Advances."

own interpretation of the mitochondrion in London, Sjöstrand attributed Palade's views partly to sloppy technique:

In order to reveal the structural geometry of a component such as the mitochondria, a rather high resolution is required. . . . The discrepancies that exist in the interpretation of the ultrastructure of mitochondria are due to differences regarding resolution and preservation of the mitochondria. When the ultrastructural components are made clearly visible, their dimensions may be measured. The value of these measurements depends on the resolution of the electron microscope as compared with the dimensions measured. . . . Describing and defining in a quantitative way the structural components observed seems an important task for the electron microscopist.[104]

The issue of preservation stems from the one difference in preparative technique between Sjöstrand and the Rockefeller workers. The Swedish group believed that it was safer to base interpretations of the living state on specimens whose cells were, as far as possible, living normally until the moment of fixation. Hence they injected live animals with fixative, cut tissue from their animals very quickly and immersed it in fixative dissolved in phosphate-buffered saline, carefully adjusted to physiological levels of total salt and pH. As Porter has recounted, the Americans had already decided this last was a nonissue, not important enough to merit serious investigation: "Palade made no attempt to employ a physiologically compatible vehicle, for he did not think it was important. Most subsequent investigators have also ignored physiological compatibility, even when selecting other buffers. By contrast, in the mid-1950s . . . Sjöstrand [and his students] advocated the use of a balanced salt solution with only minor buffering properties as the fixative vehicle."[105] Reasons for the Rockefeller group's reluctance to embrace the issue of physiological conditions of fixation are not hard to find, as we shall see below. Sjöstrand, however, thought Palade was using hypotonic fixative for the same reason he preferred embryonic tissue: with greater water content, organelles are more transparent in thickish sections. Sjöstrand privately reached the conclusion, he recalls, that Palade systematically "optimized" his technique for swollen specimens because sections did not have to be cut as thin for appropriate contrast in the final micrograph.[106]

Apart from highlighting the undeniably superior resolution of his micrographs, Sjöstrand's talk of quantitative precision referred to Palade's rough and sometimes inaccurate size estimates of cellular entities (e.g., the dimensions of mitochondrial membranes in Palade's initial report), as opposed to his own almost fanatical estimates to within a couple of Angstroms, based on many measurements from a few satisfactorily prepared mitochondrion pictures. These careful measurements were the basis of Sjöstrand's models of membrane molecular architecture. Sjöstrand urged all biological electron microscopists to follow his example of precision: "The quantitative attitude is stressed because it is necessary for the identification and classification of the different components [of structures]. . . . The electron microscope is an efficient measuring device making exact quantitative description possible, provided it is mastered to give high resolution."[107] By the rules he was advocating for the game of cytology, the electron microscopist's task was to measure and identify the molecules in cell parts, and these rules would indeed make Sjöstrand the undisputed master of the game. But there was a basic question about what the rules should be, that is, a fundamental disagreement on how well micrographs needed to be made, and how to use them properly as evidence.

Palade's paper later in the London meeting dealt with the fixation of tissues for electron microscopy. In it, he did not specifically justify his interpretation of the mitochondrion (and indeed, he did not include any mitochondria among the micrographs illustrating his contribution), but he took the opportunity to counterattack Sjöstrand's interpretive methods. Citing evidence of certain biological materials with a periodic structure analyzable by X-ray crystallography, in which the results from osmium-fixed and unfixed specimens differ, Palade cast doubt on the value of precise measurements from electron micrographs for reaching conclusions about biochemistry: "It seems, therefore, that the dimensions and spacings shown in fixed material by the electron microscope cannot be considered sufficiently true to nature to permit us to deduce the chemical composition and molecular architecture of a certain structure by finding out . . . which particular kind of molecule would best fit a given spacing."[108] Similarly, Palade riposted to Sjöstrand and Rhodin's argument that his fixative buffer was not physiological by pointing out that the true osmolarity (overall water concentration) of the cell probably differs from cell type to cell type, and was impossible to gauge accurately anyway. This deft side step construed osmolarity as a false issue, rooted in an inappropriate quest for precision on Sjöstrand's part, and a fruitless concern given available experimental technique. Of course, the value of many of the biochemical results would be impugned if it were granted that osmolarity shifts (such as inevitably accompany homogenization) irreversibly alter organelles; and moreover, if unphysiological media were seriously to affect mitochondrial structure, that would cast doubt not only on Palade's fixative but the entire enterprise of the biochemical study of mitochondria and other organelles obtained by fractionation with sucrose of any molarity. And electron microscopy was the only direct way to study mitochondrial structure. Sjöstrand's train of thought implied that the Rockefeller microscopy and fractionation biochemistry were converging on an arbitrary result.

Palade highlighted and challenged one of Sjöstrand's fundamental epistemological principles, a principle that, though not directly testable, was defensible by reference to entropic considerations and not uncommon among microscopists: "Sjöstrand is of the opinion that fixation tends, in general, to disorganize the cytoplasm so, in comparison with the situation in vivo, no structure is added, but structure may be subtracted. Accordingly the best fixative is considered to be that which leaves the specimen with a maximum of organization. Sjöstrand's view . . . supposes that there is always more order in a living than in a fixed specimen, which may not necessarily be true."[109] Palade considered it safer to rely on agreement of observed structures "with data derived from studies in cell chemistry and cell physiology," such as the work with centrifuge-purified mitochondria that indicated a containing membrane on the body. This is a highly significant respect in which Palade differed from Sjöstrand: Sjöstrand wanted to interpret his micrographs purely visually, judging fixation by the criterion of orderliness and seeking greater knowledge of molecular structure through ever-better resolution, whereas Palade, who was involved in cell fractionation himself, wanted to test micrograph interpretations against experiments on fractions. Sjöstrand's rather caustic comments after this talk, implying that Palade ought to join his students in Sweden in carefully testing all conditions of temperature and osmolarity before claiming good fixation, drew only the conciliatory (though defensive) reply that what was most significant was the increasing agreement between the two groups. Other discussants apparently did not take sides.

To summarize then, the confrontations at the 1954 London meeting highlighted a number of differences in interpretive style between Sjöstrand and Palade. To decide between conflicting microscopic images, Sjöstrand would rely on criteria internal to the micrograph and the specimen, such as resolution and section thickness, or failing that, invoke an essentially aesthetic epistemological criterion (plausible on thermodynamic grounds) to the effect that the image showing greater order in the specimen more closely reflects the specimen's original living condition.[110]

With Sjöstrand's mode of interpretation, a single perfect micrograph—even if utterly unique—was more convincing than any number of conflicting but inferior micrographs. By measuring the best micrographs, Sjöstrand would attempt to achieve his goal of deducing molecular architecture. Palade, on the other hand, refrained from addressing all questions of molecular structure by microscopy, leaving these to fractionation biochemistry, and based his interpretations of topography on a larger number of partially imperfect images (e.g., obliquely sectioned mitochondria). Moreover, Palade referred problematic questions of interpretation not to any basic epistemological principle or criterion of micrograph veracity fully under the microscopist's control. Palade would instead choose among possible interpretations of micrographs according to evidence from cell fractionation, and according to plausibility in accepted biochemical theory (e.g., Palade believed the free channel he saw in situ was not a preservation artifact because centrifuge-isolated mitochondria had it as well, and because it made sense in terms of diffusion dynamics). Information flowed from fractionation biochemistry to morphology for Palade; for Sjöstrand it only went the other way. They agreed almost completely about what constitutes a good micrograph; their disagreement concerned interpretation. The one serious difference between them over specimen preparation, Sjöstrand's insistence that microscopy was unreliable to the extent that conditions of specimen treatment were unphysiological, undercut the value of Palade's cell fractions as a criterion of microscopic structure. Because fractionation requires both homogenization (in which individual cells are broken and the contents of at least some are inevitably exposed to different osmolarities) and unphysiological media (especially sucrose), Palade could not concede this point without abandoning that outside criterion and admitting a major flaw in the biochemical approach he helped develop in Claude's Rockefeller lab. So here too the issue of interpretation was at stake.

Porter spent the rest of the summer of 1954 in Europe visiting electron microscopists and, spurred by the evidently superior quality of the competition's pictures, also looking at the latest Siemens model, which he and Palade hoped could be obtained to supplement their six-year-old EMU. Palade sent Porter some fresh thin sections to examine and urged him on: "I still believe that the Swedes have an advantage in resolution . . . but their ability to integrate their results in general cytology and in physiology remains nil. This is our main asset and it will be wonderful if, in addition, we could regain the lead in high resolution. Fly therefore to Berlin, don't get mixed up with the Russians, and bring back an 'Übermikroskop' that will show even sodium ions crossing through the membranes."[111] Palade had already stopped working directly on mitochondria, though of course they still appeared in many of his published micrographs of tissue sections. Sjöstrand, in his subsequent publications, continued occasionally to point out the noncontinuity of lamellae and mitochondrial wall when longitudinal sections of the bodies appeared in micrographs obtained from whatever tissue he happened to be studying. The specific issue was essentially left to be decided by the accumulating pictures of the community of biological electron microscopists, and most were supporting Palade's view. At Porter's standard-setting January 1956 conference (mentioned above), in which "109 investigators participated, including many from abroad [but not Sjöstrand, despite a belated invitation], and 75 papers were presented," sixteen papers were given in the section on mitochondria, and all used Palade's terminology of "cristae."[112] Of these, all five that explicitly commented on the issue of the cristae accepted Palade's model, and showed pictures consistent with the lamellae as infoldings.[113] When in September 1956 Sjöstrand organized an even bigger meeting, the first European Regional Conference on Electron Microscopy, it "proved to have attracted 370 participants from 27 countries" including the United States,[114] Porter and Palade not among them. Most of the 78 contributions on biological topics used Sjöstrand's terminology of "inner double membranes" for the mitochondrial lamellae. However, Sjöstrand himself proved willing to use Palade's terminology of "cristae" (in inverted commas),[115] and he admitted that in "isolated cases" continuity between the inner space of the "inner double membrane" and the space between the bounding membranes was apparent.[116] It is difficult to say how much the growing political clout of the Rockefeller cytologists contributed to the convincing quality of their micrographs portraying lamellar continuity with the bounding membrane as the general state of affairs; certainly, some room for varying interpretation remained, and European biochemists were not yet convinced of Palade's view even in the late 1950s.[117] There was no biochemical evidence to decide the issue. Sociological considerations will shortly be dealt with further.

In 1956 both Sjöstrand and Palade wrote review articles dealing with the structure of mitochondria, and these give us a nice indication of the dispute's status at three years. Palade, who as noted had not been pursuing the mitochondrion topic any further, gave his review as part of a major conference on enzymology and molecular biology, which in itself highlights his close connections to the biochemical world.[118] He basically repeated his earlier conclusions: the mitochondrion is bounded by two membranes and filled with cristae, which are infoldings of the inner bounding membrane. He showed some new micrographs making plain the claimed continuity between lamella and limiting membrane in mitochondria isolated by centrifugation and in situ (Plate 3.18). Although Palade could cite many recent papers that had supported his interpretation, a look at the micrographs therein shows that good pictures illustrating continuity had been forthcoming only from a few other laboratories. Porter and some Rockefeller colleagues had produced images of protozoan mitochondria in which the cristae are tubular, but nonetheless indisputably continuous with the inner membrane.[119] However, this form might be thought the evolutionarily and developmentally primitive precursor to Sjöstrand's separate plates, which made the counterexample inconclusive.[120] Because Sjöstrand's and Porter's groups were still the main sources of really sharp thin-section electron micrographs of cell structures, the opinions of both sides had considerable weight. In his conference contribution Palade again muted the conflict, which after all might reflect negatively on the value of electron microscopy in general: "The points in disagreement are decreasing in number, however. For instance, the existence of two dense lines at the periphery of mitochondrial profiles, revealed by Sjöstrand's work, was subsequently confirmed by us, and the cristae ('inner double membranes' [in Swedish terminology]) originally described as complete septa by the Swedish group, are now recognized as incomplete partitions."[121] Palade also emphasized the biochemical implications of his mitochondrial structure: the cristae could support insoluble enzymes and maximize internal surface area, the mitochondrion is a "two phase system" with an internal fluid different from the cytoplasm, and there is potentially another fluid in the "outer chamber" between the outer limiting membrane and the cristae. Palade was offering the efforts of electron microscopists like himself as a support to the biochemists, who were otherwise open to criticism that fractionation distorted cell components.

Information on mitochondrial structure has been obtained by studying mitochondria *in situ*, in their usual intracellular location. At variance with this situation, biochemical data on mitochondrial composition and activity are exclusively derived from studies of mitochondrial fractions, i.e., isolated mitochondria. If a correlation is desired between the two types of information, it becomes necessary to find out whether or not isolated mitochondria retain the structure they have inside the cell.[122]

Thus "fine structure," Rockefeller style, was pursued in tandem with biochemical fractionation studies, and must seek linkage of in situ and in vitro observations. As Palade stated, that goal was to be attained by showing, as far as possible, that mitochondria in cells and in fractions have the same structure. This could only have reassured biochemists about Palade's interpretive methods.

Sjöstrand was rather intransigent in his 1956 review, despite being able to cite only two supporting papers by workers outside his laboratory, neither of which showed micrographs quite as sharp as his own.[123] He repeated his earlier depiction of mitochondrial structure, complete with hypothesized membrane architecture, but with the subtle difference that one of the internal membrane plates in his drawing was now continuous with the inner limiting membrane, implying that septa might develop by budding from the mitochondrial envelope. The next year he published micrographs showing that such connections between septum and wall are seen, rarely, thus conceding a little ground.[124] Nonetheless, in the 1956 review Sjöstrand still held that the internal membranes "constitute almost complete septa," and that "no direct continuity between inner and outer membranes was observed" in adequate micrographs. Palade's unphysiological fixative, and rapid postmortem changes due to sluggishness in getting the tissue into the fix, might help account for Palade's observation of a free channel, Sjöstrand suggested. However, poor fixation on Palade's part could not explain the main point of contention, continuity of the lamellae with the bounding membrane, because according to Sjöstrand's principles the native state is always more orderly,[125] and a connected structure represents greater order. Therefore Sjöstrand played his strong suit, the issue of resolution and its prerequisites. The real "step down to high resolution, which here refers to a resolution of 30 Angstrom or better" was taken by himself, whereas Palade's mitochondrion studies had resolutions as bad as 200 Angstrom. Palade's sections just were not as thin. (It is also possible that the pole pieces of Sjöstrand's microscope lens happened to be unusually good.) Sjöstrand went on to discuss the ways thicker sections obscure profiles of membranes, particularly when sections are not exactly normal to them: "[Palade] interpreted tentatively the inner mitochondrial membranes as folds of the inner opaque layer of the outer membrane. This interpretation which is based on intimate contact between inner and outer mitochondrial membranes has not been supported by the detailed morphology at the site of contact as analyzed on high resolution pictures."[126] The other recent reports favoring Palade's view were dismissed still more brusquely: papers not "presented in a quantitative way have not always been considered too seriously. Science has to be based on objective quantitative estimations," insisted Sjöstrand, urging precision again. And it is true enough that few other electron microscopists met Sjöstrand's resolution standards. But not all microscopists were trying to accomplish the same thing as Sjöstrand, however much they might have admired his micrographs.

The science of "ultrastructure," as Sjöstrand saw it, should address problems of the molecular basis of life by actually looking at the molecules in good electron micrographs and measuring them. Hence Sjöstrand's preference for medial longitudinal views of mitochondria, his precisely quantitative size estimates, and his hypothesized molecular structures. Indeed, in the 1956 review he argued for a new classification of intracellular membrane structures into alpha, beta, and gamma "cytomembranes," each defined by its profile measurements and each with a hypothesized molecular architecture, discarding the endoplasmic reticulum concept as vague and obsolete (as quoted above). If one wants to make claims about the structure of molecules on the basis of measurements made from micrographs, the pictures had better be good: "[This is] the importance of high resolution electron microscopy. . . . The most fundamental and primary events in the cell that could be presented in images take place at the molecular and supra-molecular level." Those words are part of the programmatic statement that concludes Sjöstrand's 1956 review, on the proper manner and goal of "ultrastructure research." It continues: "Many old and new problems of morphology may be solved at lower resolution in establishing, for instance, *where* a certain component is formed or a certain reaction takes place. The ambition of ultrastructure research is, however, to go further and to contribute to an understanding of *how* a function may proceed. The hope that such contributions will be possible represents the really new and fascinating prospect of modern morphology."[127] This was Sjöstrand's great ambition for his microscopical molecular biology: to formulate and answer questions about molecular mechanisms. The next year, autumn 1957, the first edition of the *Journal of Ultrastructure Research* was published; Fritiof Sjöstrand was chief editor. "Ultrastructure" was not quite the same thing as "fine structure," and the journal would showcase work meeting Sjöstrand's standards of cytological electron microscopy, just as the Rockefeller circle's journal showcased its standards. Sjöstrand's standards were not simply higher than those of the Rockefeller circle, but different in kind.

Sjöstrand never really gave up his interpretation of internal membrane plates: since his student Ebba Anderssen-Cedergren's elaborate three-dimensional reconstructions in the late 1950s,[128] he has maintained that the cristae are connected to the limiting membrane only through a few narrow stalks that rarely appear in sections of good preparations, and that there is no fluid in them.[129] But the Palade model of cristae as infoldings of the inner membranes has been generally accepted since the early 1960s. For biochemists in the 1950s, not much depended on whether mitochondrial lamellae were folds or septa (although, as elaborated below, it may well have mattered *how* this issue was to be decided), because both models were consistent with the then-dominant theory that their function was to support ordered chains of enzymes. However, Palade's notion that cristae are folds of the inner bounding membrane whose interiors communicate freely with a fluid compartment between the mitochondrial membranes is essential to the chemiosmotic theory of oxidative phosphorylation, proposed by Peter Mitchell in 1961.[130] The chemiosmotic theory has held sway since the late 1960s, and if Palade's notion of cristae as liquid-filled folds is wrong, as Sjöstrand still maintains—and he has proposed an alternative model accommodating most of the same biochemical evidence—then this acclaimed theory must be false.[131] (Biochemical data from the late 1960s, dealing with the outer and inner membranes with different properties isolated from disrupted mitochondria,[132] are also reconcilable with both models.) Thus factual conclusions of today, such as the present biochemical concept of energy production, may have hinged on the outcome of prior decisions on mitochondrial structure and on the proper way, in general, to interpret electron micrographs. It is too speculative a question to answer confidently whether, if the whole community of electron cytologists had originally adopted Sjöstrand's methodology, they would ultimately have converged on today's views of mitochondrial structure and function. Certainly, Sjöstrand's continued tenacity in his own views does not offer evidence that different methods ultimately must arrive at the same conclusion. His defeat on mitochondrial structure notwithstanding, Sjöstrand retained not only his morphologically defined cytomembrane systems but his mode of interpretation; it continued to lead to controversies on substantive issues of cell structure. After his 1959 move to the University of California, Sjöstrand brought his minority school of thought to America, but it would always remain a minority.

Thus, much is at stake in the question of how and why standard methods are adopted. There is a sociological dimension to this dispute about how to interpret electron micrographs properly, and although I do not suppose that a "social constructionist" explanation of the outcome of controversy on the proper *form* of scientific knowledge can be "proved" on empirical evidence any more conclusively than the typical constructionist account of changes of the *content* of knowledge (see Introduction), a sociological account of how the Rockefeller epistemology triumphed and became standard does fit the evidence well enough. (Of course, this sort of sociological argument bears a different burden than usual because the central issue is a collective decision on standard methods rather than on the nature of the world, that is, "normative" rather than "cognitive" consensus, though the latter may follow from the former. Still, given a decision on the right question and the proper means to answer it, in this view data will generally lead to conclusions that require no further negotiations among scientists.)[133] Simply put, the Rockefeller group more successfully built an influential social network adopting and disseminating its methods, even though there was widespread acknowledgment that Sjöstrand made better micrographs. Porter's lab had already been on the circuit of the grand tour taken in the early postwar years by foreign visitors like Sjöstrand seeking the latest methods in electron microscopy, but once they began their thin sectioning, established American scientists, such as 1952–53 visitor Don Fawcett of the Harvard Medical School faculty,[134] came in an increasing stream to learn thin sectioning and to take home the skill of getting "something they could put in their microscope."[135] When Bronk started admitting graduate students to the Institute, the Laboratory of Cytology became an even greater Mecca, as noted above. Before Porter's departure for Harvard in 1961, the laboratory had already trained a large cohort, whose fondly remembered late afternoon show-and-tell of micrographs with Porter (known as "Poppa" by the students) built a shared way of looking at pictures.[136] The students imbibed Rockefeller interpretive methods along with the material practices of the Laboratory of Cytology, in a training program that launched them on distinguished careers in the electron microscopical study of the cell (see Table 1). They filled posts in America's rapidly expanding university and biomedical research sectors, naturally passing their training on to their own students and forming a network still in place today. Unlike Sjöstrand's students, they did not need to cut sections 100 Å or thinner on a regular basis to obtain results that counted as meaningful, which must have made life easier.

Furthermore, the Rockefeller and the Sjöstrand manners of micro-

TABLE I

Selected Biological Electron Microscopists Training,
Working in the Rockefeller Cytology Lab

Name	Years
Mary Bonneville	1956–61 (Ph.D., 1961)
Marylin Farquar	1958–62
Don Fawcett, M.D.	1952–53
Audrey Glauert	1953
Frances L. Kallman	1951–52
John Luft, M.D.	1956
Sanford Palay, M.D.	1953–54
George Pappas	1952–54
Lee Peachy	1956–59 (Ph.D., 1959)
Maria Rudzinska	1952–60
Peter Satir	1956–61 (Ph.D., 1961)
Albert Sedar	1953–55
J. Roberto Sotelo	1954–55
Helen Thompson	1946–48

SOURCE: Porter, "Persons Who Obtained Training from Porter or Palade in the Cytology Dept. at Rockefeller Institute 1947–1961," [1961], KPA, unlabeled carton, box "R, S, T, '56–'61." Also, annual progress reports (see notes); *American Men and Women of Science*, 9th–11th ed.

graph interpretation had different implications for both the goals and limits of the nascent discipline of electron microscopical cytology, which put the two schools in very different relationships to the discipline of biochemistry. These different relationships probably contributed greatly to the success of the Rockefeller school's network building. Porter and Palade used the electron microscope to plot biochemical reactions onto the spatial representations given by electron micrographs. But claims about the biochemical events and entities they were studying were neither generated, nor even occasionally falsified, by electron microscopy; that work of biochemical identification, characterization, and testing was done strictly by biochemists with cell fractions isolated by centrifugation. Thus the Rockefeller style insured that micrograph interpretation would not bring the morphological investigator of cellular architecture into conflict with the biochemical investigator. Sjöstrand, in contrast, used the electron microscope to establish and test biochemical facts, such as the internal structure and content of membranes, and he interpreted his micrographs according to criteria about which the electron microscopist needed to consult no ultracentrifuge (or ultracentrifuge user). These were known qualities of the micrograph and its production, such as section thickness and resolution, plus the principle that apparent order cannot be artifactual. Sjöstrand was essentially promoting a *visual* form of "molecular biology," if by this term one means a study of the way molecular structure determines biological function; he would identify and characterize macromolecules by performing geometry on electron micrographs. In the science Sjöstrand was trying to build, the electron microscope presumed a certain authority over the territory of biochemists, who were heavily invested in their "slick" new ultracentrifuges but were in a very weak position to establish for themselves that the cell components they were isolating had not been drastically altered by cell homogenization and the lengthy centrifugation protocols. On the other hand, the Rockefeller way posits a partnership with the fractionation biochemist, and a set of more modest goals for the electron microscopist that prevent conflict with the biochemist partner: mere description of topology of and associations among components in the unfractionated, in situ cell. The Rockefeller cell biologist had a metier whose definition did not entail conflict with the established biochemistry departments at institutions where electron microscopists were finding work in the later 1950s and 1960s. One might well suppose that fractionation biochemists would have made life hard for microscopists choosing Sjöstrand's way. This suggestion of alliance between fractionation biochemists and microscopists of the Rockefeller school fits with Sjöstrand's perception that biochemists have always sided against him,[137] and with the intimate articulation between biochemistry and electron microscopy established under Claude at the Rockefeller Institute in the 1940s and maintained through Porter's and Palade's work as journal editors and conference organizers. It seems confirmed in the biochemistry-inclusive character of the American Society for Cell Biology, founded in 1960 largely by Porter's circle. Sjöstrand's competing political work enlisted a smaller constituency.

And what about a rationalist counterargument to the above suggestion? Is there some clear criterion according to which the Rockefeller way of interpreting micrographs was in some absolute sense better, and more deserving to become standard for a discipline of "cell biology"— some proof establishing, Pangloss-like, that this is the best of all possible scientific worlds after all and not the product of historical contingency? The answer is not so simple as that, for instance, the epistemology promoted by Sjöstrand relied too much on one instrument (the electron microscope), whereas the Rockefeller approach was more sound just because it relied on the multiple methods of biochemistry and microscopy. For although Sjöstrand did downplay the reliability of fractionation, he relied more on other methods such as X-ray diffraction, and practiced artifact assessment within microscopy energetically. For instance, Sjö-strand would deliberately introduce error of various degrees into his pro-cedures, such as increasing lag time between tissue extraction and immer-sion in fixative or cutting increasingly thick sections, so as to recognize and compensate for errors of certain types present even under the best obtainable conditions. He also pursued many alternate fixation proce-dures, such as an early freeze-drying technique he developed for cells,[138] in a vigilant way that seems to indicate if anything greater skepticism. So this is not a matter of degrees of rigor, but of different kinds of rigor, different experimental logics. What is needed for a rationalistic dismissal of expedience as a viable explanation for the changes in epistemology that accompany the growth of all scientific research programs, it seems to me, is more than just a rehearsal of why the winners' methods make sense in their own terms; for wherever both sides in a dispute are rational, all methods can be expected to make sense in their own terms (see Introduc-tion). What would be required instead are absolute standards by which to judge these competing standards of experimental practice, each poten-tially capable of generating an impressive—though divergent—record of empirical progress. To establish the absolute superiority of today's scien-tific methods over past and alternative practices, a rational yardstick of rationalities would be necessary, a fixed measure showing how *well* we "learn how to learn."[139]

Conclusion: The Centrality of Standards for Experimental Practice

The electron microscope, together with new preparatory techniques such as methacrylate embedding, transformed the study of cell structure by providing the makings of a new experimental system on which a new dis-cipline might be founded. The new discipline, which came to be known as "cell biology" (and not "ultrastructural cytology") was of course de-voted to the study of cell components and their arrangement. But this specification of the object, plus the electron microscope as means, does not serve to define practice for that new discipline; standard experimental practices for applying the microscope to cells had also to be decided, and much else depends on such decisions. In the described conflict, decisions on material practices were not the most controversial, for the Rockefeller group and Sjöstrand prepared tissue and made electron micrographs in much the same way—the osmium fixation, plastic embedding, and thin sectioning suite of techniques that was already becoming standard by 1955. The ideal picture for both Sjöstrand and the Rockefeller workers had sharp focus, maximum resolution of detail, no holes or knife marks, and contrast showing a wide range of values from black to white across the picture field. In this highly magnified view of a section through cells, profiles of the contents were described in terms of their size, texture, shape, disposition, and apparent makeup (e.g., fibers or membranes). It was a slice that looked as much like a monochrome light micrograph of a histological section as possible, only sharper and (usually) much more magnified. That both Porter and Sjöstrand aimed to produce such a picture in the new medium was no accident: both had backgrounds in classical light microscopy—Porter in embryology and cytology, Sjöstrand in physiology and histology—so they both had worked with the classical type of picture that they managed to re-create with the new instrument. The route taken to this electron thin-section image differed: Porter, with the cultured cells that were his main object, passed through a period in which he primarily used another kind of image (much akin to Mudd's bacterial "X-rays"; see Chap. 2), whereas Sjöstrand passed through one in which he concentrated on dispersed specimens of macromolecular assemblies (such as the retinal receptors that he studied with Schmitt and portrayed in a way akin to Schmitt's; see Chap. 4).

The ways the Rockefeller group and Sjöstrand interpreted their thin section micrographs have something in common, too. Both tried to maximize the correlation of the information from electron microscopy with established knowledge of light microscopical histology, laboring to make these new kinds of pictures resemble the old sufficiently to link the two. Both relied on other instruments to evaluate the electron microscope's reliability with a given specimen; this is evident in Porter's use of darkfield and phase microscopy to show evidence of the endoplasmic reticulum in living cultured cells, and an example for Sjöstrand would be his use of X-ray diffraction to test the spacing of membrane layers observed and measured by electron microscopy.[140] Both relied to a great extent on multiple preparation procedures to determine reliability of views and the methods producing them; this was evident in Porter's viewing of cells under various fixatives by darkfield microscopy, Palade's comparison of thin sections identical except for fixation at different pH, and still more evident in Sjöstrand's practice of deliberately causing various kinds of artifacts in order to gauge those that could not be avoided. But this points to the crucial difference between the two schools: here was the sort of experimentation available to the microscopist that fit with Sjöstrand's distinctive orderliness criterion, assisting the electron microscopist to recognize artifacts by evaluation of the pictures themselves. The other strong distinction in interpretive styles lies in the relation of micrographs to evidence from fractionation biochemistry described above: where Palade and Porter would only use the electron microscope to sketch the topology of cell components, referring to biochemical experiments on ultracentrifuge-isolated cell fractions to fill in details of composition, Sjöstrand tried to determine macromolecular disposition and even to some extent composition from electron micrographs, evidently considering the results from cell disruption and fractionation so open to multiple interpretations that they could best be judged according to the best microscopy evidence from cells in situ.[141]

Mitochondrial structure, the proper way of reading micrographs, the proper role of microscopic evidence in cell biology, and the proper boundaries of that nascent discipline with those of biochemistry were simultaneously and coordinately established. Whether or not one accepts that in this particular case sociological forces might have determined the choice of standard interpretive practices by the community of electron microscopists studying cells, I hope it is quite clear that in general such choices can have serious consequences, which might in some circumstances be determining factors, for the local politics of scientific practitioners. For instance, who is to be regarded as a leader and an authority, and what the proper definition of the discipline containing the practitioners should be, both were consequences at stake in the dispute over interpretive practice described in this chapter. And at the same time that the choice of standard methods affects the politics and structure of a scientific discipline, it also can affect the content of scientific knowledge, that is, the correct understanding of the natural world. The different views of mitochondrial structure—each ultimately entailing different biochemical models of mitochondrial function—that follow from the different interpretive modes of Sjöstrand and Palade illustrate this dramatically. The structure of nature and the structure of the sciences were at stake simultaneously in the struggle over which interpretive method should be made standard, which practices "proper." Other examples of method-dependent conclusions about the natural world abound, even in the story recounted here. Light microscopists not using Janus green often saw no mitochondria at all. Electron microscopists not using Palade's fixative saw different structures. Biochemists not purifying mitochondria by the Rockefeller sucrose protocol did not obtain comparable results. All conclusions are method-dependent, but we tend to forget this until old familiar methods are overthrown (often after a bitter struggle), whether by the introduction of different methods or otherwise; then, almost inevitably, we must discard or adjust prior conclusions along with our prior methods. It might be said, without exaggerating much, that if not for standardization different researchers would be studying different worlds. For all practical purposes this is certainly the case.[142]

Science does not even begin to happen, as Dewey pointed out, until experience of the world is treated systematically.[143] And at least in fields where objects of inquiry are so heavily mediated by technique that we cannot even conceive of an unmediated representation of a phenomenon—and more and more scientific fields have this character today—the systematic methods of mediation required for science have also to be heavily standardized to maintain cohesion of the collective inquiry. If standard experimental systems for exhibiting phenomena, standard procedures for taking measurements, and standard ways of interpreting results are not maintained, different scientific workers in the same field have difficulty using each other's results, and even communicating. Consider the chaos ensuing when standardization breaks down, as in the notorious case when a cultured cell line called "HeLa" was found to have taken over certain other standard cell lines. Vast amounts of experimental results lost significance because of the failure of standardization. Results from each experiment could no longer be related to results from the others because it was no longer certain that all involved the same cell types, and knowledge in affected areas was set back many years.[144] Although Sjöstrand's survival in the field does show that some degree of heterogeneity can be tolerated by a scientific discipline, at least for a time, and not without the heavy costs associated with isolation, his story also shows what an uphill battle a methodological deviant has to fight even when he is an acknowledged master of experimental craft.[145] Practitioners without the art, or without sufficient combative spirit, have no practical recourse but the dominant and standard way. Given that not only careers and disciplines hang in the balance, but even the world as it is understood, it is no wonder that some scientists find standards worth fighting about.

Plate 1.1. Ladislaus Marton's RCA model A microscope. Reproduced from *Journal of Bacteriology* 41 (1941): 397–420, with the permission of the American Society for Microbiology.

Plate 1.2. Marton (right) demonstrating the RCA model B microscope in Camden, circa 1940. Reproduced from *Advances in Electronics and Electron Physics* supp. 16 (1985): 501–23, with the permission of Academic Press.

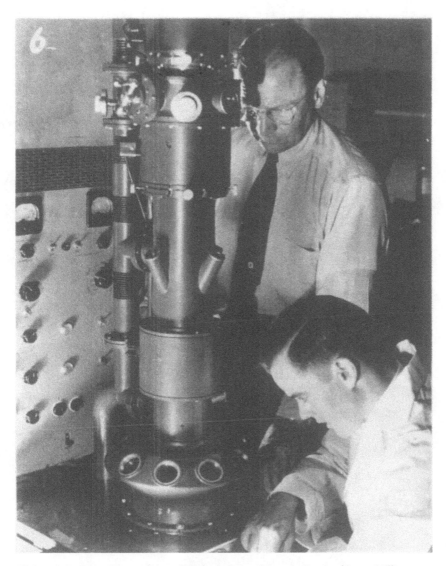

Plate 1.3. RCA publicity photo of Vladimir Zworykin, standing, and James Hillier, seated at the model B microscope, 1940. Courtesy of the David Sarnoff Research Center, Princeton.

Plate 1.4. Francis Schmitt (left) and Cecil Hall with one of MIT's RCA model B microscopes, circa 1950. Courtesy of The MIT Museum.

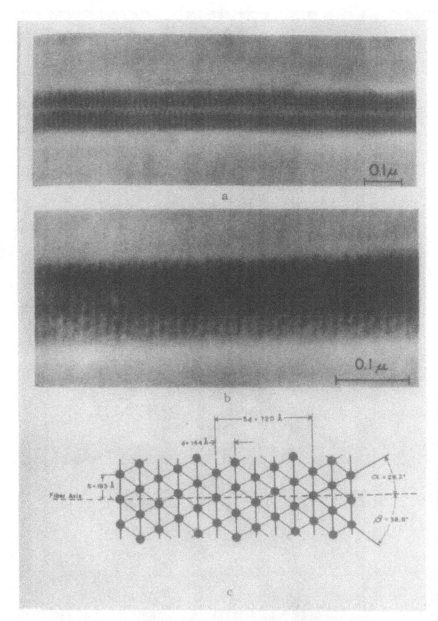

Plate 1.5. *Top and center*, electron micrographs, and *bottom*, measurements of paramyosin fibrils stained with phosphotungstate, made at MIT by Schmitt, Hall, and Jakus during the war. Reproduced from *Annals of the New York Academy of Sciences* 47 (1947): 799–812, with the permission of the Academy.

Plate 1.6. Marton and his Stanford electron microscope. Reproduced from *Stanford Alumni Review*, May 1943, by permission of *Stanford Magazine* and Stanford University Archives and Special Collections, Stanford University Libraries.

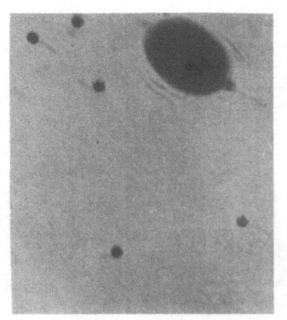

Plate 1.7. Electron micrograph of bacteriophage and a host cell, made by Marton at Stanford in collaboration with bacteriologists E. W. Schultz and P. R. Thomassen. Reproduced from *Proceedings of the Society for Experimental Biology and Medicine* 68 (1948): 451–55, with the permission of the Society.

Plate 1.8. Electron micrograph of polio virus contrasted by metal shadowing, made by Marton at Stanford in collaboration with biochemists H. S. Loring and C. E. Schwerdt. Reproduced from *Proceedings of the Society for Experimental Biology and Medicine* 68 (1948): 451–55, with the permission of the Society.

Plate 1.9. Newspaper illustration showing Zworykin and T. F. Anderson (seated) with RCA model B microscope. Reproduced from the *Philadelphia Evening Bulletin*, Feb. 13, 1941.

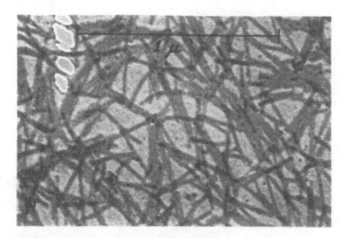

Plate 1.10. Electron micrograph of plant virus particles, made by Anderson in collaboration with Wendell Stanley. Reproduced from *Journal of Biological Chemistry* 138 (1941): 325–38, with the permission of The American Society for Biochemistry and Molecular Biology.

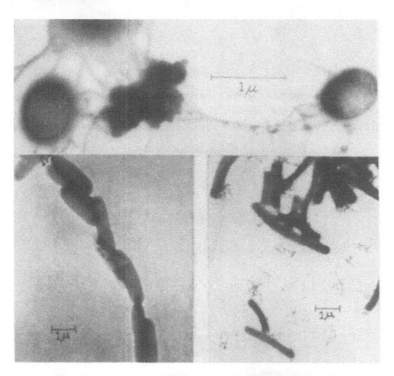

Plate 1.11. Electron micrographs of various bacterial cells, made by Anderson in collaboration with Stuart Mudd, circa 1941. Reproduced from *Journal of Experimental Medicine* 76 (1942): 103–8, with the permission of Rockefeller University Press.

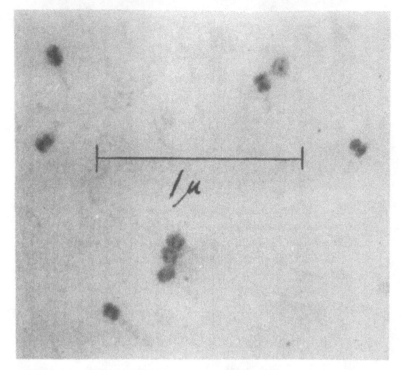

Plate 1.12. Electron micrograph of bacteriophage, made by Anderson in collaboration with Luria and Delbrück. Reproduced from *Journal of Bacteriology* 46 (1943): 57–77, with the permission of the American Society of Microbiology.

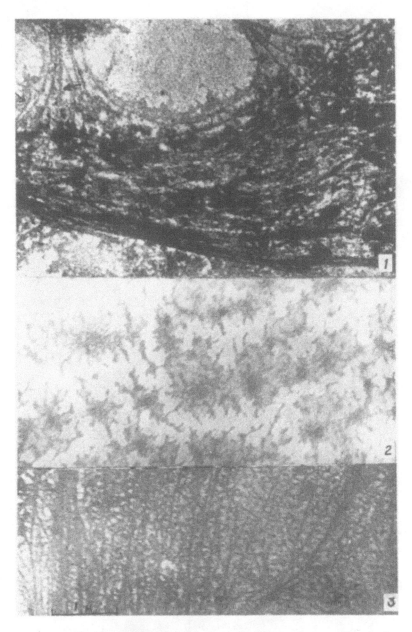

Plate 1.13. Electron micrographs of squid nerve-cell protoplasm, made by Anderson in collaboration with A. G. Richards and H. B. Steinbach. *Top*, air-dried from thick zone of smear; *middle*, air-dried from thinner zone of smear; *bottom*, frozen-dried. Reproduced from *Journal of Cellular and Comparative Physiology* 21 (1943): 129–43, with the permission of the Wistar Institute, Philadelphia.

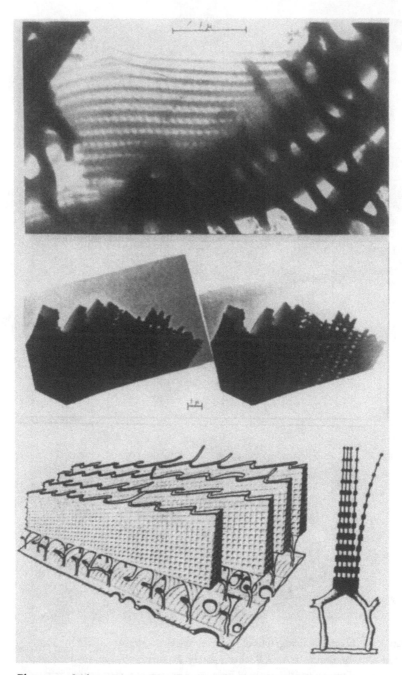

Plate 1.14. Iridescent butterfly scales. *Top*, electron micrograph; *middle*, stereo electron micrograph; *bottom*, reconstruction; made by Anderson and A. G. Richards. Reproduced from *Journal of Applied Physics* 13 (1942): 748–58, with the permission of the American Institute for Physics.

Plate 1.15. James Hillier demonstrating the RCA EMU microscope, circa 1945. Courtesy of the David Sarnoff Research Center, Princeton.

Plate 2.1. Stuart Mudd, circa 1960. Courtesy of The University of Pennsylvania Archives.

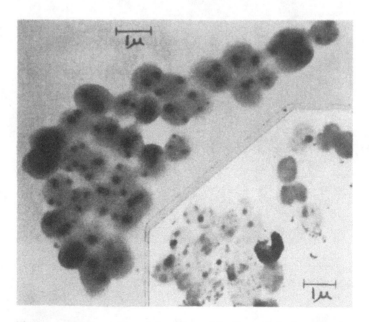

Plate 2.2. Electron micrographs of bacterial cells, made by Knaysi and Mudd, showing dark "nuclear" bodies, *left*, by 200 kv electrons; *inset right*, by clearing of cytoplasm background with hot water. Reproduced from *Journal of Bacteriology* 45 (1943): 349–59, with the permission of the American Society of Microbiology.

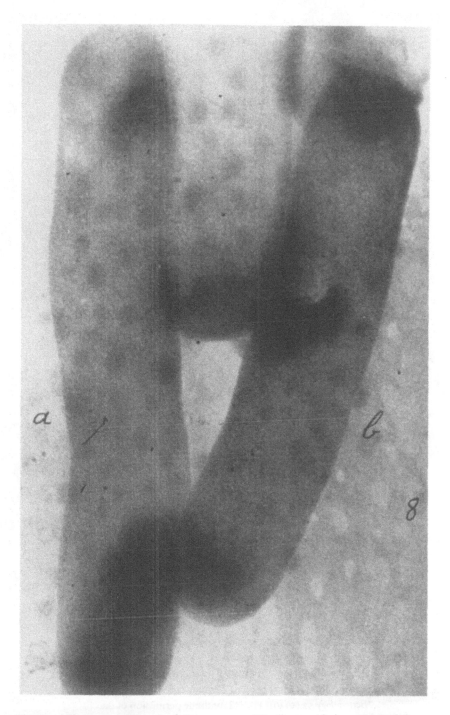

Plate 2.3. Electron micrograph of nitrogen-starved *Bacillus mycoides* bacteria, made by Knaysi and Baker, showing dark "nuclear" bodies near cell poles. Reproduced from *Journal of Bacteriology* 53 (1947): 539–53, with the permission of the American Society of Microbiology.

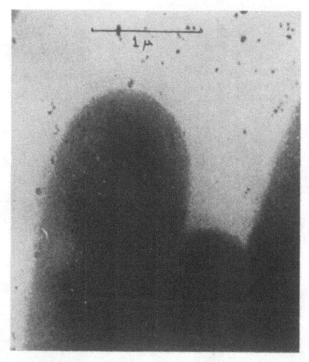

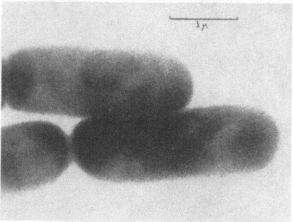

Plate 2.4. Electron micrographs of *Escherichia coli* bacteria, made by Hillier, Mudd, and Smith with a special high-contrast lens, revealing, *top*, "orientation" of protoplasmic substance; *bottom*, light "nuclear" zones. Reproduced from *Journal of Bacteriology* 57 (1949): 319–38, with the permission of the American Society of Microbiology.

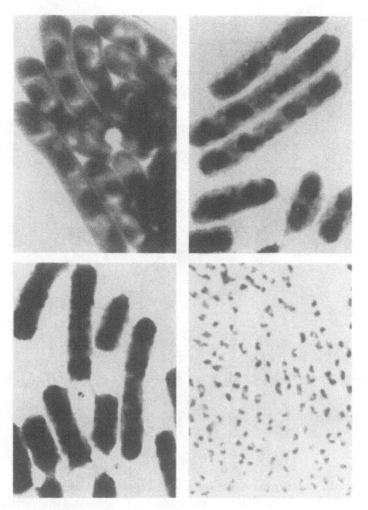

Plate 2.5. Bacteria at several stages in nuclear stain procedure for light microscopy, by Mudd and Smith. *Top left*, electron micrograph of untreated, air-dried bacteria showing light "nuclear" zones; *top right*, electron micrograph of osmium-fixed and acid-washed bacteria now showing dark "nuclear" zones; *bottom left*, electron micrograph of final osmium-fixed, acid-washed, formalin-mordanted, and fuchsin-stained bacteria showing dark "nuclear" zones; *bottom right*, light micrograph of bacteria after same final treatment. Reproduced from *Journal of Bacteriology* 59 (1950): 575–87, with the permission of the American Society of Microbiology.

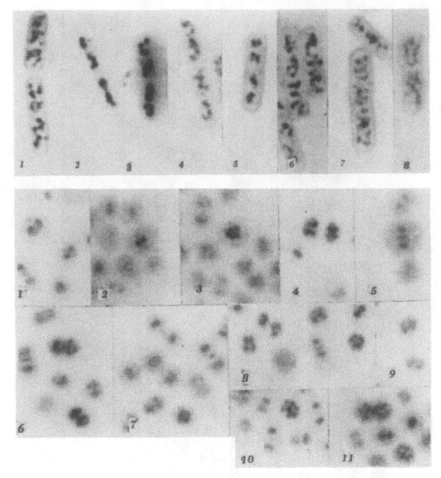

Plate 2.6. Light micrographs made by DeLamater, stained to reveal stages of "nuclear division" in, *top, Bacillus*; and *center and bottom, Micrococcus*. Reproduced with permission from *Cold Spring Harbor Symposia for Quantitative Biology* 16 (1951): 381–412, pls. 1, 2.

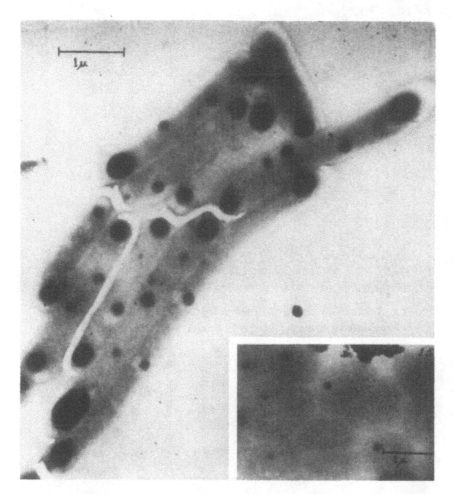

Plate 2.7. Electron micrographs by Mudd and colleagues, showing dense "mitochondria" (no "nuclei" are evident) in, *left*, *Mycobacterium thamnopheos*; and *right*, inset, *Micrococcus cryophilus*. Reproduced from *Experimental Cell Research* supp. 2 (1952): 319–43, with the permission of Academic Press.

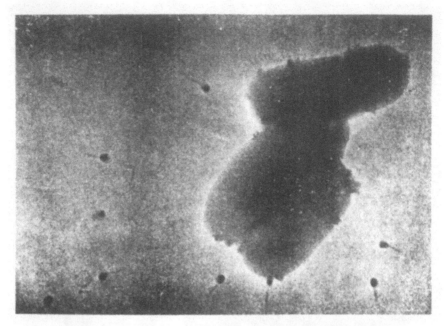

Plate 2.8. Textbook illustration of bacteriophage "swimming" toward bacteria, late 1940s. Reproduced with the permission of Chemical Publishing Company.

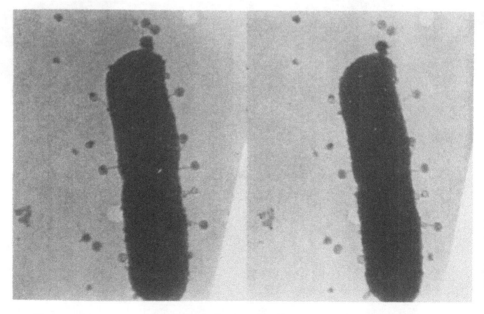

Plate 2.9. Stereo micrograph of critical point–dried bacteriophage adherent to bacterial hosts, made by Anderson; note attachment by tails. Reproduced with permission from *Cold Spring Harbor Symposia for Quantitative Biology* 18 (1953): 197–203, p. 199.

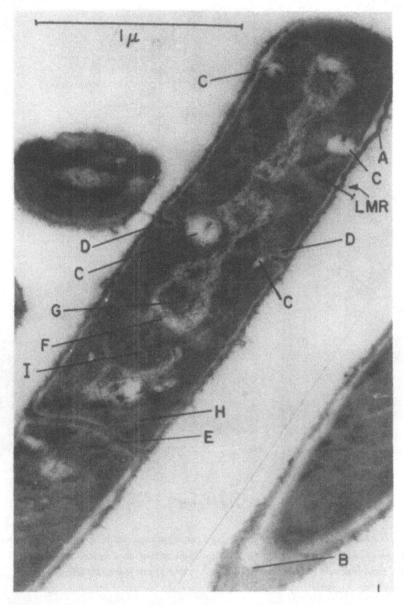

Plate 2.10. Thin-section electron micrograph of *Bacillus cereus*, made by Chapman and Hillier in 1952; *c* indicates peripheral bodies, and *LMR* indicates resolution limit of the light microscope. Reproduced from *Journal of Bacteriology* 66 (1953): 362–73, with the permission of the American Society of Microbiology.

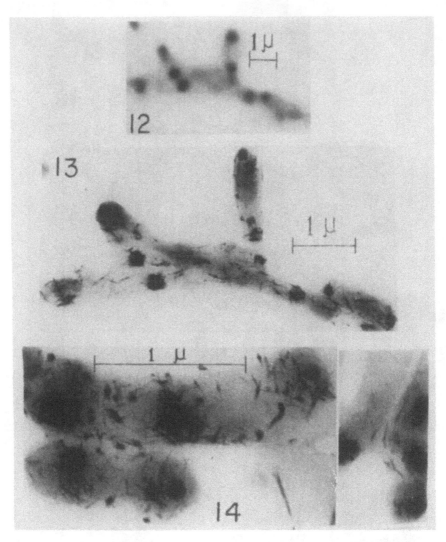

Plate 2.11. Mycobacteria grown in presence of potassium tellurite, showing dark "mito-chondrial" sites of reductive activity, by Mudd and colleagues. *Top*, light micrograph; *middle* and *bottom*, electron micrographs. Reproduced from *Journal of Bacteriology* 72 (1956): 767–83, with the permission of the American Society of Microbiology.

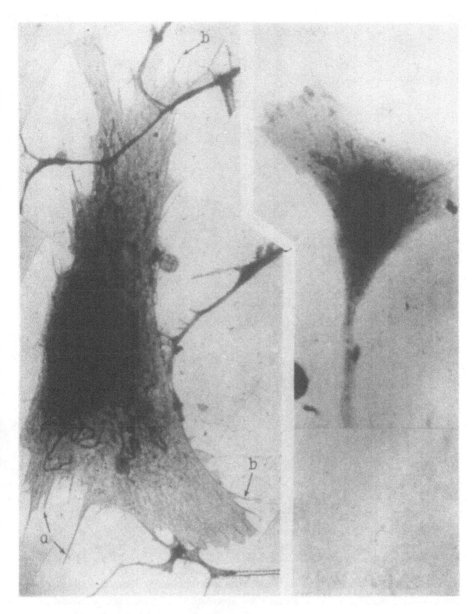

Plate 3.1. Chicken fibroblast cells grown on plastic film in tissue culture, fixed in osmium vapor, and *left*, imaged by electron microscopy; *right*, stained by the Giemsa technique and imaged by light microscopy. Reproduced from *Journal of Experimental Medicine*
81 (1945): 233–46, with the permission of Rockefeller University Press.

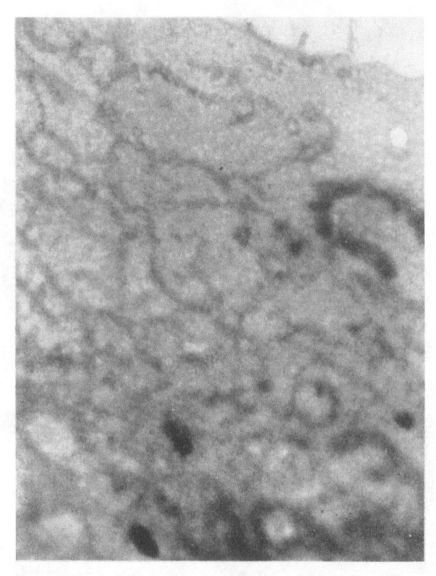

Plate 3.2. Higher magnification electron micrograph of small portion of cytoplasm in a tissue-culture cell prepared as in Plate 3.1. Reproduced from *Journal of Experimental Medicine* 81 (1945): 233–46, with the permission of Rockefeller University Press.

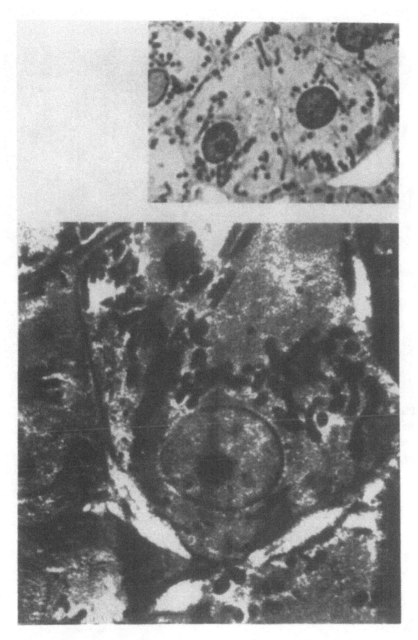

Plate 3.3. Liver tissue fixed with osmium, embedded identically in wax, and sectioned, *top*, stained and imaged by light microscopy; *bottom*, imaged by electron microscopy. Note imperfection of cell preservation in electron micrograph, invisible in light micrograph due to poor resolution. Reproduced from *Journal of Experimental Medicine* 83 (1946): 499–503, with the permission of Rockefeller University Press.

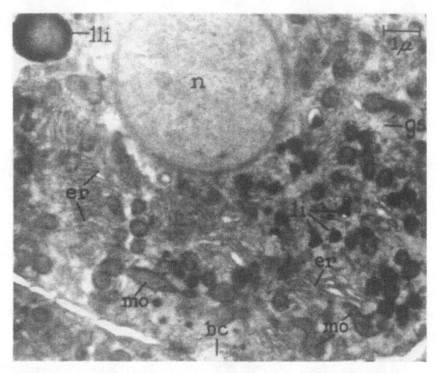

Plate 3.4. Electron micrograph of liver tissue fixed in neutral pH-buffered osmium, embedded in methacrylate, and thin sectioned. Nucleus is indicated by *n* and mitochondria by *mo*. Reproduced from *Journal of Experimental Medicine* 95 (1952): 285–97, with the permission of Rockefeller University Press.

Plate 3.5. Keith Porter (*top*) and George Palade, 1950s. Photographs courtesy of the Rockefeller Archive Center.

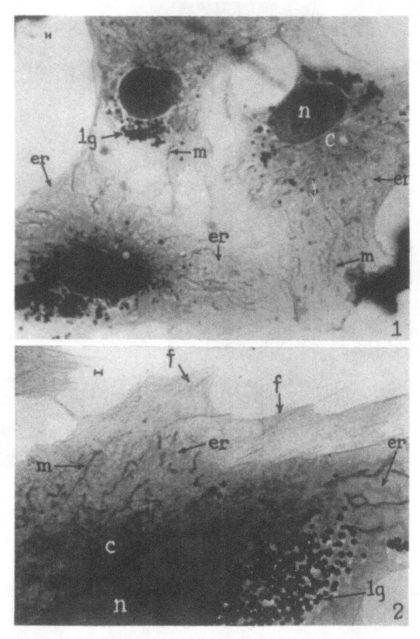

Plate 3.6. Cultured heart cells fixed with osmium vapor, *bottom*, imaged by electron microscopy; and *top*, stained with the basic dye iron-hematoxylin and imaged by light microscopy. Mitochondria are indicated by *m*, endoplasmic reticulum by *er*, and nuclei by *n*. Reproduced from *Journal of Experimental Medicine* 97 (1953): 727–50, with the permission of Rockefeller University Press.

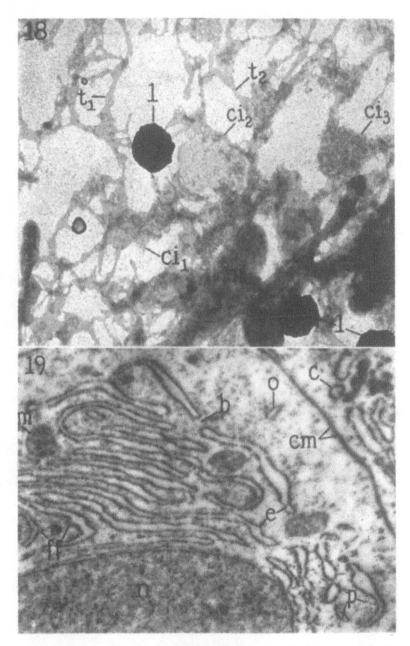

Plate 3.7. Electron micrographs of small areas of osmium-fixed rat parotid gland cells, *top*, grown in tissue culture and imaged whole; *bottom*, tissue cut from animal, embedded in methacrylate, and thin sectioned. Reproduced from *Journal of Experimental Medicine* 100 (1954): 641–56, with the permission of Rockefeller University Press.

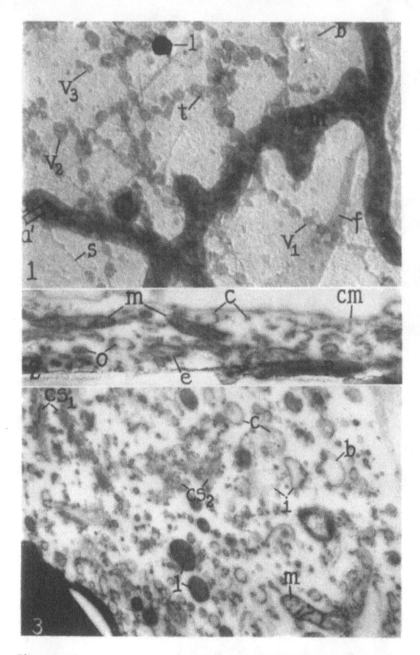

Plate 3.8. Electron micrographs of cultured, osmium-fixed chicken white blood cells, *top*, mounted whole; *middle*, embedded and thin sectioned vertically; *bottom*, embedded and thin sectioned horizontally. Approximate section thickness is indicated by *a–a′*, *top*; mitochondria indicated by *m*, and endoplasmic reticulum elements by *c, e, o, t,* and *v*. Reproduced from *Journal of Experimental Medicine* 100 (1954): 641–56, with the permission of Rockefeller University Press.

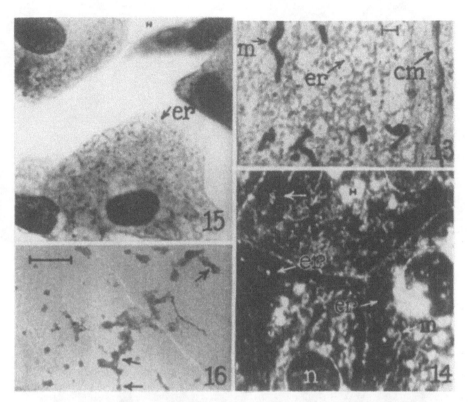

Plate 3.9. Endoplasmic reticulum *er* in, *top left*, light micrograph of cultured, osmium-fixed, and toluidine blue–stained cells; *top right*, electron micrograph of cultured, osmium-fixed cells; *bottom right*, dark-field light micrograph of living cultured cells of the same type; and *bottom left*, electron micrograph of an ultracentrifuge fraction from homogenized tissue. Reproduced from *Journal of Experimental Medicine* 97 (1953): 727–50, with the permission of Rockefeller University Press.

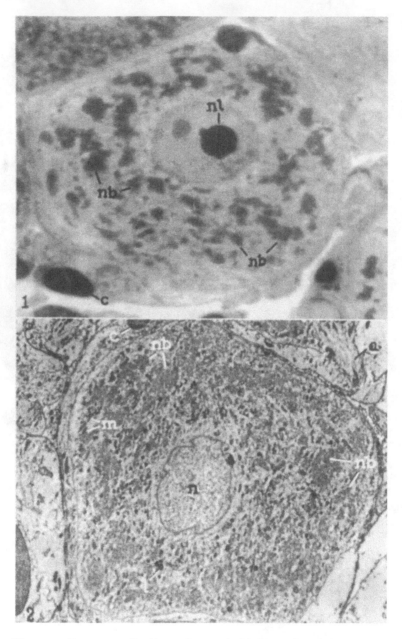

Plate 3.10. Rat neurons fixed in osmium, embedded in methacrylate, and sectioned from same block. *Top*, light micrograph of thick section stained with thionin after removal of embedment; *bottom*, low magnification electron micrograph. Note Nissl bodies *nb*. Reproduced from *Journal of Cell Biology* 1 (1955): 69–88, with the permission of Rockefeller University Press.

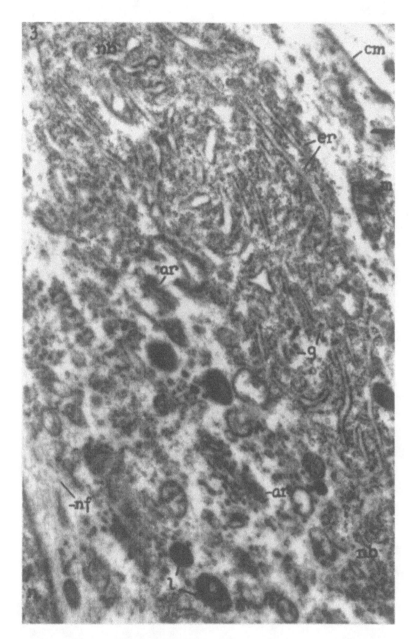

Plate 3.11. Higher magnification electron micrograph of specimen in Plate 3.10.
Note endoplasmic reticulum covered with granules *er*, and agranular reticulum
ar. Reproduced from *Journal of Cell Biology* 1 (1955): 69–88, with the permission
of Rockefeller University Press.

Plate 3.12. Fritiof Sjöstrand, circa 1960. Courtesy of Fritiof Sjöstrand.

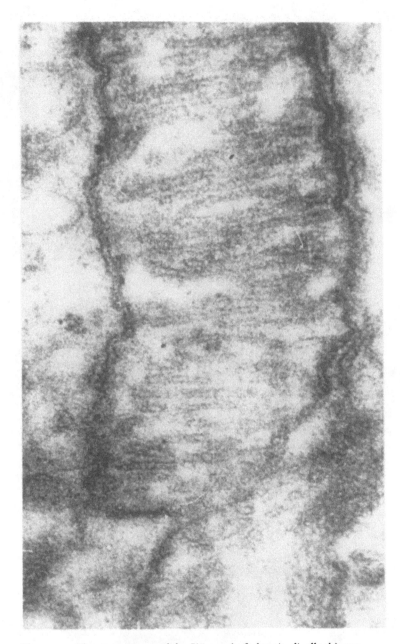

Plate 3.13. Electron micrograph by Sjöstrand of a longitudinally thin sectioned mitochondrion from guinea pig retina. Reprinted with permission from *Nature* 171 (1953): 30–32, p. 31. © 1953 Macmillan Magazines Limited. Micrograph courtesy of F. Sjöstrand.

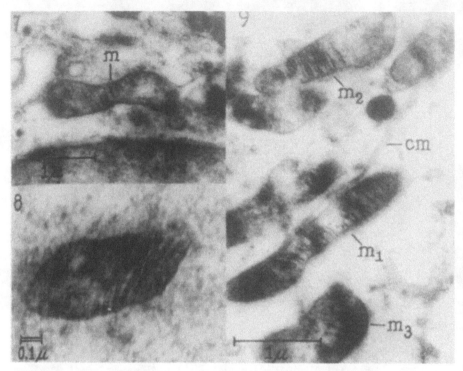

Plate 3.14. Electron micrographs of thin-sectioned rat tissue, showing mitochondria *m*. Reprinted with the permission of John Wiley and Sons, Inc., from G. E. Palade, "The Fine Structure of Mitochondria," *Anatomical Record* 114 (1952): 427–51, p. 449. © 1952 Wiley-Liss.

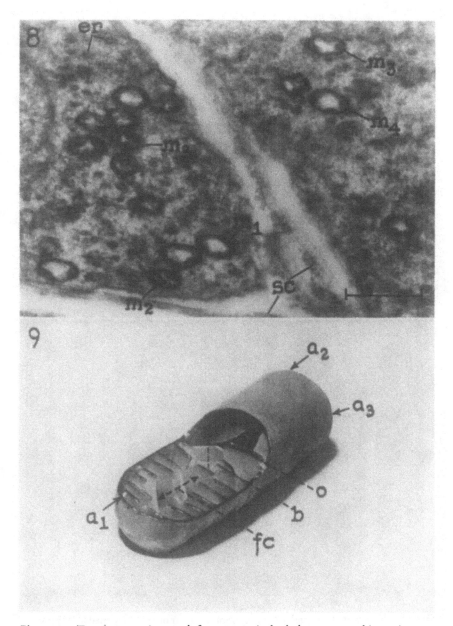

Plate 3.13. *Top*, electron micrograph from rat seminal tubule, transverse thin section, showing mitochondria *m*. Note that m_3 and m_4 show internal membranes consistent with baffled, not septate, structure. *Bottom*, wax model illustrating Palade's notion of mitochondrial structure; internal membranes are baffles, and a lengthwise free channel *fc* exists. Reproduced with permission from G. E. Palade, "An Electron Microscope Study of Mitochondrial Structure," *Journal of Histochemistry and Cytochemistry* 1 (1953): 188–11, p. 197.

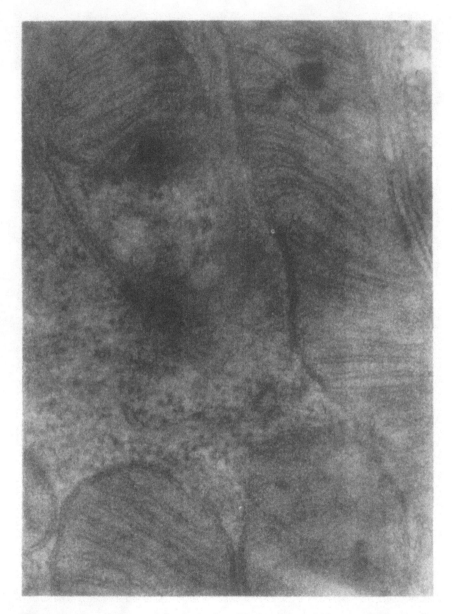

Plate 3.16. Electron micrograph of mitochondria from mouse kidney, medial longitudinal thin section. Reproduced from F. Sjöstrand and J. Rhodin, "The Ultrastructure of the Proximal Convoluted Tubules of the Mouse Kidney as Revealed by High-Resolution Electron Microscopy," *Experimental Cell Research* 4 (1953): 426–56, p. 434, with the permission of Academic Press.

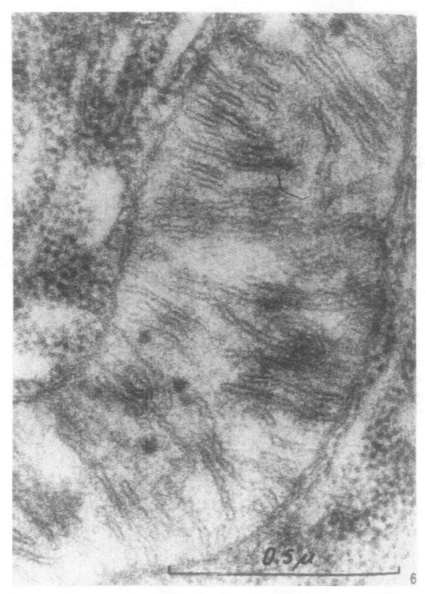

Plate 3.17. Electron micrograph of mitochondrion from mouse pancreas, medial longitudinal thin section. Reproduced from F. Sjöstrand and V. Hanzon, "Membrane Structures of Cytoplasm and Mitochondria in Exocrine Cells of Mouse Pancreas as Revealed
by High-Resolution Electron Microscopy," *Experimental Cell Research* 7 (1954): 393–
414, p. 400, with the permission of Academic Press.

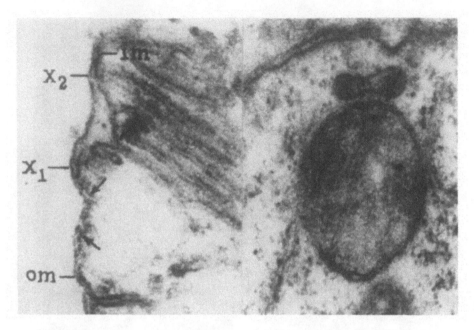

Plate 3.18. Electron micrographs of thin-sectioned rat mitochondria, showing connection (arrows) between inner bounding membrane *im* and base of cristae membranes. Reproduced with permission from G. E. Palade, "Electron Microscopy of Mitochondria and Other Cytoplasmic Structures," in *Enzymes: Units of Biological Structure and Function*, ed. O. Gaebler (New York: Academic Press, 1956), 185–215, p. 191; the right image also appeared in *Journal of Cell Biology* 1 (1955): 59–67, and is reproduced with the permission of Rockefeller University Press.

CHAPTER 4

Muscle, Nerve, and the
Iron Men of MIT

✦✦✦ [Galileo's] actual process of measuring, applied to the intuited data of experience, results, to be sure, only in empirical, inexact magnitudes and their quantities. But the art of measuring is, in itself, at the same time the art of pushing the exactness of measuring further in the direction of growing perfection. It is an art not [only] in the sense of a finished method for completing something; it is at the same time a method for improving [this very] method, again and again, through the invention of ever newer technical means, e.g., instruments. Through the relatedness of the world, as field of application, to pure mathematics, this "again and again" acquires the mathematical sense of the *in infinitum*, and thus every measurement acquires the sense of an approximation to an unattainable but ideally identical pole, namely, one of the definite mathematical idealities or, rather, one of the numerical constructions belonging to them.

—Edmund Husserl, *The Crisis of European Sciences and*
Transcendental Phenomenology

The essence of Galileo's invention of modern science, according to Husserl, was the creation of a method whereby the rich "plenum" of the world as actually sensed is abstracted into geometrically ideal forms, which can then be modeled and manipulated mathematically so as to make accurate prediction and control possible. The scientific method set in motion becomes automatic, perhaps compulsive, in demanding ever more precise measuring of sensed things in order to perpetuate cycles of its two basic movements, idealization and mathematization. According to Husserl, then, all science (though his examples all come from physics) depends on this idealization and mathematization of the world as perceived, and indeed, this twin drive has been described as the essential activity in electron micrograph interpretation.[1] No doubts need be raised about the first of these; indeed, I have just argued that idealization, as abstraction of experience according to standardized methods, is necessary to the coherence of a science of biological electron microscopy and indeed to any science. However, we have already (in Chap. 3) seen that the mode of micrograph interpretation that became standard in cell biology stopped deliberately short of the high degree of mathematization advocated by Sjöstrand. Except in his minority school, quantitative description and measurement of the forms seen in electron micrographs was excluded from the practice of cell biology in favor of an interpretive style that simply extracted topological relationships and typical shapes (*gestalten*, perhaps) of the kinds of entities, such as mitochondria and Golgi bodies and endoplasmic reticulum, found in the cell. However, there was another school of biological electron microscopy that better fits the Husserlian model of a drive toward ever greater mathematical precision, and its Galileo was Francis Schmitt (Plate 1.4), admirer of the methods of physics and ardent, self-styled "biophysicist."

The electron microscopists and experimentalists in other fields classifiable within the loose category of "biophysics"—synonymous in the 1940s with "molecular biology"—were at ground zero of the war-born changes in the scale of funding for life science (see Introduction). This was because they were positioned to benefit simultaneously from physical science support, related to the atomic bomb and the military, and from biomedical funding, related to health policy, both of which were growing explosively. Moreover, they were the sort of technophile researchers who attracted, and indeed required, large-scale patronage. Elsewhere I have described the terrific upwelling of enthusiasm for biophysics in the immediate postwar period as the "Biophysics Bubble" (on the model of the South Sea Bubble investment mania), a cultural reaction that can be regarded as a movement to compensate for the atomic bomb with a new physics of life.[2] As is widely known, the ranks of life science were swelled by physical scientists in search of new fields less intimately linked with the military and less affected by the general trend toward a corporate atmosphere in research.[3] But however important the postwar influx of physical scientists may have been to biophysics and ultimately to the rather different enterprise that today is called "molecular biology," it is necessary to remember that there already was a biophysics in America before 1945. Francis Schmitt, as one of that science's established leaders, capitalized tremendously on the Biophysics Bubble, advancing biophysics as a medically promising alternative to the deadly new physics of atoms.

Sensitive to the postatomic opportunity, he was vigorously drumming up enthusiasm for biophysics well before the postwar biomedical research boom, for instance propounding a "Manhattan Project against disease" in 1946–47. Similarly, at the December 1946 meeting of the American Association for the Advancement of Science (AAAS), where outgoing AAAS president Charles Kettering's keynote speech urged scientists to cease squabbling and find some mutually agreeable NSF proposal, Schmitt gave a talk whose conclusion, in manuscript form, reads: "If one tenth as much thought, energy, and money were devoted to the structure and properties of the living cell and its constituents [as to the atom], the results in biology and medicine might prove as spectacular as those of ~~the atomic bomb~~ nuclear physics, and perhaps more salutary from the *human* viewpoint."[4] Schmitt slightly toned down his tacit recriminations against the bomb physicists for his radio speech in a 1945–46 series organized by Warren Weaver, in which eminent scientists spoke about their work during the intermission of the popular New York Philharmonic broadcasts on Sunday afternoons. After Weaver introduced him as one of the main leaders of molecular biology, Schmitt promised the national audience that if as much support were devoted to structural research on the living cell as had been given the atom, "the results might prove highly salutary from the human point of view."[5] Schmitt did his best to cement the notion in the popular imagination that biophysics was physics for life and for humanity, as opposed to implicitly "unsalutary" or unwholesome nuclear physics. Still, whether biophysics was represented as a moral alternative to or, as some others preferred, a wholesome dividend of atomic research, biophysicists had no need to fight over why their field deserved enhanced support: biophysics benefited either way. As we shall see, Schmitt also campaigned within the scientific community to put biophysics on a permanent footing as a fully institutionalized discipline.

Schmitt and the New Biology at MIT

Francis Schmitt commanded respect for both his intellectual acuity and his erudition, not only from his loyal followers but even from those who found him a domineering or difficult person.[6] Though a comparatively young man for someone in his position of authority (he turned 40 in 1943), Schmitt maintained a formal reserve with graduate students and other juniors, and many Americans found him to have a "Germanic" stiffness that was somewhat off-putting. Still, within his MIT group he was an inspiring educator, a man of many scientific enthusiasms who was capable of communicating them well to his research students, a master of a vast range of scientific literature, and a fount of new ideas for experiments. He was also a vigorous administrator and entrepreneur, constantly writing grant proposals, networking by telephone, and traveling to meetings— even under wartime transportation restrictions. Schmitt had largely given up laboratory work in favor of an executive role by the time of his mid-1941 arrival at MIT, although summers at Woods Hole still offered him refreshing, if brief, returns to the bench.[7] Schmitt's executive flair must have been among the reasons he was chosen by the MIT administration, with major backing from Warren Weaver, to revive a stagnant life sciences department dominated by applied research in such fields as sewage engineering and food canning. The 1939 plan developed by Weaver and MIT president Karl Compton, together with graduate dean John Bunker, had been to use Rockefeller funding to restructure the university's life sciences around a new discipline of biotechnology or "Biological Engineering," which—it was hoped—would become a major new enterprise devoted to improvements in food processing, in fibers and other natural products, and in medical instrumentation, along lines analogous to MIT's phenomenally successful program in chemical engineering.[8]

As already described (see Chap. 1), Schmitt pursued two main research programs during wartime, each with its own electron microscope and technical staff: there was a sizable group of Boston doctors doing classified work with some of Schmitt's staff under his OSRD contracts for research on wound healing and on the development of artificial sutures and skin made from extrusion and spinning of purified collagen. There was also unclassified, basic research on the architecture and self-assembly of fibrous macromolecules, particularly collagen and muscle proteins, which was done mainly by Marie Jakus (then a doctoral student) and Cecil Hall (then also a doctoral student and research associate, having come with only a master's degree in physics). This wartime work by Hall and Jakus won considerable attention and respect for the electron microscopy of the Schmitt group, at least among the tightly knit crowd attending EMSA meetings in the 1940s. To some biomedical researchers in Boston, though, the electron microscope was still "Frank Schmitt's artifact machine."[9] Schmitt played an important role in winning such skeptics over to the value of the new instrument for medically oriented life science. After the war he built on his wartime connections with the local biomedical community, training many medical doctors to become researchers in his sophisticated and uniquely rigorous style, who in turn brought their MIT research ethic to the medical institutions where they found employment. Indeed, Schmitt was instrumental in shaping biomedical research as it transformed itself in postwar America. In 1947 the Massachusetts General Hospital (MGH), having just decided to invest more heavily in basic life science, invited Schmitt to join the board of directors as the sole laboratory researcher there. Schmitt helped MGH plan and execute its research reorientation, in part by putting in place a research advisory panel of local scientists.[10] This and other innovations at MGH were closely watched and widely copied at other growing medical research institutions in the immediate postwar era. But the MIT influence on medicine is another story.

At the end of the war, no American biology department was in a stronger position to exploit the new experimental technology of life science than Francis Schmitt's. MIT had a core group of half a dozen active younger biophysicists on the faculty: Schmitt's protégés Richard Bear and David Waugh, medical electronics specialist Kurt Lion, cytologist and electron microscopist H. Stanley Bennett, and John Loofbourow, the only MIT faculty member from before Schmitt's time whom Schmitt really included in his circle (Loofbourow had come a year before Schmitt, and his interests in molecular aggregation, particularly blood clotting, harmonized well). In 1947–48, after finishing his Ph.D. under Schmitt, Hall would also be officially added to the faculty. Loofbourow's wartime experience as an administrator in MIT's large and complex "rad lab" radar group made him ideally suited as Schmitt's executive officer, handling almost all of the department's routine business and leaving Schmitt free to play his preferred role as "idea man" and public figure. In late 1945, Food Technology split off as a separate department and moved to temporary buildings, giving Schmitt more room and less administrative worry, and allowing him to change the department's name to simply "Biology" (rather than "Biology and Biological Engineering") for the 1945–1946 academic year.[11] There were two electron microscopes already, and in 1946 Schmitt managed to buy a $12,500 ultracentrifuge and a $4,500 electrophoresis apparatus, both instruments that had become commercially available for the first time at the war's end. MIT biology was equipped for the future, and with demobilization, enrollment shot upward. In 1947 Weaver granted Schmitt's department $250,000 for a six-year period, which, together with the outside grants from NIH and ONR and elsewhere that Schmitt and his band showed great skill in garnering, helped the MIT biologists keep pace with their chief rivals in academic "molecular biology" after the war, Linus Pauling and George Beadle's team at Cal Tech.[12] By 1949–50, there were four electron microscopes, four X-ray diffraction units, four ultracentrifuges, four spectrophotometers, and plenty of other expensive apparatus in Schmitt's department. And at the end of 1952, the Rockefeller Foundation's largesse was redoubled with a $500,000 outright gift.[13]

Obviously, the MIT life sciences had ample resources for development of the technology-intensive, "big science" style of biology that Schmitt pioneered and disseminated. Schmitt also helped build an appropriate home for the new biology. His department's labs had become increasingly crowded with the postwar influx of students, and Schmitt prodded the MIT administration to make good on promises of a new laboratory building by compiling a list of the many generous offers he himself and his young faculty members had received from other life science departments—a list that demonstrates how much established biophysicists were in demand at American universities in the late 1940s. Planning had begun by 1946 for a tower to be shared by the biology and food technology departments, and a suitable sponsor was found in Campbell's Soup chairman John McGowan Jr. (an MIT alumnus), whose $1 million saw to it that the building was named for MIT's great food technologist, John Dorrance. Ground was broken in the autumn of 1950, and by mid-1952 Schmitt began moving his biologists into the four upper floors of the seven-story, glass- and steel-faced Dorrance Laboratories building.[14] "Modern in every sense of the word" (as an MIT magazine put it), and boasting cinder block and linoleum interiors with fluorescent lighting throughout, this "skyscraper among the Institute's educational buildings" might be regarded as emblematic of the institutionalization of a new type of biology, indeed a new model of academic science in America—the first of countless unornamented bulky structures to crop up on campuses, planted by administrations to foster the kinds of research that would capture a large share of America's postwar government support for science. Twenty-three thousand square feet of the edifice were designated for biology department research and office space, about four times the area allocated to formal teaching, which in itself is profound testimony to the orientation of the new life science Schmitt established at MIT. Electron microscopy took up about a third of the research space, and the rest was given over to other instruments for characterizing biological macro-molecules.[15] Dorrance was a house for researchers and their machines.

Learning Molecular Biology with the Iron Men

During the war Schmitt had tried to articulate a synthetic vision bringing together three agendas for his department: "physical and chemical biology," "biophysics and biochemistry" (which Schmitt saw as based more in the physical sciences and motivated less by traditional questions of life science than the first category), and "biological engineering" (which would include food and sanitation). At the war's end these fine distinctions evaporated, Schmitt behaving as if Food Technology's split from the department effectively absolved him of the charge he had been given by the MIT administration to develop an applied life science.[16] (There are indications, as we shall see, that the administration was not altogether pleased with Schmitt's abandonment of its bioengineering scheme.) Graduate students could take doctorates in either biophysics or biochemistry, the latter field at the time represented by only two faculty members. Noteworthy is the lack of genetics or any sort of zoology or botany, except what might fit under the rubric of "microbiology," "cytology," or "general physiology"; essentially, biophysics was the stock-in-trade.[17] Very similarly to Jacques Loeb, founder of general physiology (hence biophysics) in America, Schmitt envisioned a life science that would permit living organisms to be designed and engineered from first principles like other machines.[18] It was his emphasis on quantitative precision and high-level theory that set Schmitt apart, as an avatar of academic high culture, from the sanitation engineers and food technologists with whom he was supposed to cooperate, according to the MIT administration's more practical notion of "biological engineering." Schmitt's Loebian dream of an exact life science, and the centrality of the electron microscope to it, appears in a rich amalgam with the atom at the conclusion of a 1946 lecture. Schmitt's outline reads:

Analogy with atom. At close of last century physicists thought of the atom as an indivisible unit. Then came the discovery of electrons, radioactivity, the nucleus, protons, and the other atomic particles. Result is modern atomic age, electronics, etc. May we expect a similar control of living material when we know more about the molecular construction of protoplasm? No one has sufficient vision now to predict what may be possible in another 20 years as a result of the widespread and intelligent use of the electron microscope.[19]

TABLE 2

Selected Biological Electron Microscopists Training,
Working in Francis Schmitt's Group at MIT

Name	Years
H. Stanley Bennett, M.D.[a]	1945–48
Arthur R. Denues	1947–48
Arne Engstrom, M.D.	1951
Betty Geren (Uzman), M.D.[b]	1948–50
Jerome Gross, M.D.[b]	1946–48
E. Jean Hanson	1953–54
Alan J. Hodge	1950–52, 1957–60 (Ph.D., 1952)
Hugh E. Huxley	1952–54
Woutera van Iterson	1948
Marie Jakus	1941–47, 1948–51 (Ph.D., 1945)
Eduardo de Robertis, M.D.	1946–49
J. David Robertson, M.D.	1948–52 (Ph.D., 1952)
David Spiro, M.D.	1950–56 (Ph.D., 1956)
Fritiof Sjöstrand, M.D.	1947–48

SOURCE: [Schmitt], "A Review of the Program for Training in Medical Research at the Department of Biology, MIT." "Appendix D," Apr. 1954, FOS, carton 1, folder 51. Also, *American Men of Science*, 9th–11th ed.

[a]Bennett was an assistant professor (of cytology) at MIT in this period.

[b]After leaving for research posts in Boston medical centers, Geren remained affiliated with MIT as a Research Associate through the 1956–57 academic year, and Gross through 1955–56.

Schmitt's biophysics would provide exact knowledge of how protoplasm was constructed, and thus how it might be reconstructed. Technological advances as impressive as the atom bomb, and, it was hoped, much more wholesome, would come from the new generation of electron microscopists.

In the immediate postwar years, Schmitt's department was a magnet for postdocs and more senior life scientists wishing to learn how to use the electron microscope. Some came for just a few months as part of a tour of leading labs, whereas others ended up staying for years under one or another grant (see Table 2). In the 1946–47 academic year Schmitt's own lab had at least six postdoctoral visitors, and the number stayed as high through the first half of the 1950s. The biology department's population of postdocs seems to have reached an early peak in 1948–49 with at least eighteen and probably more, but when a big Commonwealth Fund grant came in 1951, the strong orientation toward postdoctoral training became an institutionalized way of life.[20] The handful of technicians was a comparatively minor part of the work force. As mentioned, many postdoctoral visitors were medical doctors preparing for a research career, and the proportion of medical doctors among MIT's biologists in training remained high to the end of the 1950s, despite a certain culture clash. Schmitt insisted that the medics take strenuous courses in mathematics and physical science as background. "Are you a physician or a scientist?" one of them remembers being asked by an MIT faculty member. "In one class, as the equations encompassed the room from blackboard to blackboard, one [other physician trainee] leaned over to me and said, 'I'd like to see that bastard do a gall bladder!'" the story continues.[21] Under the Commonwealth Fund training grant alone, by 1956 about two dozen researchers had come away with graduate degrees in biology from MIT in addition to the medical degrees with which they arrived. The caliber of these students was high (for instance, five of the eight doctors supervised for their Ph.D. degrees by Schmitt had received their M.D. degrees from Harvard Medical School); like Schmitt's other students and visiting fellows, many went on to found their own labs and rise to prominence.[22]

Schmitt's training program was designed to impart rigorous backgrounds in biology and physical science simultaneously, to avoid the usual pitfalls of interdisciplinarity. "Too often the chemist or physicist . . . without benefit of experience or background in biology . . . has either failed to realize the significance of his findings or has drawn conclusions which may apply to the partial system or model . . . but could not possibly apply to the actual biological system." Schmitt's solution for his biophysicists was that they simply had to know more: "competency in the several borderline fields in which the physicist, chemist, and biologist may work will tend to avoid both superficiality and a feeling of lack of authority which is frequently experienced by those who pioneer in these fields."[23] Serious expertise in more than one academic field is a stringent demand, but Schmitt offered such a compelling example and such force of personality that many were led to take up the challenge. As noted, medics entering the group had to take extensive remedial course work, and similarly, those trained primarily in physical science were expected to take biology courses.[24] For the advanced graduate students and postdocs, formal education was centered on a graduate seminar Schmitt began to teach in the autumn of 1948, called "Tissue Ultrastructure." Apart from Schmitt himself, and group members talking on their own areas of expertise, it featured guest lecturers from MIT and nearby institutions, and encouraged lively discussion and debate. Schmitt did not so much teach the seminar as "command it" with an astounding degree of erudition, requiring students to keep up a blistering pace of reading.[25] All these unique rigors for students came with unique opportunities, too. Schmitt would often bring international scientific luminaries to MIT for a lab tour and a seminar or two—people like William Astbury, John Randall, A. J. P. Martin, Linus Pauling, John Runnstrom, and Lawrence Bragg—so as to keep the group abreast of the most exciting new ideas and techniques worldwide.[26] There was flexibility and maximum opportunity to try all of the department's excellent equipment. Even short-term visitors had a chance to try their hands at electron microscopic investigations of specimens of their choice.

The laboratory atmosphere was enthusiastic and free-wheeling, not unlike that which Jacques Loeb had cultivated in his iconoclastic group at the University of California around 1910.[27] For instance, Schmitt recalls with relish the experiments on squid nerve that he used to organize in the early 1950s. In 1936, Schmitt had been introduced to squid as an experimental subject by the English neurophysiologist J. Z. Young,[28] and was impressed with the advantages of the long nerve fiber in the squid's mantle, which is an extraordinarily large, and thus easily handled, single cell. Ideal though the giant nerve might be for experimental manipulation, supply was a problem for the MIT group when they were not summering at the seaside. Substantial numbers of squid had to be brought, alive and in good condition, to the MIT labs housing the preparative and analytic apparatus; and because the work had to be done quickly, almost all of the lab personnel needed to be mobilized. It was the sort of adventure in teamwork that tends to breed strong esprit de corps, especially when combined with the rigorous mathematical rites of passage and the group's awareness that its training was unique and extraordinarily intensive. From the Office of Naval Research, Schmitt's sponsor in nerve work throughout the 1950s, in the spring of 1952 Schmitt borrowed a special "squid truck," and equipped it with a refrigerated transport tank for the animals.[29] Navy regulations required that the driver had to pass a stringent test involving an obstacle course, and initially the only one who could drive well enough was Betty Geren, a young doctor studying with the group. On the day of an experiment, it was arranged that Geren would go out at dawn on a fishing boat in Rhode Island and, immediately upon docking, transfer the animals to the "squid truck," telephone MIT, and then race the sloshing truck to the lab as quickly as possible. The drafted researchers and technicians of the "squid team" awaited her in a parking lot, where they hastily dissected the live squid assembly-line fashion and collected the protoplasm from the giant nerve axon (the long extension of the nerve cell body) in each animal. This "axoplasm" was then rushed, as quickly as possible to minimize postmortem degeneration, inside to the ultracentrifuge and other apparatus, where the process of purifying and visualizing the subcellular components and determining their makeup could begin. By the end of 1952, the squid experiments had become fairly routine, and the burden of labor shifted to extra technical staff, many of whom were local undergraduate students.[30] For the summers of 1952–1957 the lab averaged about 2,000 healthy squid dissected per season, from 50 or 60 squid trips—all courtesy of the Navy's interest in the microcircuitry of nerve.[31] (The nerve studies are discussed below.)

If the mantle of leadership in general physiology belongs to the scientist who best sustained the tone set by Loeb, and carried on by William Crozier at Harvard through the interwar period—what Philip Pauly has termed the field's "undiscipline"[32]—then Schmitt must be seen as Loeb's and Crozier's successor. The lively esprit de corps of Schmitt's lab group extended to social events, as evidenced by the antics at EMSA meetings that are today still vividly recalled by some participants. At the meetings the Schmitt group radiated an aura of superiority and treated most of the other biological electron microscopists with a certain disdain; Anderson and Robley Williams (see Chap. 5), both with physical science backgrounds, were among the few exceptions. Schmitt's nickname for his group, the "Iron Men," expressed in his Germanic fashion Schmitt's pride in the team's traditionally masculine scientific virtues: the acceptance of rigorous training, the energetic pursuit of certainty through daunting experimental exertions, the drive for quantitative precision, the tough-minded demand for deterministic explanations—all the sort of attributes supposedly more common among physical scientists than biologists. A favorite story is that at one EMSA meeting, the MIT group lacked liquor for a party they were throwing. Hearing happy voices from a party in the RCA suite upstairs, the "Iron Men" dispatched scouts to raid the party, some by fire escape. All returned with liberated bottles. They were a hard-drinking crowd and proud of it. Astbury, though he never worked in the MIT lab, was a good friend of Schmitt's and an honorary "Iron Man" who proved his mettle when, unflinchingly, he completed an invited lecture while finishing a water tumbler full of pure gin, with which he had been presented as a practical joke.[33] The women of the group, such as the squid truck–racing Betty Geren (who was also among the commandos of the liquor raid), seem to have mustered the necessary machismo rather than protest, despite a sense that Schmitt demanded from them an extra measure of proof of their scientific dedication.[34]

Schmitt's many lines of research represented a multifaceted "study of the love life of the molecules," as a flyer announcing one of his more casual talks quite aptly put it (illustrating, it seems likely, a lay reaction to the neologism "molecular biology," for if molecules have a biology they surely have a love life).[35] Schmitt considered the electron microscope the foremost instrument for solving the main problems of molecular biology, especially the problem of fibrous protein. "Nature apparently employs the protein fiber as the chief element in the architecture of the cell," as he said in his aforementioned national radio speech. There he lauded the electron microscope as the most "powerful tool of biophysical research." The living cell is like a factory, Schmitt offered, invoking a not very original metaphor, and biologists are far away, watching the raw materials delivered and the products coming out of the plant. Whereas the light microscope provides a pair of binoculars with which to observe the men working at machines through the windows, the electron microscope and other tools of biophysics offer admission to the factory floor and a close look at the machines.[36] In a 1947 lecture series in Boston, Schmitt fairly rhapsodized about the electron microscope, pointing to it as "the instrument best qualified to get at chromosome ultrastructure," "undoubtedly the most powerful instrument for getting at cytoplasmic ultrastructure," similarly preeminent for solving the mysteries of nerve and muscle function, and, in conclusion, "particularly significant and spectacular" in advances toward discovering the "molecular structure of protoplasm." The cover illustration of the graduate school announcement for life sciences at MIT featured an RCA EMU, very appropriately, given the electron microscope's commanding presence in Schmitt's department.[37] Of course, Schmitt also oversaw the application of other biophysical methods, such as X-ray diffraction, and the use of ultracentrifuges and chromatographic apparatus, further to characterize what the macromolecules are made of and to quantify how they are arranged. But the electron microscope was the guiding light.

Visions of Life at Schmitt's MIT: Nerve

Schmitt literally sought to *see* how protoplasm's "megamolecules" were put together. Consistent with his belief in the primacy of the protein fiber, and with the hope soon "to visualize the polypeptide chains directly" and thus to read the blueprint of the fiber, Schmitt redoubled his work on collagen at the war's end. The Schmitt lab explored the possibility of determining the detailed structure of the protein by staining it with metals that react specifically with particular amino acids, and then examining the distribution of stain in high-resolution electron micrographs to read the positions of those amino acids—an elaboration of Jakus's early success with phosphotungstate (Chap. 1).[38] The group also studied the alternate arrangements of collagen filaments into larger fibrils that could be made to form by spontaneous assembly from pure solution, under differing chemical conditions. It must have seemed an ideal experimental system to learn the general rules governing protein fiber construction.[39] The structure and function of muscle, another biologically important fibrous structure, also became an area of intense study, as we shall see. And nerve, Schmitt's mainstay subject before the war, became once more a principal focus of his lab's work. The fibers and fibrils visible with the light microscope in fixed and stained nerve preparations had once been leading candidates for "the substrate of nerve conduction," Schmitt noted, but early in the twentieth century "the ascendancy of the membrane theory together with a growing distrust of structures which can only be demonstrated after fixation caused physiologists to lose interest in morphology." The time had come, considered Schmitt in April 1949, for physiologists interested in neural conduction to turn again to the microstructure of nerve, both the neglected internal structure and in the outer membrane believed to hold the key to signal transport, using the new biophysical instrumentation.[40]

In his wartime project on the protoplasm of the squid axon, Anderson and his physiologist collaborators had decided that the several classes of fibrils and filaments they saw in the expressed nerve juice were varying aggregates of the much smaller rodlets indicated by polarized light studies on wet, fresh axoplasm (see Plate 1.13).[41] However, as noted earlier, they could not determine which of the fibrillar forms visible with the electron microscope existed in life, because appearances varied so much with preparation conditions. As Schmitt was the first to admit, the problem of the "structure of the axon of fresh nerve fibers is beset by certain fundamental indeterminacies" because an electron microscope can only observe prepared specimens, and the manipulations required by other techniques similarly required manipulations that could be expected to alter the living state.[42] Nevertheless, with Eduardo De Robertis, a medical doctor working in Schmitt's group from mid-1946 through early 1949, he pressed onward. Schmitt and De Robertis looked at dispersed "smears" of protoplasm from the nerves of amphibians, mammals, crustaceans, as well as of squid, fixed with formaldehyde in the nerve axon or freeze-dried directly after squeezing it out (without fixation). And the greatest possible care was taken to isolate the axon from surrounding material during preparation so that only axoplasm would be observed. The MIT team saw none of the structures visible to Anderson; but Anderson had rinsed his axoplasm smears in water and in salt solutions, a procedure that in the Schmitt lab caused the disappearance of the fibril structures they did see, and the appearance of aggregates more like Anderson's. Under conditions they thought best, what Schmitt and De Robertis saw in all nerve types was a kind of "indefinitely long" fibril 500–600 Å wide, which on staining with phosphotungstate had dense edges and a light core (Plate 4.1). That the fibrils are tubular seemed almost certain: elliptical outlines could sometimes be seen at fibril ends, and the appearance of a light center did not vary much with duration of beam exposure or staining, ruling out solid fibrils falsely appearing to have a hollow core from slow uptake of stain or from preferential "bleaching" of the central area. Even more "safe to conclude" was that the fibrils belonged to the axoplasm itself rather than the axon sheath or other surrounding tissue: the numbers of fibrils seen did not correlate with the abundance of connective tissue in the different kinds of nerve; the disappearance of the tubules in most rinse solutions indicated a lability inconsistent with the well-known toughness of collagen and other connective tissue components; and the familiar collagen fibers ought to be distinguished easily in the axoplasm preparations if connective tissue was present. These hollow-looking fibrils making up the nerve were dubbed "neurotubules." Moreover, the neurotubules seemed linked to nerve conduction in that they disappear from dissected frog nerve at the same time that nerve function ceases.[43] To sum up, Schmitt and De Robertis contended that their neurotubules are present in axoplasm in the living state because they were found even in nerve fixed fresh before extrusion of axoplasm, and because the neurotubules could be made to disappear in smears of unfixed axoplasm by using Anderson's methods, thus explaining their absence from his prior observations (this fit the generally accepted theory that axoplasm was a colloid delicately balanced between different aggregative states). Anderson's structures must be "disintegration products" of the MIT group's neurotubules, presumably reaggregates of the same subcomponents under artificial conditions, they concluded.

Schmitt was quite enthusiastic about these "rather startling" findings regarding neurotubules, and De Robertis announced the results at meetings in November and December 1947. "This discovery," Schmitt considered, "opens up new avenues in nerve research. The difference in density in the core as compared with extratubular material strongly suggests a polarization which has important electrophysiological implications." Thus Schmitt was thinking that the tiny tubes extend along the nerve fiber axis and play a role in nerve conduction.[44] To follow up the possible functional role of the neurotubules, Schmitt and De Robertis did more extensive work on live extracted frog nerves, and on rabbit nerves severed in the living animal. In both in vitro and in vivo cases, the neurotubules appeared by electron microscopy to degenerate over time in close correlation with the loss of nerve responsiveness. Similarly, De Robertis and Schmitt found that in vertebrate nerve freshly infected with any of several types of virus including polio, neurotubules from dispersed preparations of nerve segments show infestation with particles (either viruses themselves or the products of viral damage; Plate 4.2) at positions along the nerve fiber that correlate with the progress of the virus down the nerve, known from the possibility of isolating infective virus at the various distances from the infected nerve end.[45] Thus viruses infecting nerve seemed to invade by passing along the neurotubules. When De Robertis presented the polio findings at the 1948 EMSA meeting in Toronto, despite carefully not claiming that the particles they could see with the electron microscope were viruses themselves, the pictures created "quite a stir" and Schmitt was worried about "unwanted publicity" from the excited reporters there. Schmitt's worst fears may have been confirmed when, shortly afterwards, one of the neurotubule pictures appeared in a *Time* magazine piece altogether lacking his customary caveats on micrograph interpretation. Schmitt and De Robertis had proved that nerve "fibers were cables made up of many hollow tubes about one millionth of an inch in diameter," *Time* reported, and had displayed "remarkable pictures of the polio virus marching in orderly files" within the neurotubules.[46]

Such a close correlation of structural change with functional change, as Schmitt and De Robertis observed in both the degeneration and infection experiments, is what biologists expect of a structure causally linked to that function, so these experiments supported Schmitt's hypothesis that the neurotubules play a role in nerve conduction. Moreover, as I have argued elsewhere, monitoring the appearance of an entity under different conditions of physiological activity, in an experiment aimed at learning the entity's function, reinforces the conclusion that the entity is real and not artifactual; the real existence of the entity is presupposed in the experimental design, and the entity seemingly confirms its existence just by behaving consistently within each experimental and control group.[47] But as we will soon see, despite its auspicious beginnings, the neurotubule began to misbehave when novel specimen preparation procedures were applied, and was soon discounted as a probable artifact. The rigor and resourcefulness with which Schmitt tested his interpretations, and the candor with which he would abandon previous interpretations when it became necessary (though he is said to have been very angry with De Robertis for putting him in a position where public retraction was necessary),[48] illustrate why Schmitt was such a respected figure among electron microscopists despite his distance from the bench.

At the end of spring 1949, De Robertis left for Montevideo to head up a cytology research group, and Geren, a postdoctoral fellow in Schmitt's lab for a year already, took the lead in the structural research on nerve.[49] Even in April 1949, two months before his last neurotubule publication with De Robertis was submitted, Schmitt had already begun to have doubts about the entity.[50] Ongoing experiments with degenerating nerves turned up some intact neurotubules in nerves long past their ability to function. Moreover, on close investigation of the repeating substructural pattern along the neurotubule, evident in some favorable preparations, it became clear that the neurotubule markings were roughly equal in periodicity to markings observed in collagen fibrils (650 Å); this was highly suspicious in that collagen-rich connective tissue is always associated with the nerve sheath. The technique of viewing smears of nerve tissue made it impossible to determine from electron micrographs the original locations of the entities in the nerve, because all the parts of a short section of nerve were mixed together on the grid. To resolve the question of neurotubule localization, Geren and Schmitt turned to the newly announced method of embedding tissue in methacrylate for thin sectioning. Sections of squid, frog, and mammal nerve all showed filaments in the axon protoplasm, often with a beaded appearance, quite distinct from De Robertis's light-centered neurotubules (Plate 4.3). Light-centered fibrils of the size expected of neurotubules were found in only a few micrographs, which is not surprising, given the different preparatory methods (no phosphotungstate stain was used, for example); however, where typical-looking neurotubules were observed, they were in the connective tissue sheath (Plate 4.4) around the axon. These findings implied that the neurotubules were actually a sheath component mistakenly attributed to the internal axoplasm. And new experiments showing neurotubules in unfixed preparations before and after washing with various salts suggested that they were not unstable after all, which, together with the cross-banding pattern similar to collagen, implied that they were collagen fibers altered in some way by preparation procedures.[51] Neurotubules were abandoned as an artifact, and quickly forgotten in Schmitt's hot pursuit of the exciting new "neurofilaments." These were eventually identified with the entity that cell biologists came to call "intermediate filaments" when they appeared in a variety of tissues in addition to nerve.[52]

Joining in the Schmitt lab's broad efforts to develop better thin-sectioning technique,[53] Geren continued to devote much of her research efforts to the nerve sheath. Anatomy taught that nerve axons, whether sheathed in a fatty coat containing the substance myelin (like major nerves in vertebrates and some in invertebrates) or not, always are accompanied by cells of a distinctive type called Schwann cells, but the reason was unclear. Using a buffered osmium fixative prior to embedding and thin sectioning, not unlike Palade (see Chap. 3), Geren examined the association of Schwann cell, axon, and myelin sheath in nerves from a variety of vertebrates and invertebrates.[54] Schwann cells in some nerves, Geren and Schmitt observed, are flattened into a thin (about 0.5 μ) layer enveloping the axon, and the mitochondria of the axon are concentrated at its periphery, just beneath the membrane over which lies the Schwann cell. Given that Schwann cells, like mitochondria, are often full of intracellular membranes, and that in some cases it looked as though mitochondria protruded through the Schwann cell membrane into the axon beneath, it seemed that the mitochondria might actually be budding off the Schwann cell into the axoplasm.[55] Wrongheaded though this particular speculation might have been, the general train of thought about the intimate association between axon and Schwann cell led to one of the great successes to come from the Schmitt group in the era.

Electron microscopy by Sjöstrand, Geren, and a number of others had shown that the myelin sheath of vertebrate nerves consists of up to 30 layers of lipid (fat) and protein, each about 95 Å thick. X-ray crystallography had confirmed this and also suggested that the lipid molecules were oriented radially whereas the proteins were predominantly tangential. Studying myelinated nerves from chick embryos, Geren had found that the myelin sheath is laid down during development in progressively greater numbers of these layers, rather than by continued accretion to a fixed number of layers. In the early stages of sheath formation, the few layers enveloping the axon are thicker than in the mature nerve, with its

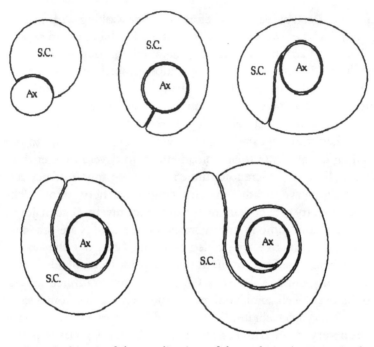

Fig. 4. Geren's theory of the myelination of the embryonic axon *Ax* through repeated enfoldings of Schwann Cell *S.C.* Drawing by author, after Geren.

many thin layers. Looking at thin sections from the earliest stages (Plate 4.5) , Geren understood the way in which the myelin sheath originates from the Schwann cell with what is remembered as a flash of inspiration.[56] The Schwann cell first envelops the axon, and meeting at the other side, flows back around in the reverse direction, thus setting down membrane layers of alternating orientation (Fig. 4). With time, the Schwann cell cytoplasm between the layers is lost, leaving just the thin layers of membrane wrapped around the nerve: the myelin sheath.[57] This theory, which soon came to be known as the "Geren jelly role model," still stands today. One reason it achieved quick acceptance, I would suggest, may be that Geren used electron microscopy to demonstrate the model in the same manner that classical histology uses light microscopy, merely showing that a topological relationship between two tissue features changes over time, using specimens at different stages of development. Geren made no claims about the functionally significant physical properties of the membrane structures on the basis of electron micrographs—though this sort of Sjö-strandian effort was not uncommon in the Schmitt group—reasoning like a standard cell biologist in this instance.

Visions of Life at Schmitt's MIT: Muscle

Nowhere did Schmitt's pronouncements that renewed attention to morphology would—with the tools of biophysics—solve longstanding problems of physiology prove more justified than in the case of muscle contraction. As an example of the style of biology Schmitt fostered, and as one of the more spectacular contributions of the electron microscope to life science, this story deserves special attention. Since the beginning of the century, the dominant theory held that a muscle fiber contracts through a conformational change in long molecules within it, which would kink or wind to assume a more compact shape and thus shorten, probably through a change in pH or, on the analogy with nerve, ion exchange.[58] From the 1920s through the 1940s, the major protein myosin and several others had been extracted from muscle, and a large amount of biochemical work accumulated around the problems of how these proteins were interlinked, and how during contraction chemical energy was related to the conformational change of muscle fibers made of these proteins. But even more than the biochemistry of energy transactions in muscle and the associated mass of confusing data, visible changes in the banding pattern of striated muscle during contraction were the greatest puzzle that theories of muscle aimed to explain. The relaxed striated muscle of vertebrates shows alternating bands with different staining and polarized light characteristics, called A (anisotropic) and I (isotropic) bands, the I band marked across the center by a dark Z line and the A by a lighter H zone (Fig. 5). During contraction, the Z line remains the same while the H disappears, and the A band grows relative to the I and apparently at its expense: it looks as though the A splits and each half moves toward its adjacent Z, impinging increasingly on the I zone until at maximum contraction this may altogether disappear. Other bands are also observable under certain conditions, and considerable attention was devoted to these as well as to the higher order architecture of the fibrils and fibril bundles making up the muscle fibers, but this level of detail will suffice to preface the present story. It was generally accepted, by the early 1940s, that there were long molecular chains containing myosin stretching great distances in muscle fibrils, and that along these chains regularly repeating complexes between myosin and other substances were responsible for the appearance of the bands seen by light microscopy.[59]

At the end of the war, it was clear to Schmitt that muscle was "scheduled for big advances," and that the electron microscope would make all

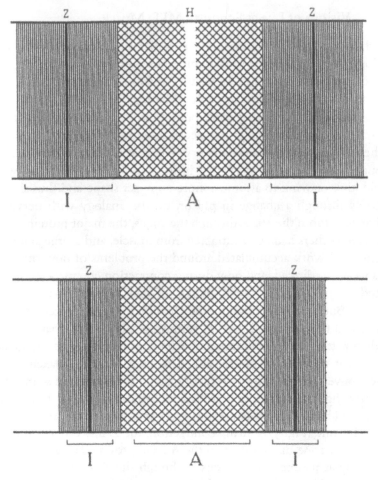

Fig. 5. Banding pattern of striated muscle in, *top*, relaxed; and, *bottom*, contracted states, showing positions of Z lines, H zones, anisotropic (A) and isotropic (I) bands. Drawing by author.

the difference in cracking the old problem of the mechanism of contraction.[60] Significant new developments included the wartime demonstration by Hungarian biochemist Albert Szent-Györgyi's group in Szeged that what had been taken as myosin protein was actually two separately purifiable proteins, one of which retained the name "myosin" and the other was called "actin." In the test tube, both proteins would reassemble from solution as filaments, either separately or in a complex if mixed. Soon after the war, Astbury showed that the X-ray diffraction pattern of muscle was nothing but the sum of the patterns of pure actin and myosin lying along the same axis, implying that both proteins exist as a parallel filament or filaments in muscle's native state. The wartime electron microscope studies of Jakus, Hall, and Schmitt had indicated that muscle contains indefinitely long "myosin filaments" stretching continuously through all the repeating bands (Plate 4.6).[61] (Schmitt was initially doubtful about the real existence of Szent-Györgyi's "actin," though by late 1946, Hall and Jakus had isolated it and shown it, in reconstituted form, to be filamentous by electron microscopy; even in the spring of 1949, Schmitt did not find Astbury's evidence that both proteins exist natively as filaments convincing.)[62] To Schmitt the micrographs were the best evidence, and they indicated that the "contractile units" undergoing the conformational shift were subunits of a single species of long polypeptide making up the filaments. The overall picture was of a bundle of long molecules, each of which could be pictured as a chain whose links could individually or en masse become knotted in response to a chemical stimulus, potentially causing great overall shortening in an instant. Although favorable preparations showed a "knotted or beaded appearance" along the fibrils as the theory of contraction by folding or knotting would predict, the appearance was not reproducible enough for Schmitt fully to endorse the actual existence of these knots in the late 1940s. The group must have tried hard to make that beaded appearance more predictable, because in some preparations the beading showed a 400 Å periodicity, which would have confirmed Bear's finding of such a periodicity by X-ray diffraction. For the moment, though, Schmitt concluded that the contractile subunits must be below the resolving power of their instrument.[63] There was puzzlement in the MIT group that the boundary between A and I bands looked much sharper than supposed from light microscopy, since the A band was attributed to the presence of an "A substance," which should have flowed into the I band given the absence of membranes between A and I. However, this puzzle in itself does not seem to have been a substantial enough anomaly to cast doubt on the single filament knotting model of contraction.

By 1950 Schmitt's group already had numerous competitors in the business of muscle structure. Among the more formidable was Szent-Györgyi himself, who had fled the new communist regime in Hungary (despite preferential treatment given the Nobelist) and come as a visitor to the NIH laboratory of biophysicist Ralph Wyckoff. The Szent-Györgyi team had found that when purified actin was made to polymerize and then viewed (after metal shadowing; see Chaps. 5, 6) with an electron microscope, the actin consisted of filaments 100 Å wide, the same width as the main filaments observed by electron microscopy of native muscle teased apart or dispersed in a blender.[64] These dispersed filaments from native muscle were seen to be indefinitely long, and to pass continuously through all bands, just as the Schmitt group had observed (though Schmitt had reported filaments of more variable width); this continuity was confirmed when the muscle was subjected to a series of washing procedures, including one known from biochemical work to dissolve myosin, which removed almost all trace of the banding pattern but left the 100 Å filaments intact. Schmitt's 400 Å periodic nodosity within the filaments was also evident to the NIH group in favorable preparations. Szent-Györgyi's interpretation was that the main filaments constituting muscle are actin complexed with myosin in the A region. Another competing group was that of thin-sectioning pioneers Daniel Pease and Richard Baker, newly installed in their electron microscope cytology lab at the University of Southern California's medical school. Pease and Baker, using their own embedding procedure (collodion and paraffin, rather than methacrylate) on both chemically fixed and freeze-dried muscle, then cutting fairly thin (0.1 μ) sections and dissolving away the embedment, produced a very different kind of picture than either shadowed or stained dispersed fibrils. Nonetheless, they too essentially confirmed Schmitt's views that filaments are continuous and that changes in banding pattern are caused by migration of accessory substances along the filaments.[65] Furthermore, an Australian group including Alan Hodge, who would soon join the Schmitt lab to complete a Ph.D. on muscle, published particularly sharp muscle micrographs in *Nature*, both shadowed and stained with phosphomolybdic acid, convincingly showing continuity of filaments and Schmitt's 400 Å periodicity.[66] Schmitt's beliefs (except actin's artifactuality) were seemingly being confirmed. Keith Porter, on the other hand, collaborating with a group of Chicago medical researchers, came to opposing conclusions. Using dispersed muscle fibrils made to contract to different degrees by chemical treatment, and afterwards fixed and shadowed, this group found no evidence that filaments extend continuously through more than one banding period, and no evidence of the 400 Å periodicity.[67] They did find, however, that the filament's relaxed thickness of 200 Å increased during contraction, probably, they thought, due to adsorption of "A substance" (myosin?) onto the actin filaments. In sum, around 1950 there was a bewildering array of conflicting interpretations of muscle ultrastructure, based on a profusion of electron micrographs made by the wide variety of preparation techniques deployed by all the rival laboratories. Among the contentious questions were the dimensions of the basic filaments, their composition (actin or myosin or both together? was there a real periodicity within the filaments?), and also their length and continuity—though the majority of researchers accepted that the filaments were indefinitely long. There was a general assumption of only one main type of filament that contracted through conformational shift.

Just as the introduction of methacrylate embedding in 1949 sent Porter and Palade into concentrated methodological research on how best to prepare and section tissue using it, so too the Schmitt lab was thrown into a period of intensive thin-sectioning experimentation. Microtomes were built, bought, and modified; glass knives were improvised with smashed bottles. Soon sections thin enough for electron microscopy were being obtained.[68] From 1950 to 1953, work in the competitive muscle field intensified and diversified, but no convergence on the controversial issues was forthcoming. The Australian group published more very high resolution micrographs of dispersed muscle fragments, stained and shadowed after treatment with various washing procedures, showing indefinitely long, continuous filaments that lost all appearance of nodosity with washing; the interpretation was in line with Szent-Györgyi's, namely, that the basic filaments are actin, bound to myosin at regular intervals.[69] Porter, working with Bennett (now moved from MIT to Washington University in St. Louis), who had recently reported a helical fine structure of muscle filaments, applied the new Rockefeller fixation and thin-sectioning protocols to muscle. Bennett and Porter found that the A band does not change in density during contraction, whereas the I band does, casting doubt on the previously accepted notion that migration of A substance into I has something to do with the mechanism.[70] Meanwhile, by the end of 1949 Szent-Györgyi began settling into his own Institute for Muscle Research, at the Marine Biological Laboratory in Woods Hole.[71] His group was joined by Delbert Philpott, an electron microscopist who stopped out of a Ph.D. program in Chicago to take up the opportunity, and in 1952 a second-hand EMU-2 microscope was purchased from RCA and installed personally by Hillier.[72] Looking for the basis of resting muscle's ability to be stretched or hyperextended, on the suspicion that whatever muscle component is responsible cannot be the same one that accounts for contraction, Szent-Györgyi and Philpott used the new thin sectioning techniques and found, once again, continuous filaments, and these they presumed to be actin only. They thought they saw special elastic segments built in to these filaments that showed evidence of a springlike spiral substructure.[73] With the introduction of thin sectioning technique, electron microscopy seemed to be making the muscle picture grow more rather than less confused.

In late 1952 Schmitt's group was joined by Hugh Huxley, followed in early 1953 by Jean Hanson, both visitors who had just received doctorates in England and who wanted to acquire electron microscopic skills at the source. Hanson, who since her 1951 doctorate had been working under J. T. Randall at the biophysics unit of Kings College in London was sent to MIT at Warren Weaver's suggestion and with a Rockefeller fellowship. Huxley, having just completed his Ph.D. in 1952 on X-ray diffraction studies of muscle, was sent by his supervisor at Cambridge, John Kendrew.[74] Using muscle fibrils prepared in glycerin by a protocol popular with biochemists, which can be made to contract slowly with the addition of a solution containing the energy-rich chemical ATP, Hanson had been using visible light to observe banding changes during contraction with the King's College unit's newly improved phase microscopy equipment.[75] Huxley's thesis work had compared X-ray diffraction patterns of muscle in relaxed and rigor mortis states (since normally contracted, nonrigor muscle gave poor patterns), confirming Astbury's findings that the major outlines of molecular structure in the muscle were not much changed in contraction, and could be interpreted as the shift with respect to one another of *distinct* myosin and actin filaments that remained parallel, rather than as the reconfiguration (folding, winding) of the protein filaments themselves. Huxley had then supposed that there were arrays of two kinds of filaments that went from a less to more ordered form of packing on contraction.[76] On arrival at MIT, he participated in the development of yet another microtome design for plastic thin sectioning, with Schmitt's graduate students Hodge and David Spiro, the former just finishing and the latter starting thesis projects on the ultrastructure of muscle. Soon Huxley had transverse (crosswise) thin sections of muscle, fixed according to Palade's recipe and stained heavily with the Schmitt lab's favorite phosphotungstate concoction, which showed that the regular hexagonal array of two kinds of filaments that Huxley expected was present in the A band, except in the H zone, which had only the thicker filaments (Plate 4.7).[77] The idea of two sets of parallel filaments that might move relative to each other was already articulated, then, but the filaments' chemical identification and their changes during contraction were uncertain. Shortly after Hanson's arrival at MIT she gave a seminar on her phase microscope work showing the changing banding pattern an individual (unstained, unfixed) muscle fiber undergoes in the course of contraction. Reportedly, there immediately followed an intense period of collaboration between Hanson and Huxley, in which a plethora of experiments were devised to correlate the changes visible by phase and electron microscopy in muscles undergoing biochemical treatments and, of course, during contraction.[78]

Their first joint experiments clinched the argument that the "A substance" was myosin and the main lattice of filaments, present in both A and I zones, was made of actin. Watching individual isolated muscle fibrils with both phase and polarized light microscopy during treatments that biochemists used to extract myosin from muscle, they were able to see the dark, anisotropic appearance of the A bands disappear before their very eyes. Electron microscopy of such extracted muscle confirmed that the thick filaments Huxley could find only in the A bands had disappeared. Thus, muscle is made of thick and thin filaments, the thick filaments of myosin present only in the A bands, the thin filaments of actin extending from the Z line to the middle region of the A band, where they were linked to the actin filaments of the next Z line by finer connector filaments across the H zone (which were as yet invisible, but necessary to explain how the fibril stayed together after myosin extraction). Though the details of this history are controversial, it seems certain that in the *Nature* publication of 1953 where they reported this work, what happens in contraction was still unclear to Hanson and Huxley: "Observations cannot at present distinguish between a genuine shortening of the filaments in the A-band, a migration of A-substance into the I-band, or a retraction of I-band filaments into the A-band."[79] According to their supporters, Huxley and Hanson had in mind this last model (in which, during contraction, the I band shrinks because the myosin filaments in the A band draw the actin filaments in), but were unwilling to argue it without stronger evidence that the myosin filaments of the A band remain unchanged. If this is the case, Hanson and Huxley must later have regretted their scruples.[80] The two went on to take phase contrast pictures of individual glycerinated muscle fibrils, which were made to contract slowly with carefully controlled amounts of ATP in solution, showing that over the range of contraction found in living muscle, the A bands remain constant in length, all shortening being at the expense of the I bands. Also, when fibrils are stretched, only the I bands change, whereas the A bands retain a fixed length (the actin filaments giving the extra length, depending on conditions of stretching, either from their connector across the H zone, or from their joint with the Z line, as Szent-Györgyi and Philpott suggested). When myosin is extracted from fibrils of various degrees of contraction, Hanson and Huxley found, no further contraction can be induced with ATP, and the entire fibril has the appearance of the I zone in an unextracted specimen, suggesting that the muscle is indeed held together by actin filaments somehow joined across the H zone.[81] Thus, it was concluded that the A bands are made of myosin filaments (the thicker ones in electron micrographs) of constant length, overlapping thin actin filaments, and both filaments must be present for contraction to occur. The hypothesis of a mechanism of contraction was that the actin filaments move along the myosin filaments, drawing the Z lines anchoring them closer together. The slack in the actin filaments would be taken up by their folding in the region of the H zone at the center of the myosin A band. This "sliding filament" model of contraction was consistent with most biochemical data on muscle, with the X-ray crystallography and electron microcopy results indicating a dual filament array, and accounted for all patterns observed by light microscopy on stained or live material viewed with special optics, establishing constant A band width during contraction.[82]

The novel model stirred up considerable excitement in Schmitt's group, and as he has recalled, "people could be seen in the hallways with interdigitated fingers of their two hands moving with respect to each other, and the investigators wondered how the two sets were linked so as to produce contractility."[83] Schmitt may have been urging Huxley and Hanson to account for this linkage between the filaments before they announced their interpretation. Schmitt augmented the intensity of the muscle scene late that autumn of 1953, when he organized a National Academy of Sciences symposium at MIT on "The Muscle Machine," bringing some of the most important biochemists in the field: Fritz Lipmann, J. T. Edsall, and, naturally, Albert Szent-Györgyi. Huxley had the privilege of presenting the micrographs from what Schmitt called the "Hanson-Huxley experiment" (probably the combined phase and electron microscopy on extracted and contracting muscle), and their sliding filament hypothesis on the mechanism of contraction, to this august crowd.[84] There is no record of the reaction, but it cannot have been unfavorable, because soon afterwards Hanson and Huxley decided to publish their sliding filament idea "without any clue as to coupling be-

tween the two sets [of filaments] caused the sliding," and Schmitt communicated the paper to *Nature*.[85] Here is an excellent example of how
Schmitt facilitated innovative work in his lab, by collecting a critical mass
of well-trained and motivated young researchers of diverse backgrounds
together with ample materials and apparatus, by making them talk and
work together in his seminars and group experiments, and by bringing
the leading lights of life science to visit as a stimulus. The alumni of
Schmitt's laboratory remember the atmosphere as electric.[86]

The young authors' elation at their accomplishments must have been
blunted when they learned that *Nature* had decided to publish, in 1954,
their piece side by side with another by a pair of physiologists from
Cambridge, A. F. Huxley (no relation) and R. Niedergerke, who had
arrived independently at much the same model based on films of contracting live muscle fibers viewed by interference microscopy.[87] This technique allowed careful measurements of changing band widths much as
Hanson's phase microscopy had done, which is to say that the Huxley and
Niedergerke paper proposed the same model, but on the basis of a subset
of the data offered by Hanson and Huxley, who were also employing
X-ray diffraction and electron microscopy. (However, the Cambridge
group had been working with living-condition muscle fibers, whereas
Hanson had used glycerin-soaked fibers, which, though they are still
capable of contraction, might conceivably show an artifactual banding
pattern.) From the tone of recollections of Randall and both Huxleys,[88]
the allocation of credit in this matter has long remained a sensitive issue
among British life scientists, one that turns on judgments of the value of
different sorts of evidence derived from the methods and apparatus belonging to different disciplines (e.g., electron versus light microscopy, the
living muscle of the physiologist versus the glycerinated muscle of the
biochemist), which therefore make it a site of interdisciplinary conflict
over status. The matter of credit need not concern this account, except
inasmuch as it shows another case in which scientists' decisions about
appropriate epistemology may be politically loaded (see Chap. 3).

At the July 1954 International Conference on Electron Microscopy in
London, where Sjöstrand and the Rockefellerians had their first direct
confrontation, the presentation of the Hanson-Huxley work was also not
without its drama. Before them in the program, their fellow Schmitt lab
alumnus A. J. Hodge, who had returned to Australia after finishing his
MIT degree, presented some high resolution thin-section electron micrographs (made with the microtome he and Huxley worked on together

to develop) of insect flight muscle before and after myosin extraction procedures. In his transverse sections Hodge saw only thick filaments, which obviously contained myosin because they disappeared with myosin extraction, joined to one another at frequent intervals by lateral cross-bridges.[89] In Hodge's preparations the thick filaments appeared to have centers of low density that, he suggested, might represent an actin core coated by denser myosin.[90] Hanson and Huxley, of course, maintained that there were parallel filaments of two distinct kinds; in addition to showing Hanson's phase microscopy evidence that the A bands only disappear on myosin extraction and retain constant length during contraction, they supported the two-filament view with Huxley's new transverse thin sections showing more sharply the compound array of large (110 Å) and small dots (40 Å) in the A band, that is, end views of the thin and thick filaments, seemingly linked by fine cross-bridges (Plate 4.8).[91] Though Hanson and Huxley had ideas now on the mechanism that makes the filaments slide—they supposed, in a paper published in 1955, that their cross-bridges might ratchet the myosin filaments along the actin—the crucial point for the attending electron microscopists was this compound array of thick and thin filaments.[92] Many were not convinced that the thin filaments were real, which left open the traditional interpretation of indefinitely long filaments of one type, the actin backbone decorated at intervals with myosin. For instance, Bennett (as noted, another Schmitt microscopy alumnus) proposed that Huxley's small dots between the thick filaments in transverse sections might better be interpreted as poorly fixed remnants of the longer cross-bridges between thick filaments that Hodge had observed, not end views of thinner filaments. Huxley countered by suggesting that although Hodge's long cross-bridges between thick filaments might be a real feature of his particular insect muscle, Hodge's pictures probably showed no thin actin filaments because he had not adequately fixed the tissue. Not surprisingly, Hodge himself agreed with Bennett, pointing out that neither X-ray diffraction nor any other data gave decisive grounds for abandoning the picture of continuous actin-containing filaments of a single class, periodically complexed with myosin.[93]

Hanson and Huxley continued their collaborative work on muscle after returning to England. At Sjöstrand's Stockholm conference in September 1956, the two presented substantial new findings to answer the objections they had encountered in 1954 and 1955, this time on the insect flight muscle that had given Hodge his different pictures. In this muscle, there is virtually no visible I band; but since in its normal function the muscle contracts only a very small percentage of its length, it would need none if the sliding-filament model were correct. Huxley and Hanson showed new, higher-resolution sections prepared as before with Palade's osmium fixative and additional phosphotungstate staining and now imaged using a new Siemens instead of an EMU. Huxley had finally made longitudinal (lengthwise) sections that clearly showed thin filaments midway between the thick, joined by frequent cross-bridges, and even hints of the hypothetical connectors joining thin filaments across the H zone (Plate 4.9). Hodge's longitudinal sections were probably too thick (as the more heavily cross-bridged appearance of his thick filaments implied) and possibly too poorly fixed, Huxley and Hanson argued, to support any claim that the thin filaments did not exist. And their transverse sections of insect muscle showed an array of small and large spots, interpreted as cross sections of the thin and thick filaments, as in their earlier work with vertebrate muscle. Moreover Huxley and Hanson now claimed (but did not show) that they had occasionally obtained transverse sections showing only thin filaments, as would be expected in an I band, thus blocking interpretation of the small dots as nothing more than cross-bridges between thick filaments.[94]

Whereas Huxley and Hanson had worked to extend their pictures and interpretations from vertebrate to insect flight muscle, Sjöstrand and his student Ebba Andersson presented a paper that pushed in just the opposite direction, effectively confirming Hodge's conventional single-filament model and extending it from insect to frog and mouse striated muscle. Using Sjöstrand's physiologically buffered fixative and the usual meticulousness in exposing tissue to the fixative as rapidly as possible and sectioning very finely, the Swedish workers produced transverse and longitudinal thin section views as sharp as Huxley and Hanson's (Plate 4.10). Sjöstrand and Andersson found only one type of filament, 40–50 Å wide in the I band of relaxed muscle like Huxley and Hanson's thin filaments, but according to the Swedes the A band consisted of thick segments along the same filaments. In contracted muscle the filaments of the A band appeared thicker in proportion to the degree of contraction; thus, it seemed probable that during contraction segments of filament converted from their thin form in the I band to a thicker and more compact form found in the A band (which would explain decrease in I band width), and that on further contraction the A band form would condense still more.[95] Thus Sjöstrand and Andersson's transverse thin sections were given diametrically opposite interpretation, despite the striking similarity in appearance to Huxley and Hanson's, vindicating the traditional theory of contraction by filament folding. The thickened filaments of the A band appeared to be joined by many cross-bridges, as Huxley and Hodge also agreed; these bridges do not go straight from filament to filament, however, but are kinked in the middle, presumably because of shrinkage of the fixed tissue. (Since the X-ray diffraction data indicated that the major filaments were more than 440 Å apart, whereas thin sections showed thick filaments less than 300 Å apart, a great deal of shrinkage during preparation for electron microscopy was indeed indicated.) Sjöstrand and Andersson argued that it might be these kinks, when several of them were viewed superimposed on one another in excessively thick sections, that gave the impression of a second set of thin filaments between the primary thicker ones. Thus the thin filaments Huxley and Hanson saw in transverse sections of A bands were explained away as artifacts of sectioning.

As for the Huxley and Hanson longitudinal sections showing thin filaments between the thick, their muscle specimens were glycerin-soaked whereas the Swedes used tissue fixed fresh; glycerin in the hands of the Swedes reportedly caused splitting or fraying of the A band filaments, which would account for these thin filaments as a simple preservation artifact. (One can imagine how biochemists accustomed to using glycerinated muscle might have greeted this suggestion by Sjöstrand!) The Swedes' own unglycerinated longitudinal sections showed continuity between thin and thick filaments at the A-I transition. Though they could not have been unaware of the phase and interference microscopic observations that A bands remain constant in length during contraction, Sjöstrand and Andersson registered faith in their own electron microscopic findings that both A and I bands are shortened in contraction—the I bands much more so. Thus they evinced greater trust in their evidence from fixed specimens than in the lower resolution light microscopic evidence, a trust defensible on grounds that there is no reason to expect that within a given preparation protocol contracted muscle will shrink especially more than the uncontracted muscle that serves as a baseline for measuring both A and I contraction. This is the same sort of reliance on evidence and epistemological criteria fully under the electron microscopist's control that was noted earlier of Sjöstrand. And true to form, from their micrographs the Swedes managed to draw elaborate inferences about the helical protein structures of the filaments and even about the supercoiling that explained the filament shortening they observed during contraction. In 1957 this paper was published in the inaugural volume of Sjöstrand's new journal, the forum for an "ultrastructural" style of his own that was becoming increasingly heterodox (Chap. 3).[96]

The sliding-filament model depended on the real existence of distinct thin filaments (at 40 Å, no larger than the resolution limit obtained by all but a handful of biological electron microscopists at the time), and the existence of two distinct sets of filaments in the whole muscle could only be established by electron microscopy. Neither the community of practitioners nor its elite were agreed on what represented the faithfully preserved state and what was artifact. In 1957 Huxley countered Sjöstrand with a paper not only plainly showing the thin filaments between the thick filaments (in the A band), and carefully explaining how all the views of continuous single filaments could have resulted from excessively thick sections of muscles made of two filament types, as Hanson and he had proposed, but also generating single-filament views like those obtained by others through deliberate degradation of his micrograph resolution (Plate 4.9).[97] This dose of his own medicine convinced Sjöstrand, with Huxley's longitudinal sections of unglycerinated muscle showing that despite poorer contrast, the thinner filaments could still be distinguished between the thick. Even so, discussion at the summer 1958 convocation of the international biophysics community in Boulder—orchestrated by Schmitt (see below)—still manifested the undecided fate of the model. All three contributors commenting extensively on muscle were ambivalent: Bennett, impressed with Huxley's "elegant" longitudinal section micrographs appears to have been leaning in the sliding filament direction, while Hodge was much more critical, pointing out that the electron microscopy was inconclusive and that even Huxley's own X-ray diffraction evidence suggested that the thin filaments are artifacts of muscle deprived of ATP (as in rigor or by glycerination), and Manuel Morales, a biochemist, considered that single filament mechanisms were compatible with existing evidence and more likely on general biochemical considerations.[98] However, these doubts about the reality of separate thin and thick filaments seem to have faded rather quickly in the face of increasingly clear thin-section electron micrographs showing their expected appearance, and by the early 1960s debate was centered around the mechanism of the sliding between the two sets of filaments rather than their mere existence. Indeed, Andrew Huxley marks Hugh Huxley's longitudinal thin sections showing distinct thin and thick filaments, shown at the 1956 Stockholm meeting and first published in 1957, as the last major landmark in establishing the sliding filament model.[99] Perhaps this case represents another in which biologists, especially those not very familiar with the huge range of views obtainable from a given specimen by electron microscopy, gave so much credence to clear pictures supporting an attractive hypothesis that they discarded viable conflicting interpretations too easily (as in the case of mitochondria, according to Sjöstrand). Perhaps; but the sliding filaments have not "misbehaved" since and have served as sound rootstock for a flourishing branch of investigation, so grounds for worry are lacking. The model is an idea that works.

Conclusion: Schmitt's Geometrical Style of Electron Microscopy

Although some in Schmitt's lab regarded Keith Porter as a competitor, presumably because of his work on muscle and collagen (see Chap. 3),[100] the MIT group was never engaged in the same enterprise of "cell biology." MIT remained a site for the application of electron microscopy to research programs in biophysics cum general physiology. Rather than delve into the complexities of the many parts of the living cell and their topological interrelations in the organism, the Schmitt lab was most successful in carrying on with the approach developed in wartime: quantitatively precise analysis of the fine structure of individual—and isolated, if not naturally occurring in pure form—cell components pursued through direct measurement of high-resolution micrographs. The electron microscopy was paralleled by X-ray crystallography, polarization optics, or other physical methods applied to similar specimens (and for these methods specimens must be isolated or pure). Mathematization of the electron microscopic image was pushed to its maximum by recourse to an ensemble of physical instruments. Theory of self-assembly was improvised from these sorts of quantitative data, by making electron micrographs at various stages of structural aggregation. It was a geometrical, even architectural, approach to the forms of isolated macromolecule complexes, of protein filaments and other simple systems, and to the disposition in space of the subcomponent molecules making up the complexes. Schmitt was after the blueprints of the machinery of life, perhaps trying to become the future "medical research man" Linus Pauling spoke of in his own Philharmonic radio talk, who "will be molecular architect . . . able to draw the atomic blueprints for promising compounds" in order to cure disease and manipulate biological processes.[101] That Schmitt at least considered constructing protoplasm in the test tube is proved by his explicit denial that he had achieved this goal, at a National Academy of Sciences biophysics symposium that he organized. "No one has 'synthesized life'; rather, we have described some of the rules of the process by which the . . . structural and chemical properties of the parts of the cell . . . produce the properties of living organisms."[102]

Such a precision engineering approach only suits, it seems with the benefit of hindsight, simple specimens having a crystalline regularity or sharply defined chemical composition very much unlike the irregular structures with which the cell biologist inevitably must deal. The muscle story surely counts as a success of Schmitt's school, because almost all the major players—Huxley, Hanson, Sjöstrand, Hodge, Bennett—had been through his lab and absorbed his influence. The specimen in question, striated muscle, was especially suitable because it had the requisite regularity (and purity, too, since muscle cells consist mostly of the filaments). The terms of the debate over muscle structure were the number of filament types (characterized by their size as well as chemical behavior with respect to staining and extraction), the disposition of the filaments relative to one another, and the geometry of the subunits composing them. These were all read quantitatively from micrographs, as Schmitt first showed how to do, and the spacing and angles obtained by this visual exercise in biological geometry were checked where possible with those from X-ray crystallography. The electron microscope played a crucial role in the first proposal of the successful sliding-filament model; moreover, the veracity of the model was settled largely on the basis of high resolution electron micrographs. Thus the right questions and material for investigation came together with Schmitt's mathematizing epistemology, leading not only to a vindication of Schmitt's method among electron microscopists, but also in general to a triumph of the electron microscope in physiology. For it is unlikely that debate on the sliding filament hypothesis could have been fully put to rest without direct visual demonstration of the two kinds of filaments and their arrangements in muscle by electron microscopy, even if it had been proposed on the basis of other evidence.

As to the question of the electron microscope's impact on biophysics, this is more difficult to address than in the case of bacteriology (Chap. 2), which maintained a continuous if evolving identity throughout the post-war decades, and likewise the case of cell biology (Chap. 3), which emerged from cytology as a distinct new discipline based on the electron microscope by the late 1950s. The electron microscope certainly contributed to biophysical fields of knowledge like nerve action and protein self-assembly, and inasmuch as standard methods for the biophysical practice of electron microscopy took shape, these were the ones developed and disseminated by Schmitt. The difficulty is that the discipline in which work such as Schmitt's was located had an indistinct and rapidly shifting definition during the postwar decades. Whereas, as noted, Schmitt's prewar and wartime work had certainly fit the model of Loebian general physiology, in the immediate postwar period Schmitt mostly called himself a "biophysicist" (an established prewar synonym for "general physiologist") along with a heterogeneous group of other biologists, including devotees of Max Delbrück's bacteriophage experimental system for molecular genetics. In this fluid period, Schmitt emerged as a national leader in the effort to found a larger and more firmly defined discipline of biophysics that would include his kind of electron microscopy, together with a variety of other approaches to the physical properties of living things. It may be that Schmitt's ambitions for biophysics were not fulfilled in the form he envisioned largely because of competing efforts at redefinition and reform of biological disciplines, among which the splitting off of many biological electron microscopists as "cell biologists" must figure prominently. Another, as I will suggest below, may be that biochemistry in the period was adopting the ideas and methods of physical chemistry that distinguished one field within Schmitt's brand of biophysics, depriving him of another constituency he envisioned as joining his reinvented discipline.

Excursus: Schmitt and the Disciplinary Politics of Biophysics

In the first postwar years, Schmitt's band of half a dozen biophysicists dominated the MIT Biology Department, as old guard sanitation faculty members left service and the food technologists seceded. Biochemistry was represented in the department only by two of the ten teaching faculty on staff at the war's end. But in 1952 the administration made a commitment, reflecting the growing stature of biochemistry as a field of basic science (as opposed to its traditional status as a service discipline for medicine), to establish a biochemistry division as a semi-autonomous unit within the biology department, with its own relatively modest but dedicated operating budget. When this decision was made, it was officially understood that a semi-autonomous biophysics division would soon be formed as well, leaving other life sciences under biology.[103] In 1952 a new junior appointment in biochemistry was made, and in the summer of

1953, under the newly recruited head of biochemistry, John M. Buchanan, the division was officially founded. When in early 1953 the administration informed Schmitt that it intended to renege on its commitment to establish the biophysics division, claiming that the cash situation did not permit adherence to the agreed timetable, the tone of Killian's letter makes it obvious that the decision must have been a serious disappointment to Schmitt. "Dear Frank," Killian wrote, "I want to explain to you why the Budget Committee reached the conclusion that we should not now go ahead with the establishment of a Division of Biophysics. . . . [T]he Committee felt it could not authorize the proposed expenditures for space changes and for new personnel which were part of your proposal. . . . Such an expenditure would most certainly appear inequitable to other departments in the light of . . . the capital expenditures which have gone into the [new] Dorrance building."[104] Limited at home, Schmitt would soon direct his efforts to build up biophysics beyond the walls of MIT.

Presumably because of some combination of frustration with the MIT administration's resistance to his plans for an autonomous biophysics division and unhappiness with executive duties for the whole biology department (also, his son and Loofbourow had met unexpected deaths in 1950–51, both of which evidently affected Schmitt very strongly),[105] in 1955 Schmitt abdicated as head of the biology department in favor of a rare "Institute Professorship," leaving him free from teaching and administrative responsibilities. The summer of 1956 saw Schmitt reorganizing his own group as a "unit" and moving it into most of the Dorrance building's fifth floor.[106] In his unit Schmitt had sufficient grant support to maintain what was in effect a miniature department: a handful of full-time postdoctoral researchers under him, including Alan Hodge, whom he brought back in 1956 as his unit's executive officer, and a couple of biochemists (freeing him from the need to arrange collaborations with the faculty biochemists downstairs). There remained a full complement of graduate students, technicians, and the network of collaborators he had built up in neighboring medical institutions.[107] Schmitt also had a full supply of his own equipment, including an EMU and the Philips electron microscope he had acquired in 1951, and several ultracentrifuges; in 1957 he was able to obtain three new electron microscopes (a Siemens Elmiskop 100, and two Hitachis, types HU-10 and HS-5, Schmitt's choices clearly illustrating how RCA had lost its lead by this time).[108] Schmitt's institutionally insulated group was able to explore his new idea that gene regulation depends on the spontaneous reaggregation of proteins and short nucleic acid molecules depending on small changes in the chemical environment, based on an analogy between the banding pattern visible in certain chromosomes and the pattern in collagen fibrils that his lab was learning to vary at will,[109] and to move ahead intensively with connective tissue and nerve. The lab's approach included not only structural studies but the sort of work in protein purification and characterization, and the physical chemistry of protein aggregation and self-assembly, that was becoming increasingly common in biochemistry departments.

Schmitt was not retiring as a champion of biophysics in 1955, but moving his leadership efforts to the national stage. The research areas constituting biophysics had been burgeoning in America since the war: not only was biophysics gaining vigor at universities where it had existed already, such as MIT and Cal Tech, but new academic and research units were springing up in prominent institutions everywhere. For instance, new departments and laboratories of biophysics were established in the first postwar decade at Yale, Chicago, Pittsburgh, Johns Hopkins, and at Ann Arbor and Berkeley too (as discussed in Chap. 5); biophysics was also established at medical institutions, including several sites in the National Institutes of Health and at the grand new medical school of the University of California at Los Angeles.[110] Given the high cultural currency of biophysics—driven, as I have argued, by post-Hiroshima nuclear fear, and no doubt later reinforced by the hydrogen bomb and Eisenhower's "Atoms for Peace" initiative—together with the rapid NIH budget expansion that resumed after the Korean war, and Schmitt's warm connections with the Bostonian medical aristocracy (indeed, Schmitt had turned down the deanship of Harvard Medical School in 1949), it comes as no great surprise that in the autumn of 1954 Ernest Allen, chief of the NIH Division of Research Grants, approached Schmitt to ask if he would organize and chair a study section to distribute funds for extramural research in biophysics. Schmitt agreed to do so, on condition that the NIH permit him great latitude to promote and organize biophysics as he saw fit.[111]

Allen assented, assigning Schmitt's friend and NIH official Irving Fuhr as the project's executive secretary, and naming three biochemists (who specialized in applying physical chemistry methods to protein structure) to the study section.[112] Schmitt also recruited a few of his own members, neurophysiologists like himself and former physicists. At the first meeting of the Biophysics Study Section, Allen put in an appearance to reaffirm NIH's commitment to Schmitt that the new study section was to take an unusually active role in "providing leadership at the national level" for biophysics, for example, by organizing and funding conferences. The main event of the meeting was a fierce wrangle, presumably with biochemists the NIH had brought into the group, over the name of the new body; by the end of the day the study section was redesignated as "Biophysics and Biophysical Chemistry," setting apart biochemical physical chemistry and making clear that it stood under the section's aegis and furthermore that this field was not to be subsumed under "biophysics." There was general agreement that research "attention should be focused chiefly at the molecular level" and be fundamental, deemphasizing forms of biophysics with a clinical slant, for instance radiobiology, which was covered under an existing study section in any case, and tissue biomechanics.[113] The venture represented by this study section was based on a partial convergence of the interests of Schmitt's circle of self-styled biophysicists and a set of biochemists committed to somewhat nontraditional approaches based in physical chemistry. The uneasiness of the alliance providing his platform would cause persistent, perhaps ultimately insurmountable, difficulties for Schmitt in his efforts to promote a new biophysics.

In the spring of 1955 the study section increased its membership in both camps, began reviewing applications for both the NIH and ONR, and began giving out grants (from the NIH, $310,000 in 1955, $1.1 million in 1956, $2 million in 1957, and increasing amounts thereafter).[114] Under Schmitt's direction, plans were also made to hold small conferences in the Midwest, the West coast, and the Eastern seaboard of the USA. At these the leaders of North American biophysics were to be invited to give informal presentations about the status of the field at their home institutions and to discuss the definition of "biophysics," its prospects, and the best strategies for its development. The first meeting was held in mid-September 1956, at Ann Arbor, Michigan. Schmitt and half a dozen other study section members sat around a table with about twenty select conferees from the region. Speaking for the NIH, Fuhr repeatedly stressed that "the grants program is very broad and flexible and probably can be used to meet many situations confronting the biophysicist." Schmitt gave something of a vade mecum into the era of big-science biomedicine for his more timid biophysical colleagues. At several points in the session, conferees needed to be reassured that their feeling "dishonest" when claiming the medical relevance of their projects, on the NIH application forms, merely reflected a "guilt complex" that ought to be overcome; the

NIH administration had utmost confidence, the biophysicists were told, that virtually all the basic biomedical research they might do would ultimately bring the kind of benefits Congress was expecting. Though there was still some worry that research would have to be redirected toward disease-related problems, a number of the conferees expressed hope that the NIH study section would provide the "prestige" needed to "make biophysics respectable," and would help when recalcitrant "university administrators have to be convinced that biophysics is an important area." Schmitt even gave hints on how to apply for the free money being offered, stressing that the study section would buy instruments, improve facilities, and finance graduate and postdoctoral personnel and major conferences, all with no strings attached. Biophysics was in "ferment" he said: the time was right to think big and to advance the field's status.[115] There was much discussion of when and how a biophysics society should be formed, but by the second conference, held in early December 1955 at Berkeley, California, it had been decided that the organization of a professional society lay beyond the NIH study section's purview (though organization of a society proceeded apace, as described below, with help from Schmitt).

At this second meeting, dominated by Schmitt's fellow wizards of grantsmanship Wendell Stanley and Linus Pauling, the NIH assurance that basic research was eminently fundable was not news, and engendered no comment. Much more discussion went to the problems of placement and recognition: finding jobs for students with new Ph.D.s in biophysics, and the relation of this to the status of the field. It was noted that students were better off with a degree in an established discipline like physiology or biochemistry, at least until more departments of biophysics opened and began hiring faculty. Linus Pauling, in particular, seemed hostile to the idea of a discipline of biophysics, suggesting at one point that "it may be that biochemistry should just absorb this field." (Of course, comfortably ensconced in Cal Tech's chemistry division and sumptuously funded to pursue work that would fall within the nascent discipline Schmitt envisioned, Pauling had nothing to gain and possibly much to lose by the development of the new discipline.) The third time Pauling raised the issue of whether there really was such a thing as biophysics, Schmitt quashed him, reminding Pauling that biochemistry had moved from a service role to true disciplinary status not long before, and asserting that biophysics was close behind on the same road.[116] Schmitt was obviously the central figure at these conferences, and he used the gatherings to promulgate his majestic vision of biophysics's destiny.

After the second meeting, the study section members communicated their impressions and recommendations to the NIH, and fault lines within the section were still evident. Biochemists Hans Neurath and J. L. Oncley said they were impressed mainly with what a heterogeneous grab bag the field represented, Oncley asserting that despite hearing what all the biophysicists were working on he still had no idea what "might be profitably called biophysics" and suggesting once more that the term be dropped from the study section's title. Schmitt fended off another name change at the next section meeting.[117] Though reservations had been expressed that the study section might be attempting to exert influence beyond the proper NIH mandate, Schmitt's fellow electron microscopist Robley Williams felt that the personal interaction at the conferences had actually counteracted suspicion (at least among the select conferees) that "we are attempting to organize and sway at a national level" and made clear that "our purpose is help and stimulation." Schmitt reiterated that the study section had been organized not just to dispense money but "to advise the NIH how it might use a portion of its large . . . resources in the development of a field which . . . lies at the heart of modern advances in molecular biology" and considered that this work—hard indeed to distinguish from "swaying at the national level"—was going well. With at least one other section member, Schmitt and Williams agreed that the third biophysics conference should go ahead, because the occasions were effective for boosting morale and giving participants a sense that they belonged to a large and growing community, and moreover, it might be bad "public relations" not to go through with the Eastern event. Williams (who must have agreed with Schmitt in declaring it useless to worry about "those who feel that Biophysics does not formally exist") was clearly comfortable with Schmitt's degree of activism, whereas the biochemists were not.[118]

The Eastern conference eventually took place in Bethesda in September, and would in due course be followed by a formal report on the state of biophysics by Schmitt. But already in March 1956 the next stage of Schmitt's project with the study section was moving forward. By arrangement with Allen, that month Schmitt addressed the NIH Health Council (a sort of board of directors consisting of outside medical experts and lay people) with a summary of the state of biophysics and what the study section had been doing. Describing the "present turmoil" of biophysics and its impending foundation as an independent discipline, Schmitt stressed most of all the need, at the present critical stage of the field's growth, for well-trained students and for fellowship support for all levels of researcher.

Shortly after the meeting Schmitt was informed that he should draw up a grant application to pursue his ideas on workshops, fellowships, and other training programs. After consultation with the full study section, in April an application was submitted, and by May Schmitt was looking for someone he could hire as an administrator of the study section's newly expanded activities.[119] Exhibiting complete confidence in Schmitt's leadership, the NIH deposited the first $150,000 installment of the grant, approved in October 1956, directly in Schmitt's personal bank account. Schmitt was principal investigator and the grant was administered from an MIT office by the man Schmitt had found to relieve him of the routine burden, Richard Bolt, an MIT physicist who had for several years been trying to develop ultrasonic imaging techniques for the brain (until his NIH funding was not renewed).[120]

For the first year of this grant (thought of by Schmitt as the "programming slush fund"),[121] through September 1957, Schmitt and the study section used it to support several researchers in career transitions and for summer training programs, to fund a small conference on the biological uses of nuclear magnetic resonance, and to underwrite a study of instrumentation needs of biophysicists nationwide. Small amounts were also spent on a grant to the newborn Biophysical Society over at least Neurath's objections that such direct meddling in disciplinary politics was unseemly for an NIH body, and on reprinting and distributing—mainly with the purpose of promoting solidarity within the biophysics community during the period of debate accompanying the Society's foundation—an article by British electrophysiologist A. V. Hill that defined biophysics in a vague and broadly inclusive way of which Schmitt approved.[122] For the second year of the programming grant, for which an additional $250,000 was allocated by the NIH, such ad hoc "catalytic" activities were continued. In addition, Schmitt organized an ambitious month-long national conference, the "Study Program in Biophysical Science at Boulder, Colorado," to take place in July and August 1958. This conference aimed deliberately to bring about a national consensus defining and setting standards of what counts as good biophysics for the attendees, and to display the new discipline's wares for scientists everywhere. Through its initial publication in the *Reviews of Modern Physics*, and later as a textbook subsidized by Schmitt's programming grant, the conference proceedings reached an estimated 10,000 libraries worldwide.[123]

It is evident from the Boulder conference's organization that, for

Schmitt at least, biophysics included the following main areas of research: physical chemistry of macromolecules, biological energy pathways, connective tissue and muscle protein structure and function, nucleic acid and protein synthesis (i.e., molecular genetics), cellular architecture, and nerve physiology/brain function. "Physiology of the mechanical type" and radiation biology apparently also counted as biophysics, though not, in contrast to the preceding areas, in the subcategory of "molecular biology."[124] In the Boulder keynote speech, Schmitt communicated his sweeping vision of where biophysics was going. It would soon mature as biochemistry's sister science:

Biophysics, which is rapidly developing to the status of a major branch of the life sciences, is following the pattern set by biochemistry. . . . Just as biochemists had to determine first the composition and structural chemistry of the complex biomolecules, so in the early development of the field, biophysicists have to determine first, with the aid of crystallographic and physicochemical methods, the detailed configuration of the molecular chains of which the macromolecules are constructed. It will then be necessary to investigate the forces between the macromolecules.[125]

Biophysicists, relying heavily on physical chemistry, will explore "the interaction properties of the macromolecules" through which they "spontaneously form supermolecular aggregates," which are the machinery of life. As their techniques and theories became more sophisticated, biophysicists will thus be able to model, predict, and engineer higher order components of living systems. Advances must come from two directions, Schmitt continued, introducing a top-down versus bottom-up distinction in a way that will hold some interest for anyone interested in the history of the term "molecular biology." Speaking of the analytical or "bottom-up" approach: "It is inevitable that such investigations must precede successful attempts to determine how the macromolecules interact to cause biological function. The analytical method has made possible most of the striking advances which, together with the parent science of general physiology, have come to be known as 'molecular biology.' "[126] Thus, Schmitt seems to understand by "molecular biology" not a discipline or a subdiscipline, but a style of work (namely, the "bottom-up" approach) within the discipline of biophysics, apparently including cellular architecture studies and some neurophysiology.[127] He went on to urge that biophysicists should not altogether abandon the synthetic project just because the analytic is initially more productive. After an overview of nineteenth-century controversies about "living molecules" meant to illustrate this point about the dangers of attributing too many vital properties to material parts rather than their arrangement in the organism, Schmitt reemphasized the theme in his concluding words: although molecular biologists and other investigators of the "bottom-up" style should explain as much as possible through automatic self-assembly of macromolecules, the holistic or systems approach must not be neglected. That is, a convenient methodological reductionism should not be allowed to lapse into an absolute metaphysical reductionism.[128] In his increasingly consuming interest in brain function, Schmitt at least would follow his own advice.

At any rate, the Boulder conference did bring together an impressive list of luminaries under the aegis of "biophysics." By the time Schmitt stepped down as chair of the NIH Biophysics and Biophysical Chemistry Study Section in December 1958—to be succeeded by his stalwart supporter in the section, Robley Williams—it was dispensing millions of dollars per year in NIH grants. Biophysics had attracted the attention of the Senate Appropriations Committee in 1958, and the special report requested by the Committee was in preparation under Schmitt's watchful eye.[129] Biophysics was still growing dramatically at universities throughout North America. And, as noted, a national Biophysical Society had formally been established, in a series of meetings beginning in early 1956 and culminating in the National Biophysics Conference of March 1957 at Ohio State University, thanks to tireless politicking by the so-called "Committee of Four," one of whom was Schmitt's brother, Otto. At the Ohio meeting, the more than 400 registrants were convinced that it was essential, as much "*professionally* as *scientifically*," to form a separate society rather than a new section of an existing organization such as the American Physiological Society. (One major reason was that biophysicists constituted a unique blend: physicists would not fit among the regular physiologists, nor the more biological biophysicists among regular applied physicists, and so forth.)[130] Robley Williams was elected the Biophysical Society's first president, Schmitt lending as much help as possible from the dignified remove of his position as NIH section chair. It seemed by the end of the 1950s that Schmitt could fairly claim a central role in successfully putting biophysics on a strong new foundation as a true discipline.[131]

But this foundation was on shifting sands. However much the NIH may have supported Schmitt, he depended in his plans for biophysics on a convergence of the interests of his circle of structural and genetic biophysicists (with backgrounds mostly in physiology or physics) with those of biochemists involved in physical chemistry of macromolecules. This leads me to a concluding speculation on one of the major factors in the failure of Schmitt and his allies to establish biophysics as an enduring discipline along the lines they had envisioned: they wanted to include the physical chemistry of proteins and nucleic acids under biophysics, and this put them in competition with biochemists who wanted to include this territory in their own discipline. Schmitt did have some physical chemistry–oriented biochemists on his side. As Case Western University's Wilfried Mommaerts saw it at the first of Schmitt's NIH conferences for the biophysical elite, "The main contact between biochemistry and biophysics will be in the area of the physical chemistry of biological systems. If I read the signs correctly, the majority of biochemists would be happy to drop this nuisance. Biophysics, however, would greatly benefit. . . . The unison with so-called physical biochemistry would greatly enrich biophysics, leaving to the orthodox biochemists a more homogeneous field of study."[132] As already intimated, though, Schmitt encountered great resistance to his discipline-building efforts as head of the NIH biophysics study section from the biochemists assigned to it. Or so it certainly appears from their repeated objections of impropriety to Schmitt's meddling in disciplinary politics, from their vote against the grant to the Biophysical Society, from the recurrent attempts to eliminate "Biophysics" from the study section's title, and so on. Conflict came plainly to the surface in those combative exchanges between Schmitt and Linus Pauling, that great champion of physical chemistry within the discipline of biochemistry.[133]

Given such signs of struggle, it seems worth considering that if biochemistry had not so readily adopted physical chemistry in the later 1950s and 1960s, Schmitt might have been able to hold the allegiance of those applying physical chemistry to biological macromolecules, and there might be a great many more departments of biophysics today (and, by the same token, possibly smaller, if not fewer, biochemistry departments). A look at textbooks of the 1970s and after gives an indication of what other disciplines the research areas included by Schmitt in "biophysics" have come to lie within. The physical chemistry and structure of macromolecules, bioenergetics, and protein self-assembly all belong to biochemistry now, as does some of gene expression, the remainder of which is central to molecular genetics. So it would indeed seem that biochemistry has expanded at the expense of biophysics as Schmitt conceived it.[134] Radiation biology has ceased to exist as an academic field, except as an element of nuclear medicine. The problem of cellular architecture is now cell biology, a discipline that incorporates methods of fractionation biochemistry, as we have seen, rather than a science that pronounces on the workings of the cell on the basis of structural evidence from electron microscopy and X-ray diffraction, as in Sjöstrand's style of ultrastructure research. (Schmitt's forceful contribution to the naming of the *Journal of Biochemical and Biophysical Cytology* can be read, in retrospect, as a move to construe what we now see as the discipline of cell biology as an *interdisciplinary* field between biophysics and biochemistry.) And the problem of nerve and brain function is roughly coextensive with what has come to be known as "neuroscience," the area to which Schmitt increasingly devoted himself in the 1960s as one the field's main organizers.[135] But throughout the 1950s, Schmitt's constellation of "biophysical" fields—radiobiology, molecular genetics, protein structure and self-assembly, much of neurophysiology, and electron microscopy of cells and other biological structures—fit together as a natural group, and appeared to many, not least of all the NIH, to be a plausible disciplinary unit. The glue that bound them all together included, near the top of its list of ingredients, that cultural reaction to the atom bomb that had made a new physics of life seem so attractive and important after the war and before Sputnik refocused the national attention on the physical sciences.

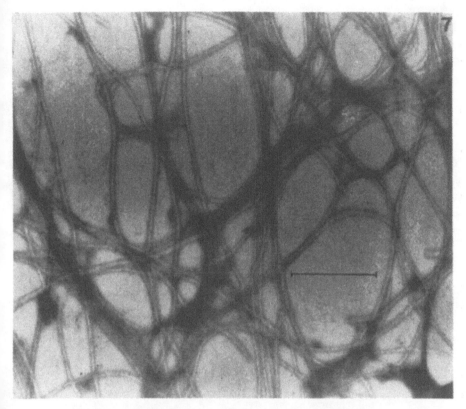

Plate 4.1. Electron micrograph of "neurotubules" from formalin-fixed, fragmented human nerve, stained with phosphotungstic acid, by De Robertis and Schmitt. Reproduced from *Journal of Cellular and Comparative Physiology* 31 (1948): 1–23, with the permission of the Wistar Institute, Philadelphia.

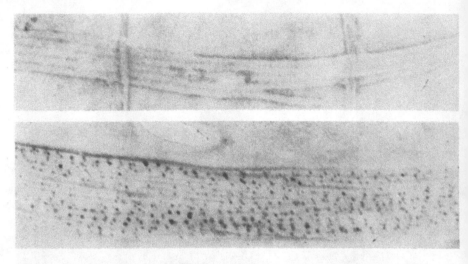

Plate 4.2. Electron micrographs of "neurotubules" from formalin-fixed, unstained, fragmented monkey nerve, by De Robertis and Schmitt, *top*, from uninfected animals; *bottom*, from polio-infected animals. Reproduced from *Journal of Experimental Medicine* 90 (1949): 283–90, with the permission of Rockefeller University Press.

FACING PAGE:

Top: Plate 4.3. Electron micrographs of a longitudinally sectioned, formalin- and osmium-fixed rat nerve fiber, by Schmitt and Geren, *top*, at low magnification; and *bottom*, at higher magnification. Note beaded "neurofilaments" in axon interior. Reproduced from *Journal of Experimental Medicine* 91 (1950) 499–504, with the permission of Rockefeller University Press.

Bottom: Plate 4.4. Electron micrograph of a longitudinally sectioned, osmium-fixed human nerve fiber, by Schmitt and Geren. Note "neurotubules" only in connective tissue sheath region. Reproduced from *Journal of Experimental Medicine* 91 (1950): 499–504, with the permission of Rockefeller University Press.

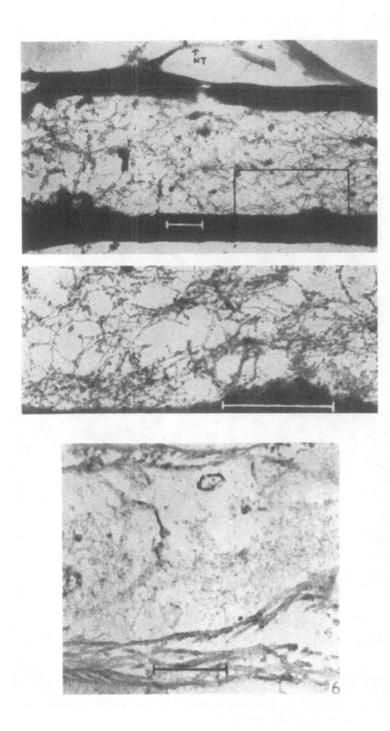

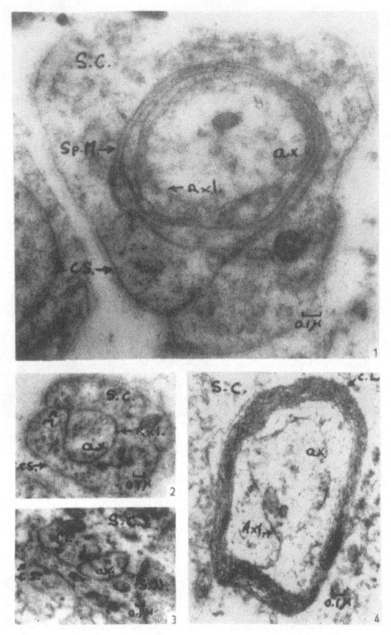

Plate 4.5. Electron micrographs of transversely thin sectioned, osmium-fixed chick embryo neurons at progressive stages of myelination, *counterclockwise from top*. Axon indicated by *ax.*, Schwann Cell by *S.C.* Reproduced from B. Geren, "The Formation from the Schwann Cell Surface of Myelin in the Peripheral Nerves of Chick Embryos," *Experimental Cell Research* 7 (1954): 558–62, p. 559, with the permission of Academic Press.

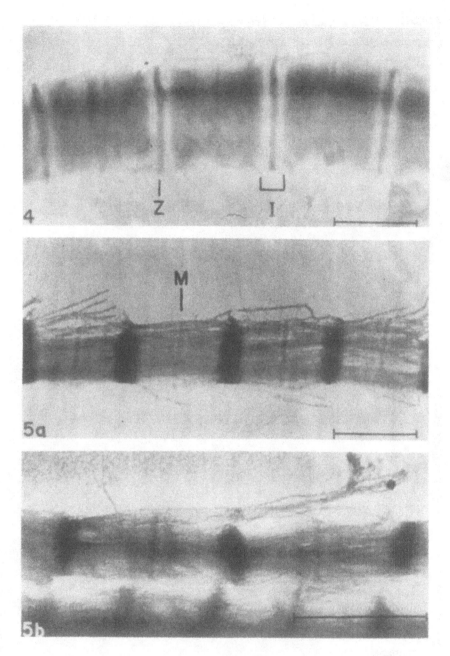

Plate 4.6. Electron micrographs of dispersed, formalin-fixed frog muscle fibrils contracted by electrical stimulation, *top*, strongly; *middle* and *bottom*, very strongly. Note disappearance of *I* zone, increase in *Z* band, and appearance of *M* band with high contraction. Reproduced with permission from C. E. Hall, M. A. Jakus, and F. O. Schmitt, "An Investigation of Cross Striation and Myosin Filaments in Muscle," *Biological Bulletin* 90 (1946): 32–50.

Plate 4.7. Electron micrographs of transversely thin sectioned, osmium-fixed, and phosphotungstate-stained rabbit muscle fibrils. Note simpler array of thicker filaments (*below, right*) where section is cut through the H band. Reproduced from H. E. Huxley, "Electron Microscopic Studies of the Organization of the Filaments in Striated Muscle," *Biochimica et Biophysica Acta* 12 (1953): 387–94, p. 389, with the permission of Elsevier.

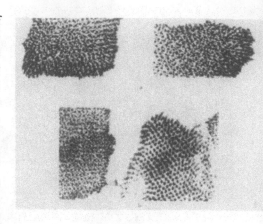

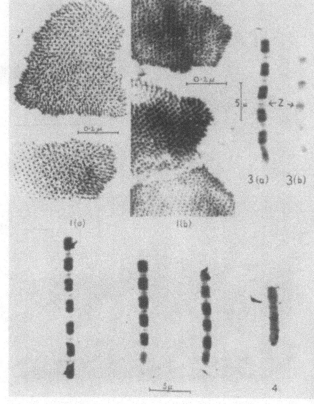

Plate 4.8. Rabbit muscle, *top left* and *top center*, osmium fixed, phosphotungstate stained, and transversely thin sectioned for electron microscopy; *top right*, before (*left*) and after (*right*) myosin extraction, imaged by phase contrast light microscopy; *bottom*, at increasing stages of contraction, imaged by phase contrast light microscopy. Note two kinds of filaments and suggestion of cross-bridges in electron micrographs; disappearance of A band with myosin extraction; and decrease of I band while A remains constant during contraction. Reproduced with permission from E. J. Hanson and H. E. Huxley, "Structural Changes in Striated Muscle During Contraction and Stretch," *Proceedings of the Third International Conference on Electron Microscopy, 1954* (London: Royal Microscopical Society, 1956), 576–80, pl. CLVII.

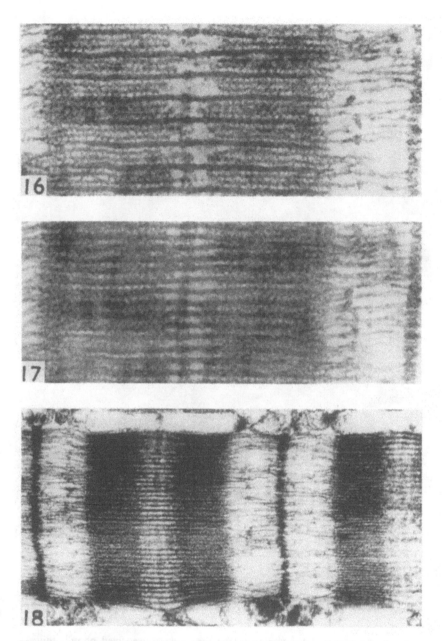

Plate 4.9. Longitudinal thin-section electron micrographs of osmium-fixed, phospho-tungstate-stained rabbit muscle, by Huxley, *top*, at high resolution; *middle*, double-printed to lower resolution; *bottom*, a thicker section of same. Note thinner class of filaments in A zone is distinct only at highest resolution. Reproduced from *Journal of Cell Biology* 3 (1957): 631–46, with the permission of Rockefeller University Press.

Plate 4.10. Longitudinal thin-section of formalin- and osmium-fixed, phosphotung-state-stained frog muscle. Note apparent continuity of thin filaments in I zone with thicker filaments in A zone. Reproduced from F. Sjöstrand and E. Andersson-Cedergren, "The Ultrastructure of Skeletal Muscle Myofilments at Various Stages of Shortening," *Journal of Ultrastructure Research* 1 (1957): 74–108, p. 81, with the permission of Academic Press.

Plate 5.1. Robley Williams at work, circa 1960. Courtesy of the Bancroft Library, University of California at Berkeley.

Plate 5.2. Wendell Stanley holding forth on viral structure, circa 1960. Courtesy of the Bancroft Library, University of California at Berkeley.

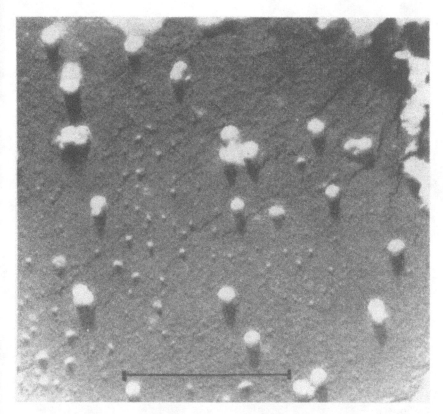

Plate 5.3. Electron micrograph of metal-shadowed influenza virus particles, by Williams and Wyckoff. Reproduced from *Proceedings of the Society for Experimental Biology and Medicine* 58 (1945): 265–70, with the permission of the Society.

Plate 5.4. Electron micrograph of metal-shadowed tobacco mosaic virus particles, by Williams and Wyckoff. Reproduced from *Proceedings of the Society for Experimental Biology and Medicine* 58 (1945): 265–70, with the permission of the Society.

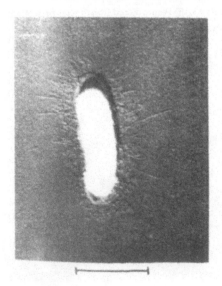

Plate 5.5. Electron micrograph of metal-shadowed bacterium, showing flagella, by Williams and Wyckoff. Reproduced from *Proceedings of the Society for Experimental Biology and Medicine* 59 (1945): 265–70, with the permission of the Society.

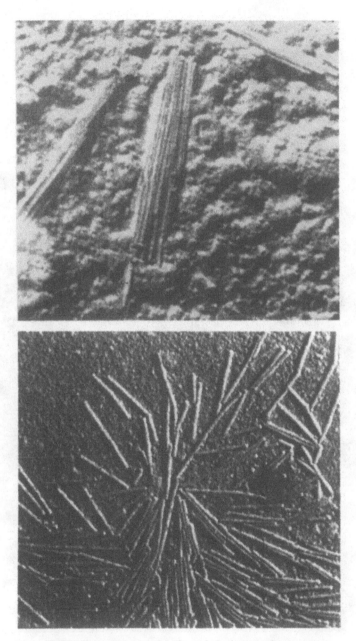

Plate 5.6. Electron micrographs of metal-shadowed tobacco mosaic virus in smears of infected plant juices, *top*, directly observed; *bottom*, after brief washing in water. Note dramatic dispersal and disaggregation of particles from minor washing. Reprinted with permission from R. C. Williams and R. Steere, "Electron Micrographic Observations of Tobacco Mosaic Virus in Crude, Undiluted Plant Juice," *Science* 109 (1949): 308–9, p. 308. © 1949, American Association for the Advancement of Science.

Plate 5.7. Electron micrograph of metal-shadowed *Escherichia coli* bacterium infected with bacteriophage. Note "honeycomb" appearance of cytoplasm, once thought to be a key to viral replication. Reproduced from R. Wyckoff, "The Electron Microscopy of Developing Bacteriophage: I. Plaques on Solid Media," *Biochimica et Biophysica Acta* 2 (1948): 27–37, p. 350, with the permission of Elsevier.

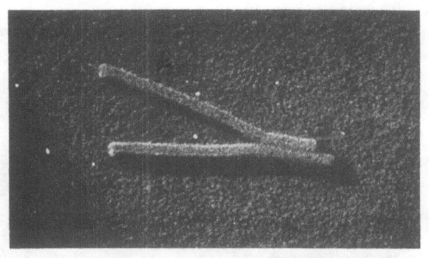

Plate 5.8. Uranium-shadowed tobacco mosaic virus at high resolution. Note similar texture on virus particles and specimen support surface. Reproduced from R. C. Williams, "High-Resolution Electron Microscopy of the Particles of Tobacco Mosaic Virus," *Biochimica et Biophysica Acta* 8 (1952): 227–44, p. 233, with the permission of Elsevier.

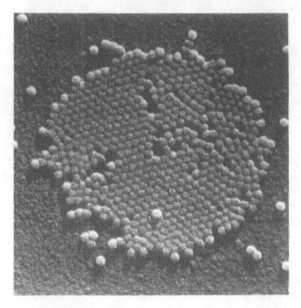

Plate 5.9. Electron micrograph of metal-shadowed, purified polio virus, made by Schwerdt and others of Stanley's group. Reproduced from *Proceedings of the Society for Experimental Biology and Medicine* 86 (1954): 310–12, with the permission of the Society.

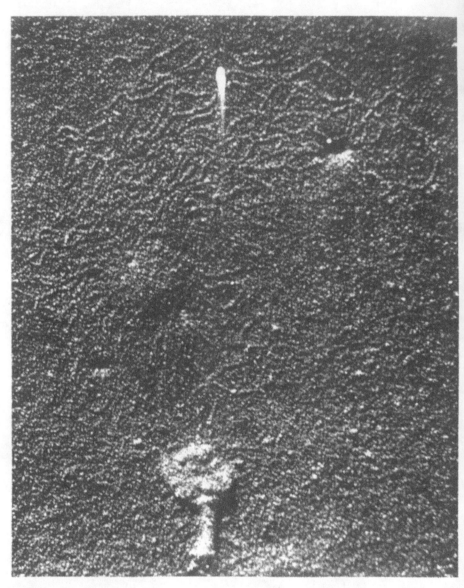

Plate 5.10. Electron micrograph of metal-shadowed DNA released from an individual ruptured bacteriophage. First published in D. Fraser and R. C. Williams, "Electron Microscopy of the Nucleic Acid Released from Individual Bacteriophage Particles," *Proceedings of the National Academy of Sciences, USA* 39 (1953): 750–52; subsequently published in and reprinted with permission from R. Williams, "The Shapes and Sizes of Purified Viruses as Determined by Electron Microscopy," *Cold Spring Harbor Symposia in Quantitative Biology* 18 (1953): 185–95, p. 192, fig. 11.

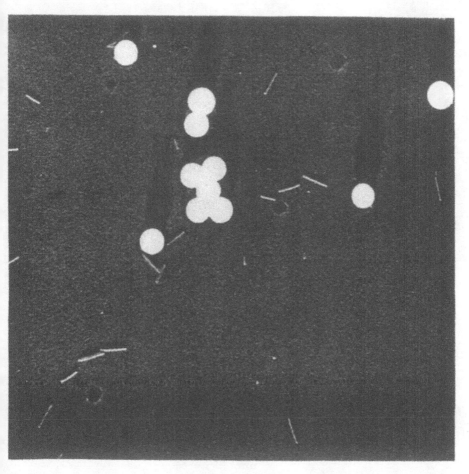

Plate 5.11. Tobacco mosaic virus and latex calibration spheres, metal shadowed. Reproduced from H. Fraenkel-Conrat, "The Chemical Basis of the Infectivity of Tobacco Mosaic Virus and Other Plant Viruses," in *The Viruses*, ed. F. M. Burnet and W. Stanley (New York: Academic Press, 1959), 1:429–57, fig. 1, with the permission of Academic Press.

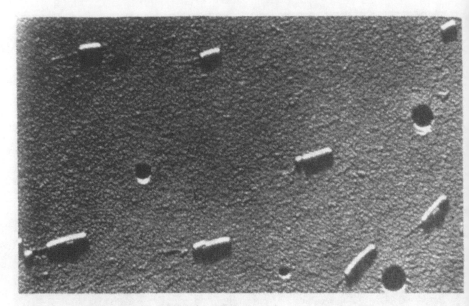

Plate 5.12. Electron micrograph of metal-shadowed tobacco mosaic virus, reconstituted from pure protein and nucleic acid, then lightly treated with detergent. Note nucleic acid strand protruding from partially degraded virus particles. Reproduced with permission from H. Fraenkel-Conrat and R. C. Williams, "Reconstitution of Active Tobacco Mosaic Virus from Its Inactive Protein and Nucleic Acid Components," *Proceedings of the National Academy of Sciences, USA* 41 (1955): 690–98.

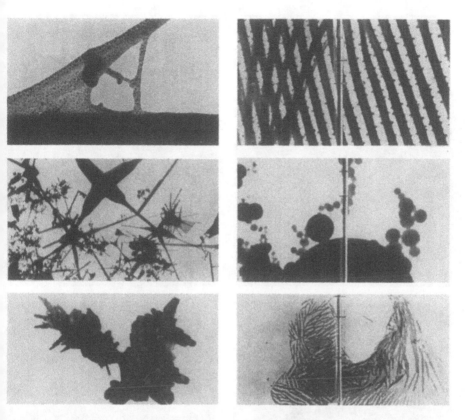

Plate 6.1. RCA publicity micrographs of various things, including face powder (*bottom left*) and mosquito "windpipe" (*top right*). Reproduced from *Life*, June 1, 1942, with the permission of the General Electric Co. and the American Cyanamid Co.

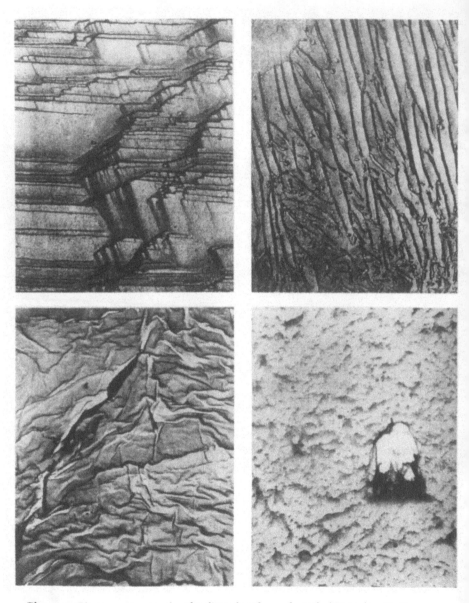

Plate 6.2. Electron micrographs of replicated surfaces of metal, skin, and tooth. Reproduced from *Life*, Oct. 10, 1945, with the permission of the Dow Chemical Co.

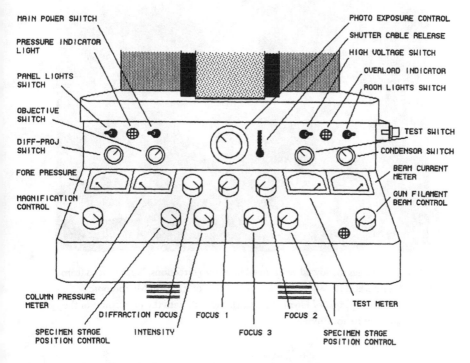

Plate 6.3. Control panel of 1952 RCA EMU microscope. Drawing by author.

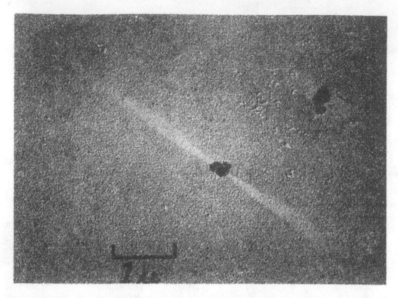

Plate 6.4. Particles shadowed with metal from two directions. Reproduced from R. C. Williams and R. Wyckoff, "The Thickness of Electron Microscopic Objects," *Journal of Applied Physics* 15 (1944): 712–16, with the permission of the American Institute for Physics.

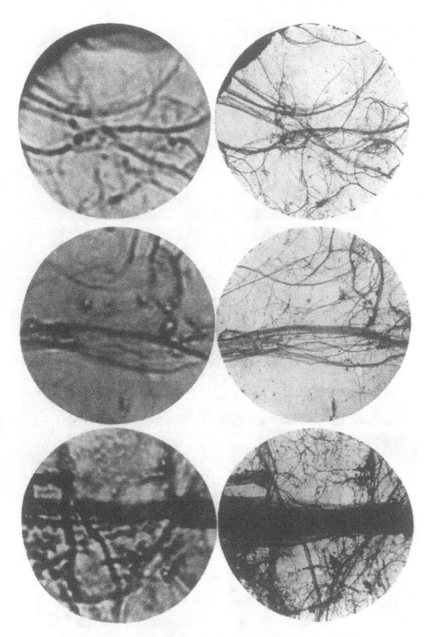

Plate 6.5. The same areas of paper specimens imaged, *left*, by light microscopy; *right*, by electron microscopy at low power. Reprinted with permission from C. J. Burton, R. Bowling Barnes, and T. G. Rochow, "The Electron Microscope: Calibration and Use at Low Magnification," *Industrial and Engineering Chemistry* 34 (1942): 1429–36, © 1942 American Chemical Society.

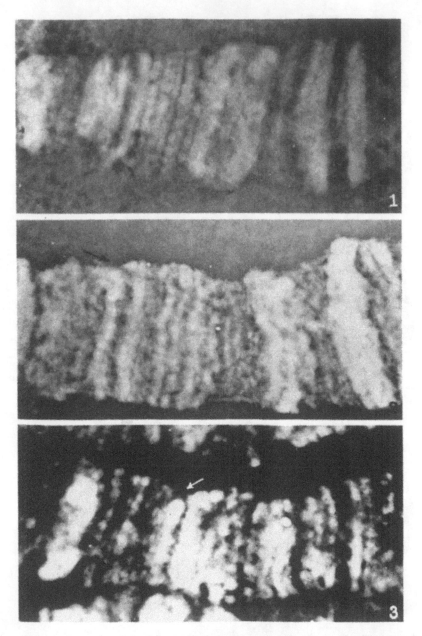

Plate 6.6. Electron micrographs of plastic replicas of fruit fly salivary gland chromosomes, by Palay and Claude. Reproduced from *Journal of Experimental Medicine* 89 (1949): 431–38, with the permission of Rockefeller University Press.

CHAPTER 5

Wendell Stanley, Robley Williams, and the Land of the Virus

❧ One arm of the pincers of praxis is therefore teleological consciousness which alters objects by virtue of the fact it grasps them differently from the outset, as the potential of the forms which are registered in the goal.

The other arm of the pincers is the unrelenting reality of practice, the ruthless logic of the things made. Workaday man is controlled and ruled over by more than merely his plans. . . . In the greater proportion of the cases he must rely on the logic of things, he has to suspend his plans either temporarily or for good, and he must yield to the course of events. . . . Man feels that whatever he is making is becoming his own. (This is so not only as regards rationalization after the event but also in cases of self-deception and self-persuasion.) In this process man produces his goals after the event so as to adapt them to a result of earlier origin. . . . Hence praxis-controlled consciousness works with a reverse teleology. It does not realize its previously elaborated goals but, on the contrary, its goals are created by the process of making.

—Miklós Almási, *The Philosophy of Appearances*

The year of the reaction to the atomic bomb, 1946, was the annus mirabilis of biophysics in America. The atomic "scientists' movement" for world government and disarmament was in full swing, and physicists were showing unprecedented interest in life science.[1] As I have described in the previous chapter (and elsewhere in more detail), biophysics, as a science symbolically capable of redeeming or counterbalancing the new physics of death with a new physics of life, was suddenly fashionable. The elite doctors and scientists of New York's Harvey Society launched the year with a lecture by former physicist Max Delbrück, who introduced his bacteriophages by explaining how remarkable their properties seemed to an "imaginary" atomic physicist studying with Niels Bohr. Also speaking to the Harvey Society at the start of the year was Stafford Warren, the Manhattan Project's chief of radiation safety at the Alamogordo and Bikini tests and soon to become the founding dean of the medical school of the University of California at Los Angeles (UCLA), who gave a description of death by radiation in Nagasaki.[2] The national airwaves were full of Manhattan Project scientists, Stafford Warren among them, pronouncing biophysics the atomic era's fount of wholesome biomedical advance, much as Schmitt did (Chap. 4).[3] In April, the American Association for the Advancement of Science meeting in St. Louis featured a special symposium, "The Philosophy of Biophysics."[4] In November, the Ninth Washington Conference in Theoretical Physics was devoted to "The Physics of Living Matter," and it brought together some of the most distinguished atomic scientists—including former Bomb-makers turned biophysics enthusiasts Leo Szilard, George Gamow, and James Franck—with those who must be counted, by virtue of their very invitation, as leading biophysics practitioners, including Schmitt and Stanley as well as Delbrück.[5] The year concluded with Nobel prizes in both chemistry and physiology or medicine going to Americans for biophysical work. H. J. Muller, already campaigning to save the human race from genetic oblivion at the hands of the atom, won the life science prize for demonstrating that radiation could cause genetic mutations. Stanley shared the chemistry prize for crystallizing viruses and thus showing that these primordial living entities could be studied by the same methods as other physical substances. Viruses, occupying the borderlands between physical and biological phenomena, represented ideal objects for biophysical inquiry: conceived as both genes and molecules (of "nucleoprotein"), they were like atoms of life, and perfect for the study of the physical mechanism of mutation.[6] Thus, Hiroshima changed the significance of virus biophysics overnight, even though it would be some time before new atomic technology brought substantial change, in the form of more readily available isotopes, to theory or practice in this field of scientific research.

Naturally, the sudden upwelling of general enthusiasm for biophysics was exploited by scientific entrepreneurs at institutions where biophysics was established (such as MIT and Cal Tech) to make their programs grow, and likewise by entrepreneurs who wanted to build new biophysics programs, such as Warren at UCLA, and the Manhattan Project circle of Franck and Szilard, who founded the Institute of Radiobiology and Biophysics at the University of Chicago.[7] Here is the first "arm of the pincers of praxis" described by Almási: scientific leaders were harnessing the new popularity of biophysics to fulfill their ambitions, to meet their consciously conceived goals of building up programs in general physiology, electron microscopy, radiation biology, and/or molecular genetics under the "biophysics" rubric. But by the "logic of things" governing the other arm of praxis, the actions of an individual may take him or her to unforeseen destinations beyond what was intended (what Dewey called the "end-in-view") in the original planning of actions.[8] Partly this displacement is social: a person's actions take on a meaning for others that is independent of the actors' goals and strategies, so the actor fulfills that socially defined role regardless of conscious intention. Thus, those scientists who were "exploiting" the biophysics bubble by packaging their research programs in that fashionable wrapper were also playing their part in upholding America's atomic politics; even Mudd, anti-McCarthy activist and promoter of Soviet-American exchange (not to mention birth control and socialized medicine), with his AEC project on the effects of radiation in bacteria (Chap. 2), and even Schmitt, proponent of a "Manhattan Project against disease" as an alternative to military research, with his Navy-funded work on the microcircuitry underlying nerve function (Chap. 4). So too, we shall see in this chapter, did virus biophysicists Robley Williams and Wendell Stanley (Plates 5.1 and 5.2) play their parts in the construction of the culture of the Cold War and its early maneuvers as they mustered support for their research programs, intention and individual politics notwithstanding; indeed Stanley, an outspoken resister of the "loyalty oath" required by the University of California and regular signer of Federation of American Scientists protests, was as liberal as Mudd.[9]

And while the social "logic of things" was working these effects, the logic of submicroscopic things was analogously displacing the conscious plans of virus researchers such as Stanley—as the virus disclosed a nature quite different from the single giant, self-assembling, machinelike enzyme made of protein they envisioned. For no scientists, even those as astute in politics and rhetoric as Stanley, realize their ends-in-view simply by force of will; rather (in a pragmatist view, at least) the world and the willfully directed ideas and methods of the inquirer meet in action, and the unpredictable results are what we know as natural things. And then, as Almási points out, by an odd "reverse teleology" whatever eventuates tends to be cast retrospectively by both actors and spectators as the goal that was being sought from the start, no matter how different from the end-in-view originally motivating action the actual result might be.[10] Thus we must be careful not to straighten history's winding path in the recounting (for instance, leaving out such digressions as tend to be absent from memoirs, like the episode I have called the biophysics bubble), especially when it seems to lead to great achievements such as the science we know today as molecular genetics.[11]

Balancing Bikini at Ann Arbor

Robley Williams, who was to become the avatar of biophysics at the University of Michigan at Ann Arbor, was an astronomer there before the war. His 1935 Ph.D. dissertation at Cornell had dealt with a new way of vacuum-depositing fine metal coatings for telescope mirrors, and during the war years Williams worked on a military contract to create coatings for, among other things, periscope mirrors. He may well have found himself drifting in his astronomical career. Also at Ann Arbor during the war as a research associate in the School of Public Health was Ralph Wyckoff, a self-described and rather well known biophysicist whose war work, like Stanley's, involved studies on the influenza virus. Wyckoff had at one time experimented extensively with the ultraviolet microscope in cytology, so he was no stranger to exotic and difficult imaging technology.[12] Wyckoff and Williams came together over an underutilized RCA model B electron microscope in the physics department. In 1944 the duo coated some of Wyckoff's purified influenza virus with metal sprayed by Williams at a known oblique angle, and then examined the virus in the electron microscope with the thought that measuring the uncoated "shadow" behind the virus particles would allow a more accurate calculation of their dimensions. The shadow-casting technique did not only that but also, by concentrating the electron-scattering effect on which electron microscopical contrast depends in the metal-coated surfaces of specimens, strikingly revealed topology and textures previously invisible in electron micrographs (because contrast is a product of mass and density across the entire depth of the specimen). The shadowed electron micrographs, which Williams and Wyckoff rapidly produced from many biological specimens, were widely received as a spectacular success, especially by fellow electron microscopists.

By the end of 1945, Williams and Wyckoff had obtained a second-hand model B just for their biological work, Williams had himself transferred from astronomy to the more tolerant physics department, and Wyckoff left for a permanent post as head of the new Section of Molecular Biophysics in the Laboratory of Physical Biology of the NIH Arthritis and Metabolic Diseases Institute—one of the many new biophysics research institutions springing up at the time. Williams was already such a convert to biophysics that September that he turned down a good job offer in astronomy on the grounds that he was happy at Ann Arbor working at "the borderline between Physics and Biology,"[13] a formula he lifted, like much of his research program itself, straight from Stanley. He must have felt an unusual opportunity to build on his early success as a fledgling life scientist. Nevertheless, Williams had a rocky start organizing his biophysics unit almost from nothing; despite recognized accomplishments in biology, it was not simple for a physicist to become a life scientist even in the first flush of postwar biophysics enthusiasm.

Williams continued to focus his research on the electron microscopy on viruses, whose attraction, with Stanley's 1946 Nobel prize, could only have grown. In 1946 he visited Stanley's lab in Princeton to learn techniques of growing and purifying plant viruses for himself. That same year Williams got a grant from the American Cancer Society (his largest at the time, but at $7,000 annually, unimpressive by postwar standards) for "Electron Microscopic Studies of Physical Properties and Growth Phenomena of Protein Macromolecules," that is, of viruses. He immediately ordered an ultracentrifuge from an instrument maker. He collaborated with bacteriologists and anatomists at Ann Arbor, and virologists at a Michigan state medical laboratory, imaging with his electron microscope the specimens provided by the biologists. One 1947 collaboration to examine chromosomes and perhaps genes themselves, with biologist William Hovanitz, ended in an acrimonious squabble when Hovanitz identified certain objects as chromosomes that Williams came to believe were contaminating bacteria. To a colleague, Williams complained that Hovanitz was the sort who thinks the "physicist is to keep the machines running and should leave the job of interpretation to the biologist."[14] In 1947 he got a small grant from the NIH to isolate and study an animal virus, and in early 1948 he went shopping for an electrophoresis apparatus to separate large proteins. As his laboratory became better equipped, and as he was able to staff it with technicians, he was able to attract graduate students and he grew less dependent on others for materials, instruments, and skills. In mid-1947, when the Institute of Radiobiology and Biophysics sounded him out about taking a job in Chicago, he felt reluctant to leave Ann Arbor with the groundwork laid, students counting on him, and the future looking brighter. In any case, Williams said, he could not leave for at least a year because of concessions from the administration and "assurances (some of them in concrete form) . . . that biophysical research was a recognized field . . . at this university."[15]

Williams still faced resistance to his plans for undergraduate courses and a doctoral program in biophysics, which became a major preoccupation for him. But eventually Williams did triumph, by representing his virus biophysics research as a biomedical dividend of the atom bomb in a local cultural context that especially valued this image. Dean of graduate studies Ralph Sawyer, who served as director of scientific experiments for the 1946 Bikini tests, had been helping devise a major new fund-raising campaign, called the Phoenix Project, to tap atomic fears and enthusiasms in order to build the sciences and especially physics at Ann Arbor. When the Regents of the University of Michigan launched it officially in May 1948, the Phoenix Project was described as a "War Memorial Center to explore the ways and means by which the potentialities of atomic energy may become a beneficent influence on the life of man."[16] (Perhaps this initiative was behind the 1947 "assurances" that prevented Williams's departure; Williams was, after all, one of the few Michigan faculty members actually prepared to do atomic biology.) One of Sawyer's early actions in his capacity as chairman of the Phoenix planning committee was to solicit Phoenix grant applications from Williams, covering experiments using the ^{32}P isotope to follow virus and chromosome replication in plants. Williams assisted Sawyer with Phoenix by giving speeches on the biological uses of atomic energy at posh trustees' luncheons on campus and at far-flung alumni clubs, and by posing with the University of Michigan president in the alumni association's brochure appealing for donations to the Phoenix Project, entitled *Michigan, the Atom and Peace*. The promoters of atomic physics needed biophysics as much as biophysicists like Williams needed the atom, and the alliance bore fruit when early in 1950 the Regents of the University of Michigan voted to establish a department of biophysics—much to the chagrin of some members of the biology department, annoyed they had not been consulted.[17] In the autumn of 1950, with this official recognition and with protein researcher Gordon Sutherland in place as his able successor as head of biophysics, and with talented young researchers such as bacteriophage geneticist Cyrus Levinthal on staff to see to the future, Williams left to join Stanley's powerful new group of virus biophysicists in California. Before merging his research program with Stanley's for good, though, William had already made considerable original contributions to the practice of virus biophysics.

Virus Structure and Reproduction in Three Dimensions

Evidently surprised by the degree of surface detail rendered visible in electron microscopy of their shadowed specimens, Williams and Wyckoff jumped to apply their metal-shadowing method to all the biological objects they could easily lay hands on. They made impressive pictures of influenza viruses from Wyckoff's own preparations, and plant viruses purified and donated by Stanley (Plates 5.3, 5.4; compare with Plate 1.12). Similarly, they quickly produced pictures of a variety of medically important bacteria, specimens of which were obtained from a bacteriologist at the Michigan Department of Health; the flagella were brought out especially sharply by shadowing (Plate 5.5). Substituting shadowing with 8 Å gold for the 80 Å of chromium they had initially favored, they used this thinner coating to obtain slightly higher resolution pictures of TMV and other plant viruses—again donated by Stanley and a Rockefeller Institute colleague—and of the large and easily purified molecule hemocyanin. Uranium soon became the favored metal for shadowing because thin coatings with it were more stable in the electron beam than gold. With an Ann Arbor anatomist, Williams even tried shadowing as a contrasting method for light microscopy of sperm and protozoa.[18] Metal shadowing was excellent for visualizing dispersed or particulate small objects like viruses: it allowed their easy resolution against the specimen support background (the relatively great scattering power of which could obscure them in the standard full depth electron image) because their prominences caught metal and left a shadow, and it sharpened images of particle exteriors and therefore permitted distinctive pictures of each virus or macromolecule. Specimens that could be obtained pure were preferable, because one would know which things in the image were the objects of interest, and virus purification with ultracentrifuges was a very active area of research at the time. But shadowing was far from ideal for specimens like bacteria or cells, because the interesting structures in their depths would be obscured by the coating that enhanced surface contrast. When the thin-sectioning methods that revolutionized research on these thicker specimens soon came, it did nothing to devalue shadowing for work on pure virus preparations (since embedding only served to mask viral structure in a contrast-obscuring plastic). Thus, in a period when viruses were fashionable objects of study and were being purified by ultracentrifugation in increasing numbers, shadow casting gave Williams technical keys to the postwar kingdom of virus biophysics at the perfect moment.

Williams tried variations of his new method in an effort to find ways of exploiting it maximally, guided by the realization that the depth-obscuring effect of shadowing made the method "a clear gain" only with small objects like viruses.[19] With help from his connections at Stanley's and other virus labs, Williams arranged to look at mutant plant virus strains for differences in both microscopic structure and infrared spectra (both of which might conceivably have shed light on the way the protein structure of genes determines their function). Naturally, conclusive results did not emerge. Together with collaborators at the Michigan Department of Health who had been trying to purify polio virus from mice and rats, Williams made an effort to improve on Marton's Stanford pictures of the dreaded disease agent (Plate 1.8). This experience with the uncertainties of identifying which was the actual virus in the menagerie of particles found in ultracentrifuge "purified" polio preparations from infected animals led Williams to advance a devastating argument, to the effect that neither he nor any other electron microscopist could claim that the virus had been visualized, even though polio particles undoubtedly were among the objects visible in published micrographs.[20] The moral was that much more careful research was needed for the deceptively easy-seeming task of imaging viruses.

Russel Steere, a graduate student nominally affiliated with botany at Ann Arbor, worked with Williams on growing and purifying plant viruses in ways involving minimum manipulation or washing; and here again, communication with Stanley seems also to have been instrumental for the acquisition of plant virology techniques. The overall goal was to visualize viruses as they occur in native plant tissues at various stages of multiplication, and so to solve the mystery of how the virus "molecule" replicates itself. Once extracted, viruses and putative replication intermediates from plant juice were contrasted by very thin shadowing with gold or uranium (Plate 5.6).[21] In this project Williams's group was following a trajectory similar to that of a number of bacteriophage researchers, including Mudd and Wyckoff, who were trying to elucidate the virus replication mechanism by direct visualization of infected host cells. Wyckoff's pictures, from his new and well-equipped NIH laboratory, of "honeycomb" and "filamentous net" structures between nascent virus particles in phage-infected bacteria generated particularly much comment and speculation as to what they indicated about virus replication (Plate 5.7).[22] These shadowed pictures from the late 1940s were essentially incommensurable with the thin-section micrographs by which phage replication came to be visualized by the late 1950s,[23] and thus became difficult if not impossible to place in the scheme of current knowledge, but for the moment they represented the focus of a vigorous research program.

Another line of research in Williams's Ann Arbor group dealt with improving electron microscopical technique, and soon led to the solution of one of the electron microscopy community's great problems: the determination of magnification in micrographs. As discussed later (Chap. 6), magnification was uncertain because no known object whose size could be accurately determined by light microscopy was small enough to serve as an internal standard in electron microscopic images. With graduate student Robert Backus, Williams found that a certain batch of latex particles obtained from the nearby Dow Chemicals facility in Shawnigan had an unusually uniform diameter in a suitable size range (2,600 Å), which was determined by various means such as optical interferometry, ultracentrifugation, and light microscopy to an accuracy of ± 6 percent (see Chap. 6). Mixed at a known concentration with a sample being prepared for electron microscopy (and then usually sprayed in fine aerosol onto grids so that the boundary of a droplet is identifiable in the electron microscope), these spherical particles served as a standard internal to each micrograph, not only of magnification but also of the concentration of other entities in the sample.[24] Backus's research projects explored the use of the spheres for biological electron microscopy. For instance, Williams and Backus were able to determine the molecular weight of plant viruses by counting the number per unit volume (given by the number of spheres in a droplet) and calculating from the dry weight and dilution. With Salvatore Luria, Williams and Backus determined the percentage of bacteriophage particles infectious in any preparation by counting the number of particles seen per unit volume and comparing with the number of plaque-forming units assayed per volume.[25] Simple though these techniques may appear, the magnification and concentration calibration available through the spheres made possible a new level of quantitative precision in experiments involving electron microscopy, and won Williams further acclaim as a biophysicist. When Williams packed up his lab to join Stanley at Berkeley in 1950, Backus and Steere followed.

Wendell Stanley's Virus Palace at Berkeley

Of all biological entrepreneurs of the immediate postwar era, Wendell Stanley capitalized on the currents of the times with the most consummate skill, building an unsurpassed biophysical and biochemical edifice at the University of California, Berkeley. Though a somewhat more affable and outgoing personality, the postwar Stanley shared with Schmitt the essential features of drive, political acuity, and salesmanship characterizing the new breed of science executive that has since become familiar in this era of big life science. Similarly, Stanley also moved quickly away from a role as bench scientist, though he certainly knew how to choose scientific personnel and to guide their research effectively.[26] In June 1946, Stanley revisited Berkeley for an honorary degree, having already spent three weeks there in 1940 as an honorary lecturer, and renewed his acquaintance with the university's president, Robert Sproul. By September, Sproul and he had begun serious discussions about his joining the Berkeley faculty. The turn of events was opportune in that Stanley's laboratory was breaking up and moving anyway: Stanley's head physical chemist, Max Lauffer, had just left Rockefeller to run the new biophysics unit at the University of Pittsburgh, which soon became a full-fledged department of biophysics, and at the end of the war it was decided that the Rockefeller Institute plant and animal pathology laboratories at Princeton were to be shut down and their staff integrated into the main facilities in New York City. Negotiations between Stanley and Sproul proceeded apace for the next few months, and in January 1947, following on the heels of his Nobel prize, Stanley was offered the job of taking over the biochemistry departments both at Berkeley and at the medical school campus in San Francisco. In addition he was to be made director of a new virus laboratory, to be entirely devoted to his creatures, the viruses, those glamorous denizens of the "borderline" betwixt life and death.[27]

Stanley accepted Sproul's offer after negotiating for such things as faculty appointments of key Rockefeller personnel he wanted to take with him. He expected that a modern new building incorporating the Berkeley biochemistry department and his virus lab would be ready by his arrival at Berkeley in late 1948, and Sproul told Stanley what he wanted to hear. The reasons for Sproul's enthusiasm are evidently complex: bio-chemistry at both Berkeley and the medical school needed new leadership and the biochemists could accept Stanley as one of their own, while at the same time Stanley's brand of biophysical work on viruses complemented the research on spectroscopy, radiation effects, and isotope studies of animal metabolism already taking place in Berkeley's medical physics division, associated with Ernest Lawrence's cyclotron lab. Together, Stanley's virus lab and the medical physics group would give Berkeley the ingredients for leadership in all aspects of biophysics. Of course, the laureate's name would also bring prestige, and his virus research was intellectually fashionable and potentially useful for agriculture. Moreover, Sproul anticipated that Stanley's entrepreneurship would bring a windfall from the foundations that had recently funded biophysics so much more generously (proportionately speaking) than conventional biochemistry—in particular, the Rockefeller Foundation.[28] And though Warren Weaver's Rockefeller program would never become a great patron, Sproul's assessment of Stanley's way with funding was not far wrong. His grant applications and speeches often opposed biophysics as the new science of life to the new atomic physics of death, to evidently good effect. He did get a $100,000 grant from Weaver to equip his new lab; but Stanley's primary initial sponsorship for the virus lab, a five-year commitment of about $100,000 per year for operating expenses, came from the National Foundation for Infantile Paralysis (NFIP) in response to a promise of medical breakthroughs (and particularly a cure for polio, in line with the NFIP charter) from basic research on plant, animal, and bacterial viruses. Stanley also managed to collect a number of smaller grants from the NIH and drug companies, including one from Lederle for work on a live polio virus vaccine, which amounted to over $50,000 per year for operations in the virus lab from 1950 to 1956.[29]

Key virus lab personnel came with Stanley from his Rockefeller Institute group in the second half of 1948, such as his right-hand man in TMV protein biochemistry, C. Arthur Knight, and the physical chemistry-savvy young ultracentrifuge expert Howard Schachman, both of whom were given regular faculty posts in biochemistry. Williams came as a full professor in autumn 1950, after Stanley finally gave up trying to recruit his friend Hillier to the electron microscopy position. Williams and Arthur Pardee, who came as a junior tenure-track biochemist in 1949 after taking a degree with Linus Pauling at Cal Tech, also got faculty posts in bio-chemistry. A permanent midlevel research staff position, without academic status, was also funded by the University. In addition to the five tenure-track positions counting Stanley's, about half a dozen faculty-level research staff positions in the virus lab were created on the "soft money" of the grants that Stanley had every reason to regard as perennial. Among the notable scientists who occupied these positions in the 1950s were animal virologist Carlton Schwerdt, and W. Dean Fraser, who (like Gunther Stent, given the less senior but permanent staff post) joined Stanley after postdoctoral training in bacteriophage research at Cal Tech with Max Delbrück. Eventually Heinz Fraenkel-Conrat, who held one such position as a research biochemist from 1952, won a sixth tenured faculty billet in 1957–58 when a virology department was officially established, absorbing the virus lab faculty. Although some of these nontenure research posts were occupied by physical scientists successfully making the postwar transition to life science via biophysics, such as Fraser and Stent (with doctorates in physical chemistry and chemistry), all of the *faculty* positions were occupied by those who were already practicing life scientists before the war's end. Of these, only Williams had actually switched from a career of basic research in physical science. Thus at Berkeley, just as at MIT and most other biophysics institutions, the main initial beneficiaries of the postwar expansion of academic opportunities in biophysics were established life scientists, not postwar converts from the physical sciences.[30]

The floor plan of the Biochemistry and Virus Laboratory (BVL) building gives a rough indication of resource and effort distribution in Stanley's institutional edifice. The roof held the virus lab's greenhouses, and the fourth and fifth stories housed the virus lab; the half-dozen faculty of the conventional biochemistry department were on the second floor, separated from Stanley's group by the plant biochemistry department on the third. The ground floor was occupied with teaching facilities.[31] Thus, judging by floor area, just as by the number of midlevel and senior staff, it seems that Stanley's virus lab biophysicists and biochemists accounted for twice the activity of the regular biochemistry department. The virus lab proper was a research unit emphasizing postdoctoral much more than graduate training; in 1954 there were twenty postdoctoral and staff researchers, in addition to the five members salaried by the University. However, doctorates were being earned there both in biochemistry and biophysics. By the time the BVL building was dedicated in October 1952, the virus lab was already impressively equipped, with a mass spectrometer for counting heavy isotopes, an automatic fraction collector for protein purification by column chromatography, two ultracentrifuges, an electrophoresis rig, a quartz spectrophotometer, and two RCA EMU-2 electron microscopes. X-ray crystallography apparatus, radioactive isotope counters, and infrared spectrometry gear were soon to come.[32] Experimentation was geared to these instruments. Knight carried on with his pre-Berkeley work on "the chemistry of mutation," mainly studying differences in the amino acid composition of various plant virus strains. Schachman developed new ultracentrifuge technique and was using it to analyze the protein components of purified virus. Pardee studied differences in the enzyme activities in normal and virus-infected cells, pursuing viral growth as a metabolic problem. Fraenkel-Conrat worked on trying to create new mutant strains by substituting chemically reactive amino acids into TMV (a clever idea, given the theory that viruses are naked genes made of protein).[33] Fraser, later joined by Stent, carried out a program that Fraser aptly described as "mostly a biochemical one, tempered, however, with Delbrükian [sic] phageology," principally the analysis of protein components of bacteriophage from various strains (after the manner of Knight's TMV research).[34] It would be fair to say that by early 1950s definitions, research in the virus lab proper was about half biophysics (counting the X-ray, ultracentrifuge, electron microscope, and molecular genetic work) and half biochemistry, though of course the distinction between the two was then vague and fluctuating. The electron microscope's role in some of these research programs is discussed below.

The Electron Microscope and Structural Virology in the BVL

In his wartime influenza virus work Stanley had occasional access to Hillier's microscopes at RCA, and in 1945 or 1946 the Rockefeller Institute in Princeton had acquired the cheaper and easier to use compact RCA model for him.[35] But only with Williams's arrival at the BVL was Stanley able, for the first time since Anderson's RCA fellowship, fully to integrate the finest possible electron microscopy into his group's research program. On joining Stanley's large Berkeley group in the second half of 1950, with its sharper division of technical labor, Williams abandoned most of the other biophysical methods he had been learning and concentrated on electron microscopy. Stanley gave Williams primary responsibility for one of the BVL's three main research programs, "An Electron

Microscopic Study on Host Cell–Virus Relationships," though he also became involved, as we shall see, in the other two: the purification and characterization of polio virus, and the study of the nature of mutation using TMV and bacteriophage.[36]

Williams, who had been using a model B all the while he was at Ann Arbor, tinkered with ways of optimizing resolution with Stanley's RCA EMUs. He worked out how to minimize vibrations, reduce thermal drift of the specimen holder, and enhance stability of the films coating specimen grids, and other small tricks, turning all his skills to making the best possible electron micrographs of TMV. The goal was to see whether any periodic surface texture could be visualized, and whether the cross-sectional shape of the virus rods could be confirmed to be hexagonal—both anticipated on the basis of X-ray crystallography to be features within the resolving power of the microscope. In viruses shattered by sonic disruption, Williams's electron micrographs gave evidence that the rods did often break into flat hexagonal plates, supporting that crystallographic conclusion. The effort to see more details of the virus surface failed, however, to show any difference between TMV virus, the supporting film, and the surfaces of other viruses, all of which had an irregularly granular appearance (Plate 5.8). Instead, Williams found disturbing evidence that his shadowing procedure only gave pictures of objects coated with metal-impregnated vacuum pump oil, generated by the specimen shadowing technique itself, which suggested that better electron micrographs would never come from it. The granular texture was an unavoidable artifact of shadowing.[37]

In a more successful effort to locate and neutralize sources of artifact, Williams also developed a freeze-drying preparative protocol for viruses that allowed for estimation of particle sizes more accurately than with ordinary air-dried preparations. The size estimates from electron microscopy of virus particles varied greatly, for a given kind of virus, from laboratory to laboratory. Even after calibration with Williams's latex spheres became widespread, the discrepancies did not completely go away, and sometimes it was even noticed that within one micrograph virus particles might have varying sizes (for instance, the particles at the edge of a clump tended to look larger than those in the middle). Perhaps virus particles were not rigid, and were capable of shrinking or spreading in the drying process. To overcome the problem, Williams worked out a simple freezing and drying method so that virus could go rapidly from a wet undamaged (i.e., infective) state to shadowing and imaging, without chemical fixation. The variation in particle sizes now disappeared, as did obvious drying effects such as shrinkage marks and wrinkling on certain viruses.[38] Because of effects stemming from surface area to volume ratios, freeze-drying is inherently quicker and easier to accomplish the smaller the entity being frozen, so it is not surprising that it proved simpler to devise a satisfactory method for the tiny animal viruses than for large specimens like cultured cells or tissue blocks.

Though Williams and the electron microscope contributed importantly to many of the BVL's research projects with such innovations, his own special program in imaging the stages of virus growth in the host cell was not the most productive. In 1953, as the dueling thin-section micrographs of Sjöstrand and the Rockefeller group were taking the spotlight, Williams moved to utilize the new thin-sectioning methods for his project on the virus–host cell relationship. The basic approach was simple: cultured cells such as the already ubiquitous HeLa line were infected with virus, allowed to continue growing for known periods, then fixed and embedded so that thin sectioning might reveal the growing virus and the internal anatomy of infected cells at various stages. There were immediate problems, though, in that it was difficult in micrographs to distinguish parts of viruses from host cell components, because the normal appearances of the latter were still largely unknown and the viral replication intermediates being sought were complete mysteries. A partial solution was the cutting of serial sections, that is, a series of thin sections from the same specimen, so that a three-dimensional reconstruction of the cell interior could be obtained. This did help identify profiles of endoplasmic reticulum and other normal cell components; however, it was quickly discovered how difficult it is to cut a number of sufficiently thin sections one after another—frequently, too many in a series had to be discarded for a decent reconstruction, so no useful information would come from a specimen. (Serial sectioning remains one of the most technically demanding operations in electron microscopy.) To improve his "batting average" by allowing the use of sections thicker than the maximum 500 Å visualizable at the EMU's 50 kv, Williams wanted a new microscope with a higher accelerating voltage. The project was covered not only under NFIP funding but also under a large NIH grant, and Williams was able to get a supplementary NIH grant to cover the purchase of the totally redesigned, $30,000 RCA model EMU-3 in 1954–55, on the grounds that the new model's 100 kv would yield the necessary penetrating power.[39] Despite the instrument, which Williams found to be an "inadequately engineered" disappointment, and the BVL presence from mid-1954 of Porter's collaborator in the tumor virus and early endoplasmic reticulum work, Frances Kallman (whose advice on Rockefeller methods was not always welcomed by Williams, she seemed to think), this BVL adventure in cell biology made no spectacular advance in the understanding of viruses or of cancer, to which the experimental design was easily adapted.[40] Williams remained primarily an artist of the metal-shadowing technique and the electron microscopy practiced in general at the BVL reflected this technical commitment.

Williams's methods of shadowgraphic electron microscopy proved crucial for the BVL's efforts to purify and characterize the polio virus. Previous work along these lines, such as Schwerdt's wartime project with Marton and Loring at Stanford, had led to pictures of a mixture of many different particles co-purifying together with the virus, all derived from the extracts of brains and other organs of infected animals (Plate 1.8). As mentioned, in 1949 Williams had punctured claims that anyone could tell, from existing micrographs like these, which of the particles seen actually was the virus. Now Bachrach and Schwerdt in the BVL were using new and elaborate chemical fractionation and centrifugation methods on this animal-cultured polio, and had obtained preparations containing just two main classes of particle, one about 100 Å in size and the other about 300 Å. It seemed likely that the larger particle was the polio virus because it was not found in control preparations from uninfected animals, but other possibilities could not be ruled out—such as the presence of two species of 100 Å particle of which one was the virus (and the larger particle species perhaps an infection by-product). What was needed was a way of accurately counting the two kinds of particles, which were only present in tiny quantities, in order to correlate concentration with biological activity. The Williams-Backus spray-drop sampling method with latex spheres would fill the bill. Using a separation cell in the ultracentrifuge, Bachrach and Schwerdt were able to divide the virus-containing (i.e., infectious) preparation into two portions, one with almost exclusively 100 Å particles and the other highly enriched in 300 Å particles and with many fewer 100 Å particles than the starting material. Then they sprayed portions of both fractions mixed with the latex spheres, shadowed, and used electron microscopy to count the concentrations of large and small particles per unit volume. Infecting animals with the two ultracentrifuge fractions showed that the virus activity was proportional to the concentration of the larger bodies, thus establishing that the polio virus was indeed the 300 Å particle.[41] The relevant data in this experiment were not high resolution portraits of the viruses, but numbers: numbers of 300 Å and 100 Å particles per unit volume in the two separation cell fractions, compared with numbers of infectious units per volume from the same fractions. Here was a form of electron microscopy that almost escaped the micrograph, the image, to a purely quantitative realm.

Here electron microscopy was only being used for counting rather than for the visualization of structural detail, but proper imaging under optimal conditions soon followed. Microscopy of polio obtained from the cleaner and bigger preparations made possible by growing the virus in tissue culture instead of whole animals, using Williams's careful shadowing and his new freeze-drying technique, confirmed that the virus particles are round and about 270 Å (the slightly larger 300 Å result from air-dried preparations presumably being a result of flattening; see Plate 5.9). Stanley must have appreciated the publicity value of these first good pictures of the dreaded polio virus, since he appended his name as author of this experiment, and the electron micrographs did indeed make newspaper headlines at the end of 1953.[42]

The 1952–53 discovery that nucleic acid, rather than protein structure, carried the genetic code, although it did slightly reduce the glamour of structural studies on viruses, seems not to have significantly hurt Stanley's enterprise. On the contrary, the BVL team was able to adapt existing experimental systems quite smoothly to what some historians have characterized as the radically new nucleic acid "paradigm."[43] Dekker, Schachman, and others, using their customary methods and equipment, shifted attention to the biochemistry and higher-order structure of nucleic acids, at one point even proposing an alternative to the Watson and Crick continuous double-strand model (namely, that DNA existed in relatively short double helical segments that might aggregate reversibly into higher-order complexes). Fraser and Williams produced some justly famous images of the DNA packed into the head of a bacteriophage (Plate 5.10); indeed, Delbrück was so impressed with these pictures he said they would "steal the show" at the 1953 Cold Spring Harbor meeting on viruses, even though it was the same meeting that featured the new Watson-Crick DNA structure.[44] Of course, Fraenkel-Conrat dropped his efforts to obtain mutant TMV by modification of proteins, because if nucleic acid is the genetic material, altering the proteins in viruses is highly unlikely to have any mutagenic effect; but he was able to adapt his TMV system to testing and elaborating the new nucleic acid theory of genetic information in ways described below. And with Knight, who had continued to study the protein makeup of various virus strains, Fraenkel-Conrat contributed to the BVL's race through 1960 to determine the entire amino acid sequence of the small protein subunit of which TMV rods are built.[45] The continuity with the BVL's prior research program on the amino acid sequence of mutant strains is obvious here. In general, so much continuity implies either that the revolutionariness of the double helix model has been exaggerated or, more plausibly, that theoretical shifts in general have a limited capacity to restructure research programs, at least in fields of experimental biology.[46]

Fraenkel-Conrat's best-known work of the period involved learning how the TMV rod could be taken apart into protein and nucleic acid components, and how it could be made to reassemble spontaneously from these elements. In this research the electron microscope, and Williams himself, again played a crucial role. Though originally Stanley had only counted the British team of Bawden and Pirie at Rothamstead experimental station as his main competition, in the early 1950s Stanley's BVL had acquired quite a few other rivals in the business of physical and chemical characterization of TMV. By 1955 biochemists and X-ray crystallographers at the BVL and in competing labs had determined that the TMV rod consists of a helical assembly of about 2,800 identical copies of the main virus protein, with a hollow core that might accommodate nucleic acid strings.[47] Initially, the Watson-Crick theory that nucleic acids had a structure that implicated the stuff as the genetic material was greeted with a certain degree of skepticism by plant virus research groups like Stanley's. For one thing, TMV contains only about 5 percent nucleic acid and 95 percent protein (as opposed to about 50 percent nucleic acid in bacteriophages); moreover, that nucleic acid is ribonucleic acid (RNA) rather than the deoxyribonucleic (DNA) found in bacteriophage and higher organisms, and modeled by Watson and Crick. Thus the double helix DNA model triggered an effort among the BVL workers studying TMV biochemistry to test it, by looking for trace quantities of DNA in the virus, and also by finding a means to determine whether the RNA or the protein of the virus—or both—was the carrier of genetic information.[48]

Because the small TMV protein had been observed to assemble spontaneously into viruslike rods, the possibility of reconstituting whole virus particles with proteins and nucleic acid presented itself. Williams and Fraenkel-Conrat accomplished this virus reconstitution with protein and

RNA purified as gently as possible from whole TMV, simply by mixing RNA and protein together and incubating under certain conditions. They found that the mixture was able to infect plants, unlike equivalent concentrations of either of the starting materials; electron microscopy on the starting materials confirmed the absence of intact virus in both. Electron microscopy on the reconstitution product showed that about 10 percent of the reassembled rods were the full 3,000 Å length of normal TMV (Plates 5.11, 5.12). The spectacular news of the reconstitution of active virus from its chemical components made an impression on the popular imagination that reached around the world. As *Punch* quipped, perfectly capturing the cultural significance of this triumph in Cold War biophysics: "two Californian [scientists] . . . 'have created life' in a test-tube. This may check the rising popular feeling that scientists are wholly preoccupied with the opposite process." Lest anyone mistake what scientific field biophysics was being contrasted with, the item carried the headline "Change from Mushrooms."[49]

To test whether hereditary characteristics really followed the nucleic acid alone, the BVL group went on to reconstitute RNA and protein from distinguishably different strains. If protein played a role in heredity, then one would expect some or all of the characteristics of the strain contributing the protein to be reproduced in progeny of the reconstituted virus. If, on the other hand, nucleic acid alone was the hereditary material, then offspring of reconstituted viruses would be entirely like the strain from which the RNA was taken. Indeed, Fraenkel-Conrat and collaborators did find a little evidence that the "protein component might slightly influence the genetic message," but concluded that nucleic acid was the main if not only hereditary determinant in TMV-like viruses, just as it had already been shown to be in bacteriophage. Moreover, the effort to determine, by testing infectivity of the supposedly pure protein and nucleic acid to be used in the reconstitution experiments, the exact amount of whole-virus contamination in these starting materials—contamination that could undermine the claim that infective virus had really been reconstituted—turned up the surprising discovery that the RNA alone can be infectious if applied to plants in high enough concentration.[50] This revelation that the RNA by itself could reproduce whole TMV when introduced into plant hosts (unfortunately for the BVL, first published by a competitor) clinched the argument that nucleic acid was the sole hereditary material in this class of virus.

The TMV reconstitution method not only permitted the protein component to be ruled out as the carrier of genetic information, but it also pointed the way to achieve the already conscious "aim . . . to produce at will a new genetic (i.e., replicating) species of molecules," that is, to find a means of actively manipulating the genes of these RNA viruses. Initially to test the locally favored hypothesis that nucleic acid existed in short segments rather than as the very long strings suggested by Watson and Crick, Fraenkel-Conrat in the mid-1950s tried "a considerable number of experiments" in which RNA from different strains was mixed together and reconstituted with protein to make active TMV particles. If the units of genetic information really were small nucleic acid segments, then some active viruses containing sequences from both strains that supplied RNA would be reconstituted, and would show themselves as disease agents that caused plant infections sharing traits from both parent strains. Unequivocal evidence for viruses of mixed character was not forthcoming, which supported the arguments from competing labs that the TMV genome was a single long piece of RNA.[51]

Still, the inability to make RNA combinations by simple mixing did not lead the BVL group to abandon efforts to produce recombinant species of RNA and introduce them efficiently into plants through reconstituted virus particles. In his talk on nucleic acid replication for Schmitt's 1958 biophysics conference in Boulder, for which most of the data discussed came from experiments with bacteriophage, Williams concluded with a promising account of the potential of reconstituted TMV as an experimental system for molecular genetics. The TMV system combined the best features of bacteriophage (wherein one assays for genetic events by investigating purifiable virus particles subsequently produced) and the best features of the bacterial transforming factor system that had initially pointed to nucleic acid as the hereditary substance (wherein one can chemically manipulate the genetic material outside the organism and then reintroduce it), said Williams.[52] The Stanley lab's heavy commitment to Tobacco Mosaic Virus as both object of inquiry and as experimental tool limited the possibilities, and simultaneously constituted the opportunities, for research pathways at the BVL. That TMV did not become the vehicle to the kind of aggressive, manipulation-oriented molecular genetics that ruled by the end of the 1960s can certainly not be attributed to any lack of vision. Rather, the "logic of things" did not favor these plans, and unforeseeable vicissitudes of fortune, such as the spontaneous recombination that permitted fine genetic mapping in bacteriophage early on, and the later discovery of enzymes to cut and rejoin DNA at specific sites (restriction enzymes), made bacteria and DNA viruses the preferred vehicle. Neither vision nor energy were lacking in the thriving TMV-centered research programs pursued throughout the 1950s at Stanley's BVL; the biochemists and structural biophysicists there had taken on as their own the goal, usually associated with Delbrück's school of molecular geneticists, of rewriting the information in the "genetic message" of living things.

Far more damaging than the double helix, and long before falling behind the times could be counted as a danger, the successful large scale testing of Salk's killed-virus polio vaccine in 1954 threw a pall over the BVL, even as Schwerdt, Williams, and other members of Stanley's group brought glory to the BVL with the pure polio virus micrographs.[53] The NFIP and Lederle both cut off virus lab funding with the answer to polio in sight. Still, Stanley's institution survived as the home of a molecular biology department chaired and, indeed, largely redesigned by Williams, changing its name conservatively to the Molecular Biology and Virus Laboratory (MBVL) in 1964, when biochemistry proper moved into its own building. Here then is one example, among a number, of a biophysics establishment from the early postwar period that later served as the infrastructural and institutional foundation for what came to be called "molecular biology."[54]

In retrospect, Stanley's postwar success cannot be attributed to the biophysical caché of structural virus research alone, but also to the very polyvalence of his virus program: Stanley's plant viruses made him interesting to plant biologists, his animal viruses gave him relevance to medical researchers and foundations, and his chemistry background allowed him to do biochemistry that was taken seriously (not that there was no strain between his virus lab people and the regular biochemists at Berkeley).[55] And this same polyvalence saved his BVL once the bottom fell out of polio. Though the virus lab had originally just dabbled in cancer research, inasmuch as some tissue culture cell lines in use were from tumors, Stanley had kept his hand in as an advocate for the relevance of virology, for instance chairing a session on basic virus replication studies at the Second National Cancer Conference in Cincinnati in 1952. Ultimately, the virus-cancer angle became Stanley's salvation. In 1955, his old NFIP patron and friend, Harry Weaver, moved to the board of directors of the American

Cancer Society, a body to which Stanley was increasingly sending grant applications for cancer research designed around tissue culture experiments.[56] The following ditty (to the tune of "Once in Love with Amy"), probably from a BVL Christmas party around 1955, beautifully illustrates the change in direction:

> Once I worked for Stanley
> He is so big and manly
> Answered all our queries,
> Full of silly theories
>
>   . . .
>
> Salk then found the vaccine,
> Now we're just relaxing
> To him goes all the money,
> We're broke and that ain't funny,
> No one knows the trouble we've seen
>
>   . . .
>
> Now he's found the answer,
> we'll all go work on cancer,
> His viruses are dandy,
> So versatile and handy,
> Now we have no worries, you see,
> Our jobs are safe until another Salk vaccine.[57]

Fortunately for Stanley's BVL, no cancer vaccine was in the offing.

So Stanley managed to justify and continue the virus lab's research programs through a change in theoretical trappings, without any radical change in their methods or objects of inquiry. Again we see, in terms of Almási's first arm of praxis, how experiment has a life of its own in the sense that experimenters generally do not change their practices as quickly as their theories, and find ways of approaching new problems with old means by repackaging the latter. Old wine in new bottles would be an appropriate metaphor, except experimentation is a productive process instead of a static product. But this functionalist logic should not rule out the possibility that those new goals, and the meanings attributed to the repackaged research by the greater culture in which science is embedded, may gradually be internalized and reshape the practices, plans, and self-images of experimenters. (Indeed, there is every indication that Stanley became a true believer in viruses as the cause of "most, if not all, cancer in man"[58] despite scant evidence.) Rather, the two processes are reciprocal, however opposite in nature they might appear. Some basic features of this reciprocal relation between experimental means and ends are explored in the next chapter.

Summary: Visualization of the Shadowy Viral Landscape

The electron microscope contributed substantially to the study of viruses in the 1940s and 1950s. On an intellectual level, electron microscopy clarified the structure of viruses and the relationships among the virus subcomponents that were being purified by biochemical methods, thus guiding purification and reconstitution experiments. Moreover, it helped quickly confirm the idea that long strings of nucleic acid are the bearers of genetic information, by showing the strings of nucleic acid contained in bacteriophage and other genetic vehicles. On the level of disciplinary formations, the electron microscope also provided one of the key methodologies, along with the ultracentrifuge for purifying viruses and in vitro culturing methods for producing them,[59] definitive of virology as it was constituted in the postwar period (in the event, outside of the nascent discipline of biophysics in which it started). Only the electron microscope could provide convincing evidence that viruses were distinct entities present in infected tissue, or purified intact therefrom despite the often violent methods used for their isolation and biochemical analysis. Without a visual criterion for viral identity and integrity, investigation of viruses by other means would have been plagued with far greater uncertainties about problems like preparative artifacts, correct identification of viral components in heterogeneous fractions, and contamination with unknown viral strains. The electron micrograph of a virus was its official, definitive portrait.

Williams developed the metal-shadowing technique to a high art for purposes of virus visualization. An inherent limitation of shadowing, recognized very early, was that its striking revelation of surface detail came at the cost of obscuring depth. Williams, unlike the technique's co inventor, Wyckoff, respected this limitation by restricting the use of shadowing to specimens of small dispersed particles. In these specimens, almost always virus particles, the outline and textures were what was distinctive and interesting; moreover, viruses were generally so small that interesting internal detail was unlikely to be resolvable without artificial contrast, even if it were present. Thus, for practical purposes these particulate specimens were already only surface, and nothing was lost in shadowing them. Where depth features had to be seen, Williams worked out ways to bring these features up to the surface and disperse them for viewing, prior to shadowing (as with the bacteriophage genome, Plate 5.10). The pictures he produced by shadowing are noteworthy in their sharp focus and full contrast range from black to white, aesthetic qualities he deliberately pursued even as he was, for purposes of maximizing information content, optimizing resolution (which depends most of all on invisible and aesthetically neutral factors, such as careful astigmatism correction and low levels of column contamination—only indirectly reflected in visible contrast range and sharpness). His electron micrographs have a strikingly realistic and intuitively accessible character, in that even an untutored viewer immediately reads them as a three dimensional surface or landscape. The subtle contrivances by which the pictures were made to read so easily as landscapes are discussed in the chapter that follows.

Williams also developed an epistemological style to go with his manner of making micrographs, and it was one that fit nicely into the integrated mix of biochemical and biophysical practices at Stanley's BVL. His hallmarks were quantification—mathematization indeed, but distinct from Sjöstrand's geometrical style of mathematization—and the identification of sources of artifact in electron microscopy that would allow artifact effects to be compensated for, even when they could not be eliminated. For instance, Williams found ways to maximize resolution in shadowed micrographs, as noted, and where features could not be resolved, he clarified whether the electron microscope yielded more or less reliable data than other instruments (as in the case of the TMV surface, where he found reason to prefer the X-ray diffraction data). He learned ways to spread nucleic acids, suddenly of greater interest at the BVL and elsewhere after Watson and Crick, on the specimen support so they could be seen by shadowing. When surface tension effects became a worry for air-dried virus specimens, Williams's simple freeze-drying method was adequate to solve the problem for the very small particulate specimens that concerned the BVL, and far simpler than the ingenious but laborious critical-point method Anderson developed for bacteria and phage (see Chap. 2). No great change in the spreading, spraying, and shadowing protocols employed by Williams's BVL school of microscopy was called for through the 1950s. Williams had put together a bundle of practices that made it relatively easy for the BVL biochemists and biophysicists, with all their diverse methods and instruments, to pick up and use the electron microscope, and he showed them how to fit the visual evidence from it with the data from their other sources.

As we have seen, with the latex sphere size standards and the aerosol sampling method he developed with Backus, the size and concentration of virus particles could be read directly and accurately from micrographs, even when quantities of specimen were so minute or dilute that no other biochemical or biophysical technique could have yielded equivalent information. Of course, this made the electron microscope very handy in the macromolecule purification procedures that so occupied BVL staff. But more than that, these techniques gave quantitative information on the sizes and distributions of entities that, in table form, could be compared with tabulated values of biological or biochemical quantities (such as infectivity, or nucleic acid and protein concentration, in milligrams per milliliter). Indeed, some of Williams's electron microscopic publications show only such tabular numbers, and include none of the microscopic images on which the numbers are based.[60] Thus, Williams developed ways to integrate the visual information from the electron microscope, through the intermediary of reduction to numerical values, with the other information about the same objects coming from other BVL workers (even the biochemists), thereby fitting his instrument epistemologically to the diverse cultural and intellectual structure of this multitechnique laboratory. The integration between visual/morphological and spatially dimensionless biochemical values was much more intimate than in the Rockefeller style of cell biology (Chap. 3), where biochemistry and morphology were plotted in essentially separate and autonomous domains and then mapped onto one another. Still, despite the drive to quantification that made possible tabular comparisons with ultracentrifuge or protein assay values, and that even occasionally banished micrographs from his publications altogether, it would not be fair to characterize Williams as an arch-Galilean mathematizer of nature implacably bent on dissolving the image into mere numbers. At least in his association with Stanley, Williams appreciated and used the image as an icon; for instance, there can be little disagreement that the entity in the magnificent pictures he made with Schwerdt (Plate 5.9) simply *is* the polio virus, to both the popular and the scientific imagination. The BVL style of electron microscopy involved abstraction in an idealizing way, then, but not the *total* mathematizing movement attributed by Husserl to all science— even on this borderland between life science and physics. Number may abide with sensuous qualities in the object of scientific knowledge.

CHAPTER 6

Through Another Looking Glass: Lived Experience and Biological Electron Microscopy

❖ Our organs are no longer instruments; on the contrary, our instruments are detachable organs. Space is no longer what it was in the *Dioptric* [of Descartes], a network of relations between objects such as would be seen by a witness to my vision or by a geometer looking over it and reconstructing it from outside. It is, rather, a space reckoned from me as the zero point or degree zero of spatiality. I do not see it according to its exterior envelope; I live it from the inside; I am immersed in it. After all, the world is all around me, not in front of me.
—Maurice Merleau-Ponty, "Eye and Mind"

It is commonly said of the Aristotelian-Scholastic physics that reigned in Europe until the scientific revolution, that its falsehood went unnoticed for so long because it borrowed strong intuitive appeal from its extension of human bodily experience to the world of bodies in general. The Scholastic physical theories mesh with what we all believe from everyday living: that projectiles slow down when they leave our hand or any other launcher, that light objects such as feathers fall less quickly than heavy stones, that the sun circles the earth. The new mathematical physics, based on contrived efforts such as those of Galileo and Descartes to see in objects only the "primary qualities" of extension abstracted from other immediately experienced properties, of course did increase science's predictive and manipulative purchase on phenomena. But history and philosophy of science has too long mistaken the mentalistic Cartesian ideology for a fair account of how science actually accomplishes what it does. What grounds are there to suppose that our present scientific understanding, despite the addition of mathematical thinking, is derived any less from the subjective, lived experience of the body's contact with the world, a world from which we could only implausibly (and with dubious motives) hope to remove our bodies for a God's-eye view, a world that, in Merleau-Ponty's colorful imagery, we find "incrusted" on our flesh from our constant immersion in it?[1] Here I will argue that preconceptions, many of which are rooted in universal bodily experience, play essential roles in the production and interpretation of experimental data, and thus enter constitutively into scientific knowledge. The advance of science depends on such preconceptions, and an understanding of their role can help answer our original question of how technology can shape science.

To paraphrase Brian Rotman, it is time to bring the body back into the picture, even in the case of mathematics.[2] The body has even been implicated in formal logic.[3] In the present case of biological electron microscopy, the problem is to discover the body, as linked to instruments, in the doing of experiments and interpreting of results. Discovering the role in experiment—sometimes productive, sometimes counterproductive, but either way fundamental and inevitable—of preconceptions of every kind (or better, following the phenomenological tradition in philosophy, "pre-understandings," the forms in which experience is given) allows a fresh return to the problem of the role of instrumentation in scientific change. Because experimental thinking is embodied action, along the way this attempt to build a bridge from pragmatism to phenomenology must pass through the experimenter's body, whatever may be the case with mathematics or logic.[4] The route I will take to the experimenter's body is somewhat indirect, and must be so because the bodily senses do not encounter the world except in a culturally prepared subject. The complex in which the body of the experimenter and his or her culture are combined needs to be disentangled.[5]

Pre-understandings have to be identified, then, and the cultural sorted from the physical where possible. Therefore I will start with mainstream culture, and show that, in the popular imagination of midcentury America, the electron microscope was regarded as an extension of the organs of perception: as an augmentation of the eyes for seeing tiny things, or even more powerfully, as a vehicle to transport the entire person to a bizarre world of the minuscule, a place whose landscapes were laden with a variety of sensed meanings to the first tourists glimpsing it. Second, I show that, notwithstanding—indeed, because of—all the skill and sophistication that scientists employing electron microscopes quickly developed, for them the microscope faded into the background and became a vehicle that projected them, like the lay public, bodily into the microscopic terrain they were engaged in exploring. In the third section, I more closely analyze the interactions between microscopist and instrument, and between microscopist and world as represented in micrographs, in order to characterize the role of the body in the understanding of microscopic space. In the fourth section, I assess the relative roles of informal bodily intuition and formal rules of inference in the case of certain knotty interpretive problems confronting microscopists, illustrating both with examples treated earlier in the text and with new material. I suggest how some of those rules themselves are founded on bodily intuitions. Up to this point the account is developed on the basis of minimal theory, but in the fifth section I attempt to recast the intuitions and the inferential rules employed by microscopists in terms of the "hermeneutical circle" proposed for experimental work by philosophers of science in the phenomenological tradition, and I try to fill in gaps to complete this reconstrual. Ultimately I employ this hermeneutical theory to address the questions raised in the first chapter about the manner in which and extent to which new instruments can be responsible for change in our scientific understanding of the world.

The Electron Microscope in the Popular Media

It is not difficult to show that, to the popular imagination, the electron microscope presented itself as a device enhancing or even substituting for the eye—a "super-eye" or "new magic eye."[6] "Man's eyes are being sharpened by a new kind of super-microscope," a 1940 *Reader's Digest* article begins, typically oblivious to the painfulness of its metaphors.[7] Some early publicity micrographs from RCA brought home the message of the new microscope's power with familiar grooming implements: an apparently sharp razor edge that featured craggy "peaks and valleys," and monstrous particles of facial talc like "great chunks of shrapnel" bristling with spikes and hooks, together exuding menace for delicate facial skin of men and women (Plate 6.1). These particular micrographs achieved especially wide circulation in the early 1940s.[8] Other things with which one has intimate bodily contact were often invoked to communicate magnifying capacities, as for example when one of the earliest book-length efforts to exploit the novel instrument stressed that "a sheet of tissue paper would appear to be about nine feet thick if seen edgewise in the electron microscope," and when James Hillier declared in his national radio broadcast on

Weaver's show that "a gnat would seem larger than a B-29 Superfortress."[9] But most frequently of all, the human frame itself furnished the yardstick of the instrument's astonishing capacities. For instance, journalists gasped that a human hair would appear, with an electron microscope, like a giant sequoia tree, and one advertisement by a chemical firm employing the instrument declared that it had "a magnification that would make a medium size man lying down, reach from New York to Saint Paul."[10] The way the body, with its parts and accessories, similarly provides the ultimate referent for scale in vernacular spatial units, such as the foot, the fathom, the hair's breadth, needs no elaboration.

Although the barbarity of Europe and the dependence of perception on customs, two of the main themes emerging from Swift's account of the outsize land of Brobdingnag, seem absent from the discourse about electron microscopy, there were distinct echoes of the human body's strange and disgusting aspect that Gulliver discovered from his close perspective. In Brobdingnag Gulliver saw the giant faces of his hosts to be horribly pitted and discolored, while *Life* found it noteworthy that the electron microscope reveals even a single skin cell to be marked by numerous deep wrinkles on its surface (Plate 6.2).[11] Similarly, the frightening cragginess of the razor blades and makeup noted above implies the deceptiveness of a seemingly smooth and handsome face. The human self-image is evidently imported to the microscopic, as an aesthetic reference point along with the body as a metric of magnification.

The electron microscope alters the experienced scale of the observer's body. He or she is made tiny by proximity with small things (when imagining standing next to a treelike hair shaft), and by the same token made huge (when stretching from New York to Minnesota). Simultaneously, the microscope adds a change of *place* to this transformation of scale, much as Galileo's telescope brought the moon out of the heavens and down to the earth. Similar revolutions were anticipated, optimistically, by one over-enthusiastic *New York Times* editor: "Give us an electron telescope, and . . . limitations will be transcended. Mars brought nearer, stars so distant that we see them by light that left them when dinosaurs still shook the Earth . . . who knows what the future holds?"[12] To the popular imagination, the minuscule features pictured by the microscope represent a new land brought near. The copy in one magazine display of electron micrographs takes for granted that microscope pictures are automatically read as landscapes, by beginning with "the . . . mountains, cliffs, furrows, and strange desert formations on these . . . pages have never been clearly seen until recently" (Plate 6.2). "Peaks and valleys" on the above mentioned razor blade indicate the same. Of course, Ladislaus Marton described passage to a whimsical territory of the minuscule with his "Alice in Electronland," that ungainly effort to win patronage through popular science (Chap. 1).[13] *Life* magazine advertised "glimpses of an unknown world," in a large-type subtitle above RCA's publicity micrographs. "Into the Invisible World" was a typical headline.[14] The type of motion said to transport one to this new land varied interestingly. *Time* readers made a "deep thrust"; *Newsweek* readers "penetrated down"; and *Life* readers took a "breathless plunge" to a place where, it seems from another piece, small things are "literally submerged from sight."[15] Depth is the dominant metaphor, but sometimes one penetrates into a murky liquid, sometimes into a thicket, sometimes to another mysterious dark place. There is an easy slide from mere perception to full relocation. But whether it is the entire person that takes a "deep dive" into the very small, or only the eyesight that probes "the secrets of the infinitesimal,"[16] the microscope projects the viewer into another realm.

Visions of the denizens and landscapes encountered in this new country were sometimes freighted with sexism, racism, and, arguably at least, colonialism. Among the sinister characters portrayed in "rogue's galleries" of germs,[17] one finds menacing "dark giants,"[18] and the "delicate" outline of the syphilis-causing organism.[19] The association of darkness with brutish large size hardly needs comment, nor the association of feminine attributes with venereal disease in the context of police mug-shots (prostitutes were being blamed for communicating syphilis, of course, but it takes two, and masculine attributes for the germ could just as easily be found).[20] And the frequency with which fibrous structures appear as a jungle, rather than a thicket or a tangle, suggests that the land of the microscopic represents a Dark Continent to a popular imagination steeped in colonialist tropes. For example, "a tiny bit of soap becomes a mass of thick-stemmed jungle underbrush."[21] All of this is, of course, the sort of cultural baggage the tourist or explorer typically brings to a foreign land. It is one indication—an unfortunate one that just happens to be especially noticeable to today's reader—of how perceptions of virtually anything can be shaped by the cultural formation of the perceiver. In any case, the viewer's experience of electron micrographs is that of a person surrounded by, or imagining travel to, an unknown and somewhat sinister country.

During the war years, the natural metaphor of a "super-eye" was strained—for seeing does not in general carry violent connotations—by journalistic efforts to maintain a martial tone. The microscope became a "powerful weapon of science."[22] "Shooting Electrons Shatter Barriers to Science March" is one headline of this ilk, while another piece begins, "Through the small window of the steel-gray console" [of the microscope], casting the instrument as a radarlike surveillance device rather than a cannon.[23] That the Germans were ahead with the electron microscope, and had been employing it in making cement for the Siegfried line and other war projects, was often repeated in the early 1940s (especially by those trying to promote electron microscopy in America).[24] For the Germans too, the stunning advances of their *Übermikroskop* had propaganda value; needless to say, the very name of the instrument in German resonates with the supposed superiority of Aryan science and technology. Thus, as already noted, in the context of wartime propaganda the RCA electron microscope was especially significant as an unclassified high-technology development with which to counter German claims of technical supremacy, and its value as a counterexample must have contributed to the media attention the device received.[25] As for hostilities with the Japanese, there surfaced increasingly sinister associations with specimens associated with tropical jungles as the Pacific war grew fiercer. The "windpipe of a mosquito larva looks like a bamboo jungle if magnified 14,000 times," according to one journalist, while "the scales of a malarial mosquito's wings resembled the skeleton of a wrecked airplane hangar" to another, confident of the audience's familiarity with war imagery at least through newsreels and photojournalism.[26]

Seeing Things

It might not seem surprising that to a lay audience, undisciplined in the scientific seeing of objects as mere distributions of matter in space, strange microscopic things still present themselves laden with meaningful textures and charged with associations springing from the sensation of bodily proximity and familiarity. Later I describe how knowledge produced by scientists working in the electron microscopic landscape similarly reflects the deep structures of spatial perception and understanding that human perceivers bring to every observation, if not the more obvious and crass cultural baggage. These deep structures, both culturally specific ones deriving from formation in a particular cultural matrix, and universal ones deriving from the human experience of living with a body and a mind of a

certain kind, have been a central issue for philosophers of the phenomenological tradition. While some have stressed the cultural component
(e.g., Habermas and late Heidegger, if not also early Heidegger), and
others have stressed the universals of experience based in specifically human embodiment (e.g., Merleau-Ponty), these ultimately are always in-
termingled, since human beings never encounter the world except embodied and culturally embedded. I have already been referring to these
bodily and cultural deep structures collectively as "pre-understandings."[27]
But before exploring these, I want to emphasize their importance. Thus, I
will first argue that the electron microscope as a focus of attention faded
into the background, and served not mainly as a key to the inference of
properties of imaged specimens on the basis of a rigorous theory of the
instrument and specimen-beam interactions (though such theory did at-
tract interest, it never developed the necessary accuracy to allow much in
the way of reliable inference), but as a vehicle to an alien landscape for its
scientific users too. Here, the only conceptual frame that need be invoked
has to do with the relations between user and microscope. As Michael
Polanyi once put it, the experimenter comes to "dwell in" his or her
instrument. More recently, Don Ihde has spelled out this basic idea in
some detail, locating human relations with technology along a continuum
of decreasing consciousness and increasing familiarity, from "alterity" to
"embodiment." At the former extreme, the user of a technology experiences it as alien or "other," usually in a way that impairs easy interaction
with the world by means of the device; contrived, formalized inference is
needed to understand anything of the world on the other side. At the
latter extreme, the instrument becomes so familiar it effectively disappears
and becomes an organic extension of the user's sense organs and limbs,
permitting an easy interaction with the world through the now "transparent" device. The user is "embodied" in the technology.[28] Later I argue
that at least some of the rules of inference are, like the intuitions mediating
the more informal and direct grasp of the world available to an experimenter embodied in the apparatus, also rooted in the body, countering
the possible suggestions that their exclusive use in experiment might be
preferable, or that embodied forms of thinking are avoidable. For now, the
only claim is that electron microscopists developed a strong embodiment
relation with their microscopes.

Electron microscopists spent and still spend much time with their
instruments, most of it alone. Microscopes are almost always installed in
basement rooms to avoid building vibrations, and usually the rooms are

closetlike, so the lights can be extinguished for viewing the screen without interfering with the work of colleagues, and so the activities of others will not disturb tweezer-handling of tiny specimen grids or thin sections. In 1943, one well-maintained model B gave about 36 hours per week of operating time for a 44-week academic year (idle periods rather than servicing presumably account for most of the remaining weeks).[29] Perhaps it was typical for a single busy microscopist to use a microscope about half that time, or eighteen operating hours a week, sitting in the dark, watching the phosphor screen and operating the controls. Add to that the many hours spent working on micrographs in a photo lab, and on the microscope itself to keep it functioning well, and one realizes that an electron microscopist is somebody who spends most of the working day in a dark room (even should there be a separate technician handling routine maintenance or routine photography), generally alone but sometimes with a mentor or trainee, and most of it shut in with the big machine itself. Electron microscopists tend to feel a personal tie with their instruments, always ensuring that nobody tries to use it whom they have not individually instructed and observed in action, and sometimes they are possessive beyond what is necessary to protect the device from harm. Many electron microscopists, including the main characters in this study, set aside one microscope as their "personal" instrument.[30] This possessiveness bespeaks the intimacy that develops between microscopist and microscope.

Through daily practice learning about and working in the diminutive world made present by the microscope, the early electron microscopists underwent changes. As they familiarized themselves with its controls, the instrument became decreasingly refractive to their wishes, and increasingly predictable in performance. Microscopists in training recapitulate all this today. They learn how to find the focal plane by choosing a particle of dirt or a hole in the specimen support, and then adjust lens power carefully back and forth, thus making the image pass in and out focus in both directions. From subtle qualities of the image, they learn to recognize when poor focus is due to charging from column contamination, and when contrast could be improved by using a different accelerating voltage or another aperture. They learn how to move around within the specimen by turning the specimen holder control knobs, and so forth. That is, as they become good at using it, microscopists begin increasingly to experience the instrument as an extension of their sense organs; in Ihde's terms, they enter into an embodiment relation with the microscope in that it becomes a "transparent" window or vehicle for their experience of the world. It is as if the microscope really is a part of the microscopist, like eyeglasses or any other good prosthetic of which one is unconscious—in the circumscribed context of the instrument's darkened laboratory lair, of course.

And when the instrument malfunctions in a way that the microscopist cannot easily fix or does not understand, vision through the microscope becomes clouded, impairing the embodiment relation and thus highlighting the machine's former transparency. The instrument's identity as not-self becomes evident, its alien quality intrudes and demands that attention be focused on it, rather than on what one was doing with it. Since embodiment is by no means restricted to imaging technology, this is familiar ground to those who have experienced upsetting failures of their computers, cars, or other trusted appliances. Of course, not much of our technology is so good that, even when it is functioning normally, we are permitted to forget its mediating role completely. Most devices have imperfect transparency, aspects that chafe and limit the embodiment relation. In the 1940s, though electron microscopists were usually grateful to get their hands on a working instrument at all, a frequent complaint about the RCA model B was the height above the seat at which several of its knobs were placed. As one former user of the model B put it, gesturing vividly: "You had to be an Orangutan, because you had controls up here . . . and down here."[31] Indeed, the operators most comfortable and skilled with the instrument tended to be exceptionally tall, like Cecil Hall (see Plate 1.4) and Robley Williams. Bicycle-type chains run over the high controls and down to positions near the seated operator's hands ameliorated this problem for some.

When RCA's successor model EMU came along in 1944, it was warmly received despite certain imperfections that tended to keep its resolving power lower than that of the model B, in large measure because of its more convenient placement of controls. Robert Holley, an industrial designer versed in the ergonomic design philosophy that became popular with the military in wartime, contributed greatly to the EMU's ease of use.[32] To the large cohort of electron microscopists trained right after the war in America, the EMU was simply *the* electron microscope, and its operation was second nature. George Chapman, certainly one of the most accomplished biological electron microscopist of this generation, in 1994 still employed a 1950-vintage EMU for his research, despite the enormous trouble involved in keeping the relic running with parts cannibalized from junked units. He had naturally tried a number of more sophisticated microscopes in the course of his long career, but he was most comfortable with the EMU and still felt that it responded best to his needs.[33] A cartoon drawn by Cecil Hall and apparently shown at the 1951 EMSA meeting in Philadelphia illustrates just how transparent the EMU had become to its more proficient users by this time (Frontispiece). The friendly, if eccentric, personality of the instrument demonstrates that it is sometimes still perceived to some extent as an other, a mediator for the operator rather than a simple extension of the sense organs. But the EMU's alterity is not intimidating or obtrusive enough to disturb transparency of embodiment—except perhaps when it is in a bad mood. The cuckoo clock and the hand crank like that of an obsolete car, which is a transformation of the vacuum valve control on early versions of the EMU, emphasize a klugelike, eccentric, and (literally) cranky character.

In any case, transparency to the user is indicated by the absence of the dozens of controls and dials of the real EMU (compare Plate 6.3). Only three remain, indicating the areas in which the instrument still presented some opacity, that is, difficulty or concern. Focus is almost self-explanatory: as with any microscope, the operator always strives for the sharpest image, but there are a host of potential menaces to optimal focus, such as charging in the column due to contamination, fluctuating beam voltage or lens current, astigmatism, and so forth. Vacuum was another obtrusive worry (as it still is today), and very often beyond the control of the microscopist. After inserting a specimen, he or she always has to wait in suspense for the column to pump down sufficiently for the electron beam to be switched on. If the vacuum is bad, contamination and charging problems increase, meaning the nuisance of partially disassembling and cleaning the column, and also the arduous search for leaks; and if it turns out that the pumps simply are not pulling hard enough, this means at best the nasty job of changing pump oil, and worse, forebodings of major delays implied in visits from service personnel who may or may not have the necessary parts on hand. And what is the third dial, "Picture Control"? The "Picture Control" is meant to be understood as a knob, because it is drawn like the focus, and microscopists are always turning a knob to adjust focus. This is the cartoon's main punch line, for the knob refers to nothing on the real EMU's control panel. Perhaps all microscopists wish there were such a knob, or even just an indicator, and what makes the cartoon funny is the recognition of the impossibility of realizing it in hardware. What makes a picture good is a big issue—one addressed at various points throughout this book, in the contexts of the research programs and researchers establishing the standards by which pictures were judged as good or bad, significant or meaningless.

Picturing Things

Thus, the electron microscopists established a relationship of embodiment with their instrument, so the popular notions of "super eye" or "magic eye" were not terribly far off the mark. The microscope became, within limits, an extension of the operator in his or her interactions with the minuscule; we might say, to use currently fashionable language, the microscope became a prosthetic sense organ and microscopists become cyborgs—as do eyeglass wearers and video game players, strictly speaking.[34] (Since virtually all the American electron microscopists had RCA instruments in the 1940s and for much of the 1950s, their tacit and intimate understandings of their art must have been quite uniform, which may have contributed to the remarkable cohesiveness of the EMSA community during its early years.) At this juncture I want to argue that electron microscopists, like the general populace, experienced themselves transported by the instrument to an alien landscape. It would be easy to cite examples of statements by electron microscopists, often for popular consumption, portraying the electron microscopic as a new land. For instance, in an interview George Palade poetically described the early days of biological electron microscopy as the "opening of a huge . . . uncharted province, of a huge equivalent of a forbidden city or forbidden continent."[35] But even the first electron microscopists came to the microscopic world not just to gawk like tourists, but to get a job done, and the more important point to establish is that in their work they in fact treated the electron microscopic as a terrain.

One especially clear case in point, showing not only that electron micrographs could be read as landscapes but also that they could be deliberately designed to be read that way, is the metal-shadowing method of contrasting introduced by Ralph Wyckoff and Robley Williams in 1944. Bacteriologist Carl Robinow was surely not alone in perceiving the images thus obtained as "moonscapes" (see, e.g., Plate 5.7).[36] Indeed, the idea for the technique first arose from the familiarity of its inventors, one an amateur and the other a professional stargazer, with the astronomical practice of measuring the size of lunar mountains from the shadows cast.[37] This way of seeing was communicated to the nonastronomical audience by invoking common bodily experience; shadow-casting microscopy reveals topography more clearly "just as an observer on a mountain-top or in a plane sees many details of a landscape at sun-rise or sunset, which would be invisible to him at noon," explained Williams.[38] However familiar this sort of image may have seemed, it was contrived by novel means: the specimens were coated with a thin layer of metal atoms by spraying them from a low angle onto the specimen, so that with knowledge of the angle one can use the length of the "shadow" where a feature has blocked metal deposition onto the surrounding support to determine the height of that feature. The actual electron beam "illumination" is at "noon," not from the direction of metal deposition. But these technicalities fade into the background except when needed for measuring, and indeed are designed not to intrude on the intuitive perception of these micrographs as everyday landscapes, as two more contrivances involved in the production of so-called "electron shadow micrographs" illustrate. Williams and Wyckoff had initially used two shadowing directions in order to gauge feature size more accurately (Plate 6.4), but changed almost immediately to single-direction shadowing for what Williams simply described as "photogenic purposes,"[39] despite a potentially serious loss of accuracy due to specimen films on the grid having locally varying slopes, often far off horizontal. Moreover, except in the first publication, these images were (and still are) always presented as negative prints, produced by the extra step of printing from a negative made from the original plate negative. If they were printed from the plate like other electron micrographs, the ground would be black and the shadow white, thus conflicting with the visual intuitions derived from everyday human seeing even more than the dual-shadow picture—which would presumably be preferred if we happened to live on a planet with two suns. Aesthetics and intuition here take precedence over both inferential accuracy and practical efficiency; presumably the cognitive advantages outweigh these drawbacks.

Conventions of pictorial representation are clearly involved here, along with fundamental visual intuitions, in that most commonly the micrographs are oriented with light apparently above the picture plane and left, so that prominences in the specimen seem to come forward just as in classical European painting (as Françoise Bastide has pointed out).[40] Actually, when these micrographs first appeared their orientation was rather haphazard, despite explicit recognition in 1945 that the orientation of a shadow image determined whether features seem to be depressions or prominences (this optical effect was brought to general attention by newspaper misprints of aerial photographs from the war, where craters appeared as mounds).[41] So this convention was only weakly obeyed at first. Other pictorial conventions were adopted in the making of a standard electron micrographic visual language, many drawn from existing scientific imaging traditions—as already indicated by the astronomical origin of shadowgraphs, by the deliberate efforts of cytologists to adapt the traditional medium of the histological section (see Chap. 3), and by Mudd's efforts to adapt traditional bacteriological staining to electron microscopy (see Chap. 2). But one would like to go further and locate fundamental visual intuitions, such as those that might make all of the various different types of micrographs read as landscapes (in a less specific sense than Williams's shadowgraphs) into which the microscopist finds him or herself bodily projected.

I will now point to some specific analogies, beyond those already mentioned in the context of "mapping" (see Introduction), between electron micrograph and map, and electron microscopist and cartographer, both in terms of representational conventions and of underlying intuitions in the visual understanding of space. The implications for maps will not be pursued here.[42] Consider the seemingly simple one (or one-tenth) micron scale bar that typically calibrates an electron micrograph's magnification, playing just the same role as the scale in a map legend. The scale bar allows the micrograph to be related to others the way maps are related to each other; without a measure of scale, one could not tell whether one representation referred to features contained in, overlapping, adjacent to, or containing the features depicted in another. One might confuse a human hair with the bristle of a louse in that hair. Thus, among the major preoccupations of the first EMSA meeting, magnification figured quite prominently: how to determine it accurately, and how to reduce it so that comparison with light micrographs would be possible. The latter was a problem because the RCA model B had no low power mode. (Note how many micrographs Porter had to use in making his low-power montage, Plate 3.1, which was reduced *photographically*.) And determining magnification was still more problematic because even slight alterations in specimen positioning on the model B could change it radically, making repeats of observations (returns to the same place, on the analogy) very uncertain. Moreover, uncertainty was exacerbated by popular excitement over electron microscopy and by spontaneous replication of micrographs, as R. B. Barnes complained at that 1942 meeting, from the point of view of an industrial chemist:

The president of my company and the presidents and vice-presidents of your companies like to hear about 100,000 diameters. They like to hear about 60,000 diameters, and those slogans of publishing, such as the magnification of a human hair which is too large to put under the electron microscope. . . . We have . . . realized that although . . . particle size or magnification was subject to question, still anyone who obtained a copy of that photograph—and I am sure you have experienced what we have, that it is the hardest job in the world to keep a copy of your own photographs for yourself—is extremely prone to take that 30,000 magnification we wrote on the back, and next thing you know there is an article in print with pages of mathematical calculations based on the little thing we happened to write on the back.[43]

James Hillier had a similar complaint about what he called his "numerous experiences in escaping pictures," as did Robley Williams; it seems Walter Benjamin's thoughts on the circulation of artwork in the age of mechanical reproduction would also apply to the work of science.[44] Barnes went on to describe his fruitless efforts to find some object large enough to measure accurately with a light microscope, but small enough to serve to calibrate magnification in electron micrographs. As Barnes despairingly concluded, "we are convinced that you cannot use a light microscope to measure the dimension of a particle so small that it is, in turn, suitable for calibration in the electron microscope." The magnification ranges of the two instruments are so different as to make the two sorts of images quite literally incommensurable (i.e., "not measurable by a common standard")! The solution he developed was to use silica or nitrocellulose replicas of a 30,000-line diffraction grating, calibrated by spectrometry. Though widely adopted, this method was not ideal in that one still had to change specimens to insert the standard, which might change magnification with the movement of the specimen holder.

The 1948 discovery by Williams and Backus of their remarkably uniform batch of Dow Chemical latex particles (see Chap. 5) provided the era's definitive solution to the magnification problem. At first the investigators attempted to determine the absolute size of these latex spheres simply by making electron micrographs of them resting on diffraction grating replicas, but soon discovered that the grating replicas themselves showed greater dimensional variability than the particles. They ultimately determined size of the spheres by measuring them, stuck to fine glass fibers, in electron micrographs in comparison to lengths along the fibers that had been gauged with a calibrated light microscope. The value obtained for the spheres' diameter was 2,590 Å. Given this known diameter, not only could objects in images containing the spheres be measured and the image magnification calculated, but if the specimen was metal-shadowed the true local angle of shadowing could also be obtained from length of the shadows belonging to the spheres. Eight years later, despite values ranging from 2,520 to 2,730 Å having been obtained by a number of other researchers using light scattering, X-ray diffraction, and cen-trifugation methods (each of which entailed its own uncertainties), Wil-liams stood by his original microscopical determination as the most likely true diameter. Regardless, accepting the most deviant reported value as an indicator of the maximum possible error, Williams's figure was accurate to better than 6 percent. What mattered was not so much the exact size of the latex spheres, but that everyone used them so that sizes of things vi-sualized in different laboratories would be comparable. The latex spheres were greeted with terrific enthusiasm and were distributed gratis to EMSA members, who sprinkled them freely on their specimens. Thus, by multiplying the use of this material metric convention, a uniform stan-dard of magnification was established; it was an "internal" standard with respect both to the image and the community of electron microscopists.[45]

Once a metric of magnification is fixed, one is in a position not only to relate scales in different views but to equalize them. Only now does one begin to tackle the problem of whether a specific thing or place in a light image is the same thing or place in an electron image of the same speci-men, that is, the practical question of how to shuttle between light and electron pictures of the microscopic world, locating particular corre-sponding features in the two spaces. After all, one cannot recognize a thing in an optical field of view as a thing in an electron field without knowing that the things are the same size. At the 1942 EMSA meeting Barnes described how he encountered this second problem of orienta-tion, recognizing known regions and features, together with scale adjust-ment when doing some research on paper manufacture.

We took pictures at reasonably high magnification, at 12,000, and enlarged to something like 36,000. We obtained some pretty pictures with extremely good resolution. We were faced with a severe problem. How were people to know what it was supposed to look like? They couldn't tell us what was good or bad in the paper industry. They wanted to see this in a manner they were accustomed to seeing. We had that come up so many times that we got into an argument with ourselves about cellulose, whether it was fibrous or has particles. We couldn't see any particles under the electron microscope; but then we would take the same identical specimen with the optical microscope and see particles. One man started us out on low magnification when he said, "I won't believe a thing you tell me until you convince me your photograph is the same identical piece of material." We realized we had to lower the magnification until it overlapped thoroughly.[46]

It was only possible to obtain low magnification, comparable with that of a light microscope, on the model B by disabling one of the magnetic lenses, as Barnes and his team finally discovered. Then they viewed the same piece of paper mounted on a grid first with a light microscope at its highest magnification, and then with their electron microscope at the lowest. Here we can see that a crucial stage in reading electron micrographs is the identification of what the microscopists themselves call *landmarks*.

Recall the compass on Hall's EMU (Frontispiece): the issue is quite literally that of orientation. Barnes used features of the specimen support and the edges of the paper specimen to identify particular fields in light and electron microscopic views of similar magnification, then looked for distinctive features within them. Because in this case he could locate identical objects in the self-same specimen photographed by the two methods, he was able to use the objects themselves as magnification standards and to reprint at equal scale and similar orientation, presenting optical and electron micrographs side by side so as to enable thorough mapping between one and the other (Plate 6.5). Barnes had, of course, already done this cartographic work in making his montage. He invited the audience to recapitulate a part of the work in reading the images: "In a set of photographs of this type, it is so easy to orient yourself. Every little mark in the optical image is found to be a filament or a crossing of a filament or piece of wax or some foreign material."[47] Thus Barnes was able to convince his skeptical paper chemists to forget the particles they saw by light microscopy in cellulose fibers. If you have ever used a map to "orient yourself" outdoors you will recognize that Barnes was employing the same procedure. You look about for distinctive features that you expect the map to represent, then you attend to the map to find where you would have to be to see what you see. There may be cycles of this orientation process, where you attend again to your surroundings in search of features that look distinctive on the map, trying to find the map in the world just as you tried to find the world in the map. Finally you succeed, and declare, much like Barnes, that "you found yourself on the map." Just so, electron microscopists locate themselves virtually in the minuscule landscape, mentally positioning and moving the body within the image, passing from landmark to landmark. Wendell Stanley, ingenious publicist in general and the top popularizer of the virus in particular, knew how to bring the layperson through this procedure by placing himself on stage amidst hugely blown up electron micrographs and model viruses, literally moving and pointing in what, to the audience, was the microscopic field (Plate 5.2).

Thus, the electron micrograph is read in a way that fundamentally resembles the way a map is read. And a map, as Poincaré observed, is an abstract model of possible perceptions and actions, linked inextricably to the body's native spatial axes.[48] At the level of convention there are other striking similarities, beyond the noted scale bar in the legend: marks and letters adorning micrographs have counterparts in the names and symbols on a map, and there is sometimes the equivalent of a windrose, too (e.g., where tissue section micrographs bear arrows or similar indicators of orientation relative to some anatomical feature in situ). And beneath the level of explicit convention, in the realm of deeply ingrained pre-understanding, map users' and microscopists' understanding of space both depend on an implicit, usually moving body as a standpoint or reference point (Merleau-Ponty's "zero point"). This raises the issue of the extent to which unquestioned visual codes and native spatial intuitions might be imported, along with the body, into microscopic landscapes in the reading of micrographs. The deeper the visual pre-understandings, the more we will share and the less we will tend to question them, thus the harder they will be to detect; but an instance on the borderline between native and deep-cultural is the "magnification series" (see, e.g., Plate 4.3), a device that imparts spatial information in a way that can no doubt be described as rational, but nonetheless depends on major tacit ingredients. High-magnification electron micrographs are presented with lower-magnification views including the same fields, with an arrow or box marking the big picture to indicate the area depicted in the close-up. The indicator means "you are here." Arguably, such a trope assumes the viewer not only to have familiarity with conventions like park signs, but also to have a specifically human frame. The zoom-in effect preserves the viewer's orientation for the high-magnification view, fixing the attitude of the viewer's virtual surrogate while changing the relative scale of the body. That is, the viewer fixes the gaze on the object and advances steadily toward it, virtually speaking.[49] In his presentations to popular audiences, Keith Porter used to reproduce this effect in temporal rather than spatial series (and thus bypassed the park sign convention), rapidly displaying a series of slides of an object at progressively increasing light and electron magnifications so the viewers would maintain a sense of where they were when they got to the most magnified views.[50] Natural though it may seem to human viewers, some intelligent beings might well find this practice incomprehensible (or at least counterintuitive), such as dolphins or other creatures with laterally positioned eyes. They do not look in the direction they move, so they cannot experience visual approach.[51] Again, the microscopist moves about in microscopic space the way a human being moves through a space bodily surrounding him or her. But what of the contents of that space? Do these also bear the imprint of the human body?

Intuition, Inference, and Interpretation in Electron Microscopy

Even though there may be universals in the manner in which space and spatial representations are understood, rooted in the human body, there will always be a learned and culturally relative component, too. For instance, the way Euclidian-Cartesian space reads effortlessly to modern Westerners can be interpreted as a result of the training of and constant reimposition of such space on the eye by our carpentered environment, rather than as evidence that space essentially has such a uniform character and that the mind is preadapted to it. Euclidian-Cartesian space is just one of the possible ways in which the human mind—indeed even a given individual—can learn to read the world;[52] this originally highly contrived abstraction, a triumph of Greek philosophy, has at length been mass-produced for popular consumption. Now, it is obvious that science takes precautions to identify and compensate for prejudices and other pre-understandings rooted in personal bias and the cultural context. In the standard view, intuitions are productive in the local exploratory context of discovery and conjecture; but the testing and justification that follow, involving debate and competition amongst the entire scientific community, prevent mistaken conclusions that may have come from intuitions from passing into accredited scientific knowledge.[53] Prejudices are supposed to be purified out of scientific knowledge, and for the most part with great efficiency, according to standard philosophy of science. But although this purification process may counteract bias rooted in specific cultural backgrounds, one would not expect prejudices that are shared by all members of the scientific community, because rooted in the body or in cultural universals, to be affected. Indeed, I will contend that the manner in which scientific knowledge is tested and justified depends on just the same bodily intuitions, even if sometimes these are formalistically elaborated into rules of inference, that first shape experience into comprehensible form. Thus when there are major advances in the methods of a science (e.g., a Bachelardian epistemic break), these may arise from learning a novel way of deploying native intuitions in embodied inquiry that transcends shortcomings of prior ways, but not from a transcendence of the human body itself. If we are content with scientific knowledge that makes sense only for creatures such as ourselves and nothing more absolute, science stands to lose nothing by this argument.

At least some strategies for evaluating experimental data, formalizable though they may be, are based in bodily intuitions and may not be independently justifiable (although, like every practice, they justify themselves through their working, their utility). Let us consider two of the most important of Anderson's rules for testing the validity of micrograph interpretation (see Chap. 1), rules fundamental to the epistemology of the biological electron microscopist and, in more sophisticated incarnations, still in constant use. The first rule holds that an appearance, if not artifactual, should be consistently present as a feature of similar specimens prepared for electron microscopy in different ways. The second, conversely, holds that a feature observed by electron microscope, if not artifactual, should be detected in the same specimen when studied by different instruments.[54] These both reflect basic strategies from everyday life by which people assess functional dependence, deciding what is the cause or who is to blame: we race the engine to see where the worrisome mechanical sound is coming from, or we try another pair of glasses to see if the newsprint really is smudgy. But the source is deeper still than this "common sense" logic of adult life. These are the same basic methods the infant uses to establish its sensorimotor self, to learn its physical boundaries and acquire the foundation of motor skills: what the infant can make move with little effort—and what feeds back changed sensation (the sight of its arms moving, the feeling of contact with objects)—belongs to its body, while what does not simply change when the infant tries to change it must be not-self, world. Conceptions of causality, substance, and space all emerge at this early moment when reflexive gropings give way to sensorimotor projects.[55] The convincing quality of epistemological moves like Anderson's must flow from their universal, instinctive basis. But how much stock ought we to put in them? Hume and Kant are among those who have found that the leap from correlation to cause is a brute psychological fact beyond justification; the reverse leap, from noncorrelation with effort to causal independence, seems no less dubious.

And regardless of how we may judge Piaget's argument that the historical stages of physics mirror the succession of physical understandings reached in individual maturation, we can still agree with the basic insight of his "genetic epistemology" that humans are endowed with a limited, instinctive epistemological repertoire, which is constantly reutilized in all stages of cognitive development.[56] What matters here is that these interpretive moves are born with a subject embodied in a particular way, and are shaped in a nursery environment already structured by social practice and material culture. They are an instantiation at a higher level of a primitive instinct, one proven adequate—so far, and strictly for creatures such as ourselves—only inasmuch as the instinct has not generally been fatal to its bearers in the course of human evolution (thus "evolutionary epistemology" provides no more certification than pragmatism, and indeed less, because the latter speaks of adequacy for intelligent practice and not merely in survival).[57] If mathematical logic itself is similarly rooted in the body and the instincts, as has been argued recently by linguist George Lakoff and philosopher Mark Johnson among others,[58] then ultimate distinctions between formal epistemology and intuition become even less tenable.

Furthermore, strict reliance on such secure—or at least instinctive and inevitable—formalist rules such as these two of Anderson's frequently fails to yield an unambiguous interpretation of the state of affairs in the microscopic world. When two preparation methods yield different appearances of a certain feature in a specimen, we are confident that one of the appearances must represent a preparation artifact. But which one? And could not both be artifacts? This was the problem Anderson faced squarely when, confronted with the assortment of views he obtained from various preparations of nerve protoplasm, he admitted that none was really "true."[59] And even when all preparation methods yield appearances that agree, how can one know with any degree of assurance whether agreement might be due to coincidentally similar artifacts, and whether the next innovation in preparative method might not yield a radically different view? For a pragmatic biologist, this indeterminate admixture of method of inquiry in object of knowledge need not be an insuperable obstacle; as Porter has recently put it (and in a truly pragmatist way): "it didn't matter so much whether the thing was there in exactly the same form we were seeing it. If we had a technique that revealed something, we described the something in terms of that technique."[60]

Conscientiousness consists in minimizing uncertainty, so multiple specimen preparation methods for electron microscopy are compared, and so also are observations of the same specimen with the light microscope or nonvisual instruments. Agreement is seldom precise. For instance, X-ray crystallography on nerve tissue before and after osmium fixation showed a difference in membrane thickness, which suggested that electron microscopy is unreliable for gauging this feature.[61] But the spacing of repeated structures like membrane leaves is the *only* kind of specimen feature accessible to both X-ray diffraction and electron microscopy, so such comparison helps in the interpretation of no other features in the micrograph. When comparison is possible and there are discrepancies, it is often debatable which instrument gives the truer picture—as when Chapman and Hillier claimed that the light microscopic observation of a membrane preceding septum formation in bacterial division was an artifact, explicable by poor resolution of the "peripheral bodies" they found clustered near the nascent septum (Chap. 3).[62] When comparison is impossible, the image obtained by the most common method of observation tends to dominate; this is why appearances like Wyckoff's virus "honeycomb" fade into obscurity without conclusive explanation, with the decline in popularity of the method that generates the appearance. In general, precisely to the extent that observation methods are different, it is uncertain how to map between different observation types of the same specimen and thus, even to begin comparing what different methods are saying. Mudd identified the mitochondria he found in bacteria by light microscopical cytochemistry with the dense bodies he found with the electron microscope, but he was unable to convince Chapman and Hillier that the "peripheral bodies" in their thin sections were the self-same entities. All were following Anderson's rules, but in different manners fitted to their different research programs and schools of electron microscopy. How, when bare rules are not enough, to decide?

A variety of interpretive practices is always in use.[63] Rigorous theoretical treatments of instrument action and instrument-specimen interactions are not infrequently attempted in order to render some of the possible interpretations more and less plausible, but seldom to any great effect. The necessary calculations are often intractable because of the complexity of biological specimens, or inaccurate because of questionable assumptions made to simplify calculations or to cover for inadequate knowledge of the specimen. Aesthetics ultimately play a major role in the electron microscopist's decisions about which pictures are trustworthy (not as an alternative to "rational" methods but as a supplement to, or even a component of them—see Introduction), and how best to reduce indeterminacy of interpretation, as even multiple-instrument enthusiast Francis Schmitt freely admitted: "in some cases [judgment] becomes as much artistic as scientific. . . . Gestalt and the interrelationship of parts are by long odds the preeminent factors in the judgment of [preparation] quality."[64] I have earlier mentioned a few common aesthetic standards, such as backgrounds maximally clear of debris, and the maximal range of darkness values in micrographs. The reference to appearances with good "gestalts" must be taken seriously, too. And another aesthetic guideline to which Schmitt alludes above—"the interrelationship of parts"—and made explicit by Sjöstrand in his criterion that an artifactual state cannot appear more orderly than the native, is still more inextricably linked to judgments about the truth value of a picture (as opposed to simple presentational impact): greater connections between depicted parts are more likely to represent the orderly, and therefore true, condition of the specimen. But the proper limits of this aesthetic logic were highly controversial, despite its defensibility on thermodynamic grounds (see Chap. 3). In general, it is controversial which interpretive strategies should count as "proper," and as we have seen, *how* to read given micrographs is often decided at the same time it is decided *what* they are depicting. That is, the correct assortment of empirical rules, theoretical inferences, aesthetic guidelines, and other means for settling any given issue is an open question, and one settled—usually locally, though there are exceptionally influential cases with far-reaching ramifications like that of the mitochondrion discussed in Chapter 3—coextensively and simultaneously with the decision on the state of affairs in the microscopic world.

The point is not that science is fallible (and who ever claimed it to be infallible?) because we may have to resort to intuitions, nor that science is doubtful because it resorts to intuitions that tend to lead us astray—as undoubtedly they may, for instance in the case of the tadpole-phage (see Chap. 2). The point is that intuitive pre-understandings function everywhere in the most reliable processes of data acquisition and interpretation and even, as noted above, in the formal means used to test alternative interpretations. In the interest of examining further how preconceptions (or pre-understandings) are epistemologically productive and necessary, let us consider the replica, a type of specimen that is, after all, purely artificial, and thus in many ways presents a greater interpretive challenge even than elaborately prepared biological specimens. Replicas have played a very important role in electron microscopy. Because the electron beam does not penetrate dense material, the transmission electron microscope is unusable for the direct imaging of surface features on metal and other dense objects. This type of specimen was of great concern to a variety of industrial users. One could assess the type and extent of wear on machine parts, for instance, only if one could examine the microfeatures of these metal or ceramic surfaces. Several ways to solve this problem, by replicating a specimen's surface in a material thin enough for electron microscopy, were already circulating widely at the first EMSA meeting of 1942. Both the concept and the technique were quite simple: the specimen is cleaned, coated with a plastic resin, and after hardening the coating is removed by some gentle process such as adhesion to cellophane tape. Alternatively, a mineral coat could be applied chemically or electrochemically, and then freed by dissolving the original in a solvent to which the replica material is insensitive. Also achieving some popularity during the war was a two-step nondestructive process, in which thermoplastic was pressure-molded to a specimen, then removed and coated with a thin film of silica, after which the plastic cast was dissolved away.[65] To gauge the third dimension from images, a stereo pair of micrographs of a replica could be taken and examined with a commercial aerial survey contour mapper (again showing that the image was, quite literally, read as a landscape).[66] Alternatively, the intensities in the transmission image of a replica could be taken as a quantitative index of replica thickness, hence elevation in the original surface—but this exercise in formal theory was crippled by uncertainties and never became reliable. After metal shadowing was introduced, the simple one-step plastic replica process became the most popular because elevation could be measured fairly reliably from shadow length. Shadowed replicas resemble terrain especially strongly (as *Life* noted, above; see Plate 6.2) and, by printing directly from the negative plate without the usual step of making a second negative of the negative, the physically negative features of the one-step shadowed replica would read in the intuitively "correct" way. That is, what were prominences in the original metal surface would be crevasses in the replica that caught little metal on shadowing, thus would be zones transparent to electrons and leave dark areas on the microscope plate, which when printed would appear light and therefore prominent. So, perhaps ironically, a replica's extra difference from an original cancels one of the extra mediations needed to make its shadowed image read as if immediately present.

But even if the surface features in a replica can be measured accurately enough, how can one be certain that the replica represents the original faithfully, since the original can never be observed? The limitations of the capacity of formal (and formalizable) interpretive modes to answer this question are severe, bringing the role of more direct and informal intuitions to the forefront. At the first EMSA meeting there was keen discussion, in connection with the epic struggle to calibrate magnification, of the making and testing of multiple replica diffraction gratings from a metal original which had been hand-machined by the master of the art, Johns Hopkins physicist R. W. Wood. Optical analysis gave results for each replica grating that varied over a certain range, depending on the material used to make replicas, but the average indicated that the number of lines per inch on the original was exactly as Wood had claimed.[67] However, the accurate replication of line density says nothing about the faithfulness with which smaller features of the metal original are reproduced. One can microscopically examine an original's surface illuminated from above or obliquely, and compare light micrographs thus obtained with the electron micrographs of replicas, reasoning that because large (optically resolvable) features in the replica resemble those in the original, the smaller features in the replica that were revealed by electron microscopy also faithfully represent features of the original too small for light imaging.[68] But questions still arise about whether submicroscopic features might be replica artifacts, especially when these small features were not anticipated; for instance, mysterious tiny "spots" appeared in vinyl replicas and nobody could say whether these reflected the structure of the original metal, the structure of vinyl, or neither (recall here Bridgman's worries, noted in the Introduction, about calibration beyond the calibrating instrument's range).[69] No replica method is ideal, it was quickly recognized; all one can do is to compare multiple replica methods when confronted with doubtful appearances, despite the limits of this strategy,[70] in the hopes of identifying as artifactual any suspect features (bubbles, wrinkles, spots) that are not reproduced similarly in all replicas.

Preconceived notions and anticipations of what the specimen ought to look like, even though it has never been seen at such a high resolution, play a role in evaluating the faithfulness of replicas, as can notions of what likely artifacts should look like. These preconceptions can be founded in either theory or "common sense" (i.e., intuitions and pre-understandings, both native and culturally programmed), or in some combination. For instance, when investigating whether exposure of bacteria to vacuum causes dramatic artifacts, Hillier and Richard Baker at RCA found the same degree of wrinkling on replicas of air-dried bacteria as on air-dried bacteria themselves in the electron microscope, and less wrinkling in bacteria treated with fixative before drying; thus they concluded that wrinkles are caused by air-drying rather than vacuum, and were faithfully represented in their replicas, and that fixation reduces shrinkage and wrinkling.[71] The preconceptions that drying is related to wrinkling, and that similar wrinkles are unlikely to appear coincidentally through both drying and replication (say, through shrinkage of the plastic film after removal from the specimen), are what makes this experiment convincing. And theory plays its role here as the grounds for supposing that live bacteria have taut surfaces like balloons and are not wrinkled—for only the electron microscope, which cannot examine live bacteria, is capable of detecting wrinkles on microbes directly. In general, features such as wrinkles or smooth and round gaps that look like bubbles can be recognized as replication artifacts because they look like the kind of wrinkles and bubbles one encounters in substances such as plastic or wax in everyday life. Here it is not a question that observations may be theory-laden; indeed one needs no theory of wrinkling or bubbling to recognize these phenomena. What is at work is the questionable assumption that everyday experience of the macroscopic is a trustworthy guide to a remote dimensional range. When in doubt, one might consult physical theory—where there is theory available. But when calculations concerning viscosity and surface phenomena were done, they "proved" one-step vinyl replicas to be much worse than improvements in visualization technique later showed them to be.[72] As noted above, unknowns about the specimen and preparative process often make physical theory unreliable or useless, as in this example. One thinks also of the many elaborate and unproductive theoretical treatments pointing in bizarre directions for ultramicrotome design (Chap. 3). Where theory disagrees with experience, so much the worse for theory.

It can even be argued that because multiple methods applied according to Anderson's rules so seldom agree closely in any obvious way, interpretation is impossible without the guidance of preconceived concerns and hypotheses, obtained with or without the help of theory (and when formal theory does play a major role, we are back on the philosopher's familiar ground of "theory-laden observations"). For example, with vinyl replicas of cells and chromosomes (Plate 6.6), Albert Claude was able to reach no conclusion about his specimens that had not been previously suggested from work with light microscopes and other instruments. He limited interpretation to supporting prior results—despite acknowledging that all previous observations had been almost entirely uninformative, and that his pictures confirmed none of the extant models of chromosome structure! Here the lack of preconceptions about whatever mysterious chromosomal structure might explain gene action seems to have been positively crippling.[73] An obvious inverse case, showing how intuitive preconceptions, derived without benefit of anything recognizable as theory, can drive a specific interpretation that later is vindicated, can be cited in Porter's early reading of the endoplasmic reticulum as canals and vesicles instead of the filaments that others and even he himself were seeing (Chap. 3). But this case does not involve replicas. Perhaps the degree of uncertainty surrounding both biological specimens and replica methods made the combination of the two inaccessible to productive intuition. Indeed, replicas of biological specimens were almost entirely unproductive in the period of this study (though new methods introduced in the 1960s and after would change that).

Hermeneutics and Microscopical Experimentation

What have we learned, then, in all of these reflections on how the science produced by electron microscopy grows out of the embodiment of the experimentalists in their instruments during interaction with the world? I have argued that intuitions—including both instinctive and learned pre-understandings as well as more conscious and deliberate patterns of thought from everyday life—play a role alongside formal inferential rules in the interpretation of microscopic evidence. Indeed, I have contended that the latter are generally insufficient by themselves, and in any event are at least partly derived from the former. A similar case can presumably be made for many if not all other experimental fields. The perceived is the physically lived in the course of active work, objects of scientific knowledge not excepted. That interposition of elaborate experimental technology does not alter this situation is unsurprising, for instruments (just like simpler tools) are extensions of the body and its way of making sense of the world. If my discussion of the intuitive so far has been vague in places, the fault is not entirely my own, for philosophers of science have provided an impoverished vocabulary for such matters (interested, as they typically have been, only with distinguishing "rational" from "irrational"). But there are happy exceptions. In particular, Patrick Heelan has advanced a "first-person" account of experimentation founded on the primacy of perception, as an alternative to an account that preserves the traditional notion of a scientific observer who, ideally, would be disembodied, testing hypotheses about the world "objectively" from outside it somehow, through data interpreted only by mathematics or logically rigorous inferential chains. In Heelan's phenomenological account, an object of scientific knowledge is known as the invariant of a set of perceptual profiles, each of which is generated by an interaction of embodied experimenter and world. Theory is that body of mathematics (and other conceptual constructs) which allows the scientist to understand the different profiles as interconversions of the same invariant, and whose purpose is nothing other than this codification of perceptual performances; thus theory is always derived from and relative to the embodied practices of experiment. Scientific knowledge therefore entails an inseparable admixture of the purposeful subject and collective conventions of experimental practice, in the object of knowledge. This view seems to offer new foundations for a philosophy of science capable of treating personalities and cultures as constructive elements of sound scientific knowledge, rather than as menaces to or corruptions of a Cartesian purity that is at any rate unattainable. "The trail of the human serpent," as James observed, is "over everything" (see epigraph, Chap. 1). However, to despair immediately on this account would be premature and, arguably, misanthropic as well.

My own sympathies are certainly with Heelan; pragmatism (especially Dewey's version of it) always stressed embodied and culturally embedded action as the root of reliable knowledge, and it seems to me that this shared emphasis points to a fruitful bridge between phenomenology and pragmatism. In this study I have been trying to develop, in the specific case of one area of scientific practice, the sort of phenomenologically "thickened" project in history of perception recommended but never executed in the (in Joseph Margolis's words) empirically "thin pragmatisms" of philosophers from Peirce to Quine (and beyond).[74] But the reason for introducing a phenomenological framework at this late stage is not to endorse it, but to explore whether one of its more promising corollaries can be applied to the problems raised in Chapter 1 about the capacity of new instruments to change scientific knowledge. This corollary is Heelan's suggested adaptation of the hermeneutic circle to scientific experimentation. The classical hermeneutical circle treats the role played by pre-understandings in a reader's grasp of a text, which could, of course, never be understood if it were altogether alien and novel. These pre-understandings are broken down into the *Vorsicht*, or the figurative language shared by text and reader, the *Vorgriff*, or the preliminary sense of what the text is about, and the *Vorhabe*, or the skills and practices employed to read the text. To a great extent the reader derives from the text what he or she is prepared to accept, and returns in circular fashion to the same condition as before the reading—only displaced a little, to the degree that something new is learned, so that the circle describes a more or less open spiral. Now it should be clear how, if hermeneutic theory could be applied to the analysis of experimentation, it might help us reach an answer to the original question about how instruments can bring novelty in science. For this question can now be rephrased as one about the degree to which the interpretive circle of scientific experimentation is open, and the way new technology figures in opening it.

For Heelan, the *Vorsicht* in experiment consists of the descriptive language current in an investigative field; for instance, in the field of particle physics, the mathematical schemas of quantum dynamics. The *Vorgriff* consists of preconceptions such as the hypotheses the scientists in a field expect to see substantiated or not, on the basis of established observations and theory, in the material investigated in any new experimental situation; for instance, the properties anticipated of new particles in an accelerator experiment. And the *Vorhabe* includes the body of skills and practices for displaying and interpreting phenomena proper to an investigative field; for instance, the design and operation of particle detectors, data recording devices, and event interpretation techniques used in an accelerator experiment.

It is evident that new instruments and methods, such as the electron microscope and preparative techniques for biological specimens that this book has been devoted to describing fall under this last category, the *Vorhabe*. The *Vorsicht* for biological electron microscopy can be located in the descriptive language common to modern biomedical fields, aptly characterized by Foucault as flatly visual and "folded in on itself."[75] A nice example of such language, and the way it was used in grappling with interpretive problems, comes from a transcript of a discussion between Porter and Peyton Rous on micrographs of some cells from a rabbit tumor in 1951:

[If the putative carcinogenic virus has a certain size,] the virus in the x3,600 pictures of Porter will yield only minute dots. These pictures contain a vast number of such dots, but nowhere any significant special aggregates. At x7,200 the dots would be ⅜ mm across, or perhaps ¾ mm if hydrated. Again no special aggregates of such dots can be seen, though on pictures . . . some radial arrangement of dots of this size can be seen in the cytoplasm of the cell at the middle of the picture. The arrangement is radial to the nucleus, and the dots are almost evenly spaced, as if they were structural elements. . . .

Question: . . . . Are any of the cells normal cells? What about the dust-like particles almost evenly distributed in the cytoplasm and nucleus . . . ? Not merely an occasional pap[illoma] cell, but every cell should show the virus.

In sum: No special hordes of particles of even size, such as might be virus, are visible anywhere, but instead innumerable particles of a great range of size.[76]

Though many new terms were introduced, such as Porter's "endoplasmic reticulum" and Schmitt's "neurofilaments," the basic character of this biomedical language, this *Vorsicht*, was unchanged by the electron microscope and adapted easily to the new visual medium. Finally, the *Vorgriff* relevant to biological electron microscopy can be located in the preconceptions about the constitution of living things motivating experimenters and from which their research questions emerge, discussed in detail in the course of each chapter. In the above quotation the *Vorgriff* can be identified as the expectations about viruses: that they should appear in every cancer cell, as "hordes" of uniform particles in irregular aggregates.

Let us now see how this hermeneutical theorization of the electron microscope's impact unfolds in terms of cases already discussed, bracketing the issue of the basically unchanging *Vorsicht* except to the extent that we acknowledge that new names for entities and new descriptive terms such as "electron-dense" were introduced and incorporated into the existing vocabulary. In all the fields I have discussed, the electron microscope was initially employed in ways closely resembling the way those fields used the light microscope, to make pictures resembling the light micrographs then in use; think, for instance, of Mudd's bacteriological preparations, or of Porter's whole cell mounts (recalling that originally Porter cultured cells in order to observe, by light microscopy, their transformation to malignant or developmentally different forms). Similarly, many preparation techniques were adopted straight from light microscopic practice in biology, like osmium fixation, phosphotungstate staining, or the early and unsuccessful wax embedment and removal procedures for thin sectioning. It would thus be fair to say that among biologists the electron microscope was initially embedded in the same web of material practices as, and was expected to perform similarly to, the light microscope. The electron microscope began with the same *Vorhabe* as the light microscope. There was indeed some relatively minor initial variation in this *Vorhabe* among the users discussed here, as evident in the differences in experimental protocol and pictorial style pointed out in each chapter; but this variation can be attributed to the different ways the users were already employing light microscopy in their different research programs.

By contrast, the set of ideas about the nature of specimens and their expected appearances used to frame experimental questions, and to interpret the micrographs resulting from those experiments, varied highly from biological subfield to subfield. Thus the *Vorgriff*, like the *Vorhabe*, came initially from the particular experimental traditions adopting the electron microscope. For instance, the general physiologist Schmitt wanted to know about macromolecular self-assembly and conformational shifts. To test the then-dominant conformation-change theory of muscle contraction, he designed experiments to look at muscle fibrils in different states of contraction, stained selectively so that the pattern of staining (representing the position of specific, repeating sites in the macromolecule) might be seen to change in some geometrically regular fashion. Therefore Schmitt was looking in his pictures for conformational shifts that would account for contraction (see Chaps. 1, 4). The bacteriologist Mudd wanted to know about the organs of microbes and their biochemical function. Therefore Mudd tried to identify them by looking for organized bodies within bacteria, and eventually by treating living bacteria with reagents that would leave electron-dense indications of a particular biochemical activity in an electron micrograph of a whole-mounted bacterium (see Chap. 2). The cytologists Porter and Palade wanted to learn about granular bodies in the cytoplasm that might influence development. Therefore they looked in their pictures of cultured cells for the small dense bodies that corresponded to the nucleic acid–rich microsome fraction that they had found by centrifuging tissue homogenates (see Chap. 3). So we see that the problems initially posed for investigation by the electron microscope, the experimental design and the manner in which the evidence would be evaluated as meaningful, already presupposed what a significant result would look like. And moreover, it must be noted, all these electron microscopists quickly found evidence of the type they were seeking: Mudd found his "vesicular" nuclei and oxidative granules, Schmitt his periodically "knotted or beaded" fibrous muscle molecules, and Porter his plasmagenelike "growth granules." As observed in Chapter 1, the electron microscope did not initially overturn many expectations about the nature of living things. Such expectations informing experimental design derived from a pre-existing *Vorgriff* that varied with each biological subfield.

How then can new technology open this deeply conservative, self-fulfilling, and apparently closed hermeneutic cycle, in which experiments are designed, specimens prepared, findings interpreted, and pictures described all in pre-established manners? The electron microscope picture, for all its similarity with the light micrograph, looks a little different, and showed its first devotees a little more than what they anticipated. Porter and Claude could see not only their expected "microsome" granules but also a meshwork of strands, tubes, and sacs in a cytoplasmic ground substance they expected to be a uniform gel. Schmitt's postdoctoral fellows saw more than one kind of filament depending on how they sliced the muscle. Anderson and Luria could see differentiated tadpoles where they expected to see just a regular geometrical solid like Stanley's plant viruses. These early users also became accustomed to the new kind of micrograph, learning how to find their way around in it and what to expect therein. At the same time, they grew so accustomed to the instrument that produced the micrographs that it began to fade into the background and almost became as transparent as a pair of eyeglasses. That is, the *Vorhabe* is subtly shifted by the new technology, as users become embodied in it and improvise new practices especially for the new kind of picture they look forward to making. Thus when they returned to their inquiry and design the next experiments, the electron microscopists were looking for something not only new, but something that could only be sought—even conceived—with an electron microscope. Now Porter looked at membrane topology and for a correlation between the reticulum and the basophilic staining property, because the latter was already correlated with protein secretion, and a system of conduits made functional sense in the context of secretion. Now Anderson worried about why phage have the structure they do, and looked for ways to escape the surface tension forces—so tremendous at the tiny dimensional scale he was exploring—that obscure the mechanism of phage mobility, attachment, and infection. The *Vorgriff* of the microscopist is changed, progressively displaced as he or she develops a novel *Vorhabe* with the new instrument and the new kind of pictures it produces. Each time around the hermeneutic circle brings a growing displacement, and new experimental fields begin to evolve. The four different electron microscopic schools described in this book's central chapters each represent a different experimental tradition, divergently evolving, presumably due in large measure to origins in different biological fields, but all starting with Anderson's RCA-NRC hermeneutic practices (and specifically from the *Vorhabe* he founded). At later stages, divergence presumably may be countered by some cross-fertilization between traditions.

Although experimental technology brings change, its power to do so is multiply constrained. If the pictures or other data (profiles, in Heelan's terms) produced by a new device are so alien that no existing interpretive methods can manage them, and no relation can be drawn between the new data and the existing data concerning similar specimens, the device produces an unrecognizable signal and is not even a candidate for adoption by an existing science. The medical thermograph is perhaps a good example of an instrument that failed because its data were too strange.[77] Complete incommensurability at the level of practice heralds not scientific revolution through a novel technique, but simply its doom to rejection. And if the evidence a new device produces can be related to known objects, but can only be interpreted in such a way that it conflicts with accepted wisdom based on traditional methods, a new instrument is highly unlikely to be adopted—as the RCA-NRC committee overseeing the electron microscope's introduction seems to have been well aware (see Chap. 1). Thus, a new piece of experimental technology is deemed acceptable when it confirms most of what is already regarded as known, that is, when interpretive practices are found, such as Anderson's rules (which, it is worth reiterating, were somewhat, but not radically, innovative), allowing it to be "calibrated" against established knowledge. And not only must a new device prove unthreatening cognitively, but it also must comfortably fit the established culture and political structure of a field. For instance, the radio telescope was made an astronomical instrument not by astronomers, but by radar engineers after the Second World War working to exploit their technical expertise and continue their existing modus vivendi by shifting research projects. The radio telescope required new techniques and a different way of life from those customary among optical astronomers. Radio astronomy is still largely a separate discipline.[78] The same sort of social dynamics can be used to explain why the electron microscope was initially taken up most enthusiastically by biophysicists, a group of technophile life scientists both marginal and lacking in disciplinary coherence at the time, and thus lacking a deeply ingrained, established modus vivendi. It can also explain why the application of the instrument in cytology led to the formation of a new discipline (cell biology). Cytology and histology were established, low-technology medical fields, set in a century's worth of traditional ways. It is not surprising that they should not be open to adopting or understanding the new techniques readily, thus requiring a schism of the electron cytologists into their own discipline (evidence supporting this hypothesis can be found in the Rockefeller cytology group's difficulty getting work published in journals catering to traditional pathology, cytology, and histology, which led to the founding of their own journal; see Chap. 3). And we have seen that the particular way in which the electron microscope was made a cornerstone of this new discipline of cell biology had much to do with the political structure and disciplinary ecology of the medical sciences.

So new techniques must be assimilable to both paradigm and habitus, or both cognitively and culturally, because the hermeneutic circle into which they must fit comprises both sorts of components. But where does this formula leave us? It is a simplification, though (it seems to me) an accurate and helpful one, and we can say more. What in Chapter 1 I vaguely described as "conservative forces" affecting whether, and in what manner, a new device becomes an experimental instrument, turn out to be a complex and heterogeneous set of factors that channel experimental practice: some are rooted in our native cognitive capacities; some in the historically based intellectual and technical resources experimenters must work with in extending the pattern of knowledge generation in their specific fields; some in the local cultural patterns of laboratories and universities and other institutions of knowledge; some in the personalities of scientists and the politics within a discipline; some in the larger political arena where state funding, cultural agendas, and the competition among scientific fields for resources and status affects who gets to use what technology. All of these contingencies constrain and structure instrument usage. A change in any of these factors could mean a change in which instruments of science are accepted and how a community of practitioners considers them properly utilized for exploring the world. And any such change in manner of instrument usage might translate into a different content of the scientific knowledge derived from the instrument— a different construal of the natural world (as in the case of the mitochondrion, or the bacterial mesosome).[79] Knowledge of nature is relative to the method, history, and curiosity of the knower, just like understanding of literary productions. But such contingency does not constitute grounds to regard current scientific knowledge as simply arbitrary. The essence of a phenomenon, revealed by science at the core of the welter of profiles in which it manifests itself, and then tested scrupulously against the methods that skeptics might use to challenge the interpretation, retains an inter-subjective validity as the best that knowers like ourselves can attain at the moment. However, we are in no position to say what the world would look like to other creatures, or to human scientists with a very different history and technology. There are no empirical grounds for hope that we can transcend our humanity and speak about the way the world is in any absolute, body-independent sense. Any project to establish a relationship of correspondence between things as they are thought to be and as they "really" and absolutely are must remain a speculatively metaphysical project, given the contingencies of embodiment and history.

Of course, the analogy between interpretation of data from scientific experimentation and interpretation of literary productions is imperfect. There is no Author of Nature officially countenanced by contemporary science, and imagining a Gadamerian "fusion of horizons" with a nonanthropomorphic universe leads in an odd direction, perhaps even stranger than Schelling's Romantic project of reading the inner laws of nature through intuitive reflection.[80] Moreover, scientists are in many ways more like the authors than the audience of a book (or theatrical production);[81] the experimentalist interprets an *interactive* medium. Nonetheless, there are sufficient similarities in the positions of the interpreters of artistic production and experimental evidence, struggling to make sense of an over-rich mass of information and faced with choices among interpretive alternatives that can in differing ways reduce but never eliminate indeterminacies, to make the hermeneutic analogy fruitful. And the hope of learning about oneself, and of breaking free from entrenched ways of thinking, can similarly ennoble the scientific and artistic efforts to understand. As Paul Ricoeur has put it, the hermeneutic struggle to interpret the strange "enlarges the horizon of my own self-understanding."[82] As the novel conditions under which phenomena present themselves to the experimenter change, by altered technique or instrumentation, we become aware of the limitations under which things had previously been known; and we can become aware not only of the technical conditions, but also, if we are lucky and sufficiently reflective on our prior credulities, of still more deeply rooted conditions of our knowledge. In the projects of apprehending art and of interpreting scientific experiment alike, the hermeneutic circle engaging the inquirer is broken open and expanded by new media of perception, which thus become true media of thought about both our world and our selves.

Reference Matter

Notes

The following abbreviations are used throughout the Notes:

CK   Karl Compton–James Killian presidential papers, Massachusetts Institute of Technology Archives, collection AC4.

FOS   Francis Schmitt papers, Massachusetts Institute of Technology Archives, collection MC 154.

FSOH   Francis Schmitt Oral History collections, Massachusetts Institute of Technology Archives, collection MC 226.

KPA   Keith Porter papers, unsorted, at University Libraries Archives, the University of Colorado at Boulder.

KPN   Keith Porter papers, unsorted, not yet donated to library (in Aug. 1993).

MMB   Medical Microbiology administrative papers, University of Pennsylvania Archives, Philadelphia, Pennsylvania.

NMAH   National Museum of American History Archive Collection, Smithsonian Institution, Washington, D.C.

RF   Rockefeller Foundation Archives at the Rockefeller Archive Center, Tarrytown, New York.

RW   Robley C. Williams papers, unsorted, Archives of the University of California, Bancroft Library, Berkeley, California, collection 73/7 c.

SUA-DT   Donald Tresidder presidential papers, Stanford University Archives, Stanford, California.

SUA-RLW   Ray Lymon Wilbur presidential papers, Stanford University Archives, Stanford, California.

TFA   Thomas F. Anderson papers, unsorted, American Philosophical Society Library, Philadelphia, Pennsylvania.

WMS   Wendell M. Stanley papers, Archives of the University of California, Bancroft Library, Berkeley, California, collection 78/18 c.

Introduction

1. Bush, 3. The relationship between what actually happened and what Bush wanted is complex, however; see Reingold, *American Style*, chap. 13.

2. Reingold, *American Style*, chap. 13, and idem, "Science and Government." Cf. Kevles, *Physicists*, chaps. 20–25.

3. Penick et al., secs. 2–3; Sapolsky.

4. Strickland, *Politics*; and idem, *Story*.

5. Kuhn.

6. Van Helden and Hankins; Geison and Holmes; Galison, *How Experiments End*; Heilbron and Seidel; Foreman.

7. See Turnbull and Stokes; Rheinberger, *Experiment, Differenz*; Frank.

8. On scale, see Misa; on stages, see Hughes.

9. Rasmussen, "Biophysics Bubble."

10. On general physiology, see Pauly, *Controlling Life*; and idem, "General Physiology."

11. Text from *Rockefeller Foundation Annual Report*, 198–99. That a speech of approximately this content was given by Weaver at the April 1933 meeting is established on the basis of archival material by Kay; see "Cooperative Individualism," 54–64.

12. *Rockefeller Foundation Annual Report*, preface, "Members, Committees, and Officers"; concerning trustees, see *Who's Who in America 1934–1935*.

13. For a new view of the political interests of the Rockefeller circles, see Ferguson.

14. Weaver, "Molecular Biology."

15. Yoxen, "Giving Life."

16. Abir-Am, "Discourse."

17. See, for example, Kay, *Molecular Vision*; Keller, "Physics"; Kohler, *Partners*; Zallen.

18. For recent work, see Smith and Marx.

19. There are a few classic and honorable exceptions to this generalization, e.g., much of Temkin's oeuvre, and Hessen's Marxist argument that Newtonian physics was driven by the questions of interest to the rising bourgeoisie.

20. Jardine, *Scenes*, esp. chap. 3.

21. Hacking, *Representing*, vii. Priority for this sentiment probably should go to De Solla Price.

22. A. Franklin, *Neglect*; and idem, *Right or Wrong*; Giere; Hacking, *Representing*; Kosso, *Observability*.

23. Hacking, *Representing*, 186–209; A. Franklin, *Neglect*, chap. 6. See Wimsatt.

24. A. Franklin, *Neglect*, chap. 6; idem, *Right or Wrong*, 104; Kosso, "Science and Objectivity"; idem, *Observability*, chap. 4.

25. Rasmussen, "Facts."

26. Jardine, *Fortunes*, chaps. 4, 7; idem, *Scenes*, chap. 6. Also see A. Franklin, *Neglect*; and idem, *Right or Wrong*.

27. Bridgman.

28. To think here in a Bridgmanian vein, with Hasok Chang. On Bridgman's pragmatist bent, see Walter.

29. Jardine, *Scenes*; and idem, *Fortunes*; see also Cartwright; Galison, "History."

30. Collins, *Changing Order*; Pinch.

31. Latour and Woolgar; Knorr-Cetina and Mulkay; Lynch, *Art and Artifact*.

32. Lynch, *Art and Artifact*.

33. Ibid. Also cf. Amann and Knorr-Cetina.

34. Ihde, *Technology*.

35. Heelan, "After Experiment"; idem, *Space Perception*.

36. Crease.   37. E.g., Misak, Solomon.

38. Wilson.   39. Schivelbusch.

40. Hook, 75. In one of his moods, Ian Hacking appears to recommend a strong pragmatist or operationalist approach; cf. "Self-Vindication."

41. See Latour, "Drawing"; Rorty, "Anti-Representationalist View"; Tibbets; Toulmin, chap. 4; Turnbull; Wood.

42. Borges.

43. For differences between Rheinberger's views of representational spaces and mine, see Rheinberger, "From Microsomes to Ribosomes," and Rasmussen, "Mitochondrial Structure."

44. Contrary to common opinion among philosophers; see Hacking, *Representing*, 186–209.

45. Crease; Latour, *Science in Action*, introduction.

46. For a fine example of mutually inconsistent accounts of science supported empirically (though by different standards) by the same set of evidence, see Collins, "Strong Confirmation," and A. Franklin, "How to Avoid." Differing premises affecting where the burden of proof should lie in the empirical deconstruction or defense of traditional claims of science's authority have not received adequate examination. Some realists entirely refuse the burden, putting all faith in scientific orthodoxy ; see, e.g., Laudan. It could be argued that, ironically, the skeptical attitude of the "strong program" harmonizes more with the original skeptical spirit of the scientific revolution than the credulous attitude of such realists.

In addition to such complications arising from basic unresolved issues, I suspect that much of the heat in this controversy derives from mirror-image efforts on the part of sociologists, who have tools for studying conflict and consensus, and analytic philosophers, who have tools for studying formal logic, to hypostasize a methodologically convenient reduction of their subject matter, science. Hence traditional presumptions that science is reducible to logical elements, versus newer attitudes that science is "nothing but" social authority.

47. Chalmers, *Science*; Dupré.

48. Shapere. Cf. Bachelard; Tiles; Chalmers, *What Is This Thing*.

49. As I attempted before to spark, with "Facts." However, all that some critics seem to be capable of reading in that piece is a claim that particular scientific conclusions were ill warranted; see, e.g., Culp. Whether particular conclusions are warranted, and even whether the use of given epistemological moves (especially when described at such a high level of abstraction that they are empty of practical content) is warranted in some cases, are beside the point. The point is that conclusions reached in the course of science are founded (or "warranted") not just on different evidence, but on changing ways of evaluating evidence; so, can these choices and changes in epistemology be judged better or worse, and if so, by what standards? For instance, given conflicting conclusions that are both "robust," but in different ways, on what basis are we entitled to choose between forms of "robustness"?

50. E.g., Lynch and Edgerton. Blame might perhaps be laid at the door of Feyerabend's triumphant antirationalism.

51. See Jardine, *Scenes*, chap. 10. Crease gives aesthetics a prominent place in his phenomenological account of experiment as theater.

52. Hillman and Sartory; Hillman, "Artefacts."

53. Hillman and Sartory; Hillman, "Artefacts"; Hillman, "Towards a Classification."

54. Hillman and Sartory, 6.

55. For knowledge at the time covered in this study, see Burton and Kohl; Hall, *Introduction*; V. Zworykin et al.

56. Kunkle.

57. Marton, *Early History*; Qing; Reisner, "Early History"; Ruska. Also see the contributions in Hawkes. On wartime biological electron microscopy in Germany, see Rajewsky and Schon.

58. Grubb and Keller.

59. Johnson and Cantino.

60. Bracegirdle, chap. 4.

Chapter 1

1. For instance, radio astronomy is to a large extent (and in its early days, to a greater extent) a science built around one instrument, the radio telescope. It seems the discipline was founded not by practitioners of optical astronomy—a discipline nearly as dependent on its main instrument (the light telescope)—but by physicists with a prior commitment to the radar technology involved; see Edge and Mulkay.

2. Burian has used the term *domestication* for the process in which existing techniques and instruments are imported and adapted to a laboratory's "local culture." My use here designates a similar process, except on a larger scale (in that it governs acceptance for whole disciplines, rather than particular labs), and in a deeper sense (in that a machine cannot be said to constitute an instrument at all before domestication).

3. The biographical information about Marton comes mainly from Süss-kind.

4. Marton, "Electron Microscopy of Biological Objects."

5. See introduction by Denis Gabor in Marton, *Early History*.

6. An apocryphal story has it that Zworykin convinced Sarnoff that there was a market for the electron microscope by cajoling one of his friends at Amtorg, the Russian trade office, to inquire as to the price and numbers available of the machines; see Reisner, "Early History." In light of the interpretation given RCA's involvement with microscopes in this chapter, this tale might have originated as deliberately disinformative advertising, although Sarnoff was not necessarily party to Zworykin's scheme in its early phases.

7. Ibid.; Newberry, "Early Years." Cf. Kunkle.

8. Sobel, 122–67; Abramson, 108–225, 248–72; Barnouw, chap. 2; Federal Communications Commission, *Sixth Annual Report*, 70–72.

9. Marton, "A New Electron Microscope"; Reisner, "Early History"; Süss-kind.

10. Mudd to Rubin Borasky, Nov. 25, 1959, NMAH, unsorted Rubin Borasky Papers, carton 3. Zworykin and Polevitzky had a romantic relationship that probably included the entire period covered here (James Hillier, interview with author, Princeton, N.J., Feb. 1992, and personal communication).

11. Polevitzky, Mudd, and Stanley provided specimens that Marton included in his paper "The Electron Microscope: A New Tool." It was evidently Hillier who actually made these first TMV images (Hillier interview). On Stanley, see Kay, "Stanley's Crystallization." On genes and viruses in the 1930s, see Ravin.

12. Hillier interview; Reisner, "Early History." The model B price excluded photographic and other expensive ancillary equipment. In 1938, $35,000 was sufficient to install a sizable cyclotron; see Kohler, *Partners*, 375.

13. H. M. Miller memo of phone call with Marton, Dec. 10, 1940, RF group 1.1, series 205, box 10, folder 138.

14. K. T. Compton to Weaver, Dec. 18, 1940, RF group 1.1, series 224, box 4, folder 33. See Kohler, *Partners*, 316–21; also Schmitt, *Search*, chaps. 7–9. On Loeb, see Pauly, *Controlling Life*.

15. Weaver to Compton, Dec. 26, 1940; Schmitt to F. B. Hanson, Jan. 7, 1941; both in RF group 1.1, series 224, box 4, folder 33. McBain to Wilbur, Jan. 10, 1941, SUA-RLW box 114, "Rockefeller" folder. McBain to Marton, Oct. 4, 1940, and McBain to Marton, Oct. 25, 1940; both in NMAH, unsorted

L. Marton papers, carton 7, folder "Photographs. Correspondence, 1940–41." Apparently the collaboration began when Marton broached the possibility to McBain and some other chemists during a visit to California.

16. D. Pease to McBain, Jan. 14, 1941; S. E. Raffel to McBain, Jan. 15, 1941; C. Clifton to McBain, Jan. 16, 1941; E. W. Schultz to McBain, Jan. 17, 1941; all attached to "Project for the Development of the Electron Microscope at Stanford University," May 24, 1941; RF group 1.1, series 205, box 10, folder 138. Quote from McBain to Wilbur, Jan. 10, 1941, SUA-RLW box 114, "Rockefeller" folder.

17. Hillier interview.

18. Hanson memo of phone call with Marton, Feb. 19, 1941, RF group 1.1, series 205, box 10, folder 138. Marton had improved his academic credentials while at RCA by teaching classes in electron microscopy as a visiting professor in Physics at the University of Pennsylvania, 1938–39, and again in the summer of 1941 at the University of Michigan; see "Curriculum Vitae" attached to "Project for the Development." An academic career was apparently one of Marton's life-long ambitions (Charles Süsskind, personal communication).

19. Miller memo of phone call with Schmitt, Jan. 31, 1941, RF group 1.1, series 205, box 10, folder 138.

20. McBain to Wilbur, Feb. 17, 1941, SUA-RLW box 114, "Rockefeller" folder. "Report Upon the Interest Taken in the Electron Microscope at Stanford University, Feb. 1941"; Weaver to Hanson, Feb. 28, 1941 (quotation); both in RF group 1.1, series 205, box 10, folder 138.

21. E. Hawley. On this ethos in science, see Kargon; Kay, "Cooperative Individualism"; and Kohler, Partners, chaps. 4–5, 8–11. On Rockefeller Foundation research funding practices, see Kohler, Partners, chaps. 10–12.

22. See Bugos; Kevles, "George Ellery Hale."

23. Although both proposals are undated, their order is clear from a section intended to "amplify the arguments given by the first memorandum" in one of them; RF group 1.1, series 205, box 10, folder 138.

24. "Project for the Development."

25. Kohler, Partners, 156–58, 387–88; Seidel.

26. Both preoccupations of Cal Tech's Linus Pauling, one of Weaver's favorites, who with Rockefeller help was feverishly trying to make synthetic sera. See Pauling and Campbell; and Kay, "Pauling's Immunochemistry."

27. "Resolution RF41060," June 20, 1941, RF group 1.1, series 205, box 10, folder 138. Hanson to Wilbur, May 29, 1941, and Wilbur to Hanson, May 30, 1941; both in SUA-RLW box 114, "Rockefeller" folder.

28. Hanson memo of phone call with Schmitt, Dec. 30–31, 1940, RF group 1.1, series 224, box 4, folder 33. Schmitt to Hanson, Jan. 7, 1941; Bunker to Weaver, Feb. 1, 1941; Hanson diary entry, Feb. 3, 1941; Bunker to Weaver, Feb. 14, 1941; Hanson memo of conversation with Schmitt, Feb. 22, 1941; all in RF group 1.1, series 224, box 4, folder 34.

29. Hanson memo of conversation with Schmitt, Feb. 22, 1941, RF group 1.1, series 224, box 4, folder 34. Schmitt's objection to Marton as an academic colleague had something unspecified to do with Marton's wife. Claire Marton evidently irritated some Americans by boasting of her husband's heroic genius at the expense of his colleagues (James Hillier, personal communication); this refusal even to pay lip service to the American cooperative ethos in science must have diminished Marton's attractiveness as a colleague.

30. Weaver to Bunker, Feb. 28, 1941, RF group 1.1, series 224, box 4, folder 34. Schmitt to Hanson, May 22, 1941; Hanson memo of (phone ?) conversation with Compton, Apr. 2, 1941; Bunker to Hanson, Apr. 24, 1941; all in RF group 1.1, series 224, box 4, folder 35. On Toronto, see Hall, "Recollections."

31. Bunker to Hanson, Feb. 25, 1941, RF group 1.1, series 224, box 4, folder 35. The simultaneity of this draft with the similar report from Stanford, which followed a conversation between McBain and Weaver (see above), coupled with Weaver's early decisiveness, strongly suggest that Weaver took the initiative and solicited these proposals from a conviction that biological electron microscopy should be pushed.

32. MIT grant proposal, May 17, 1941, RF group 1.1, series 224, box 4, folder 35. Cf. Schmitt, Search, 125.

33. MIT grant proposal, May 17, 1941, "Appendix A," RF group 1.1, series 224, box 4, folder 35.

34. Resolution RF41059, June 20, 1941, RF group 1.1, series 224, box 4, folder 33. Hanson diary entry on interviews with Schmitt at Woods Hole, Mass., July and Aug. 1941, RF group 1.1, series 224, box 4, folder 35.

35. Schmitt to Hanson, Jan. 14, 1942, RF group 1.1, series 224, box 4, folder 36. Marie Jakus, telephone interview with author, Feb. 8, 1994.

36. Schmitt, Search, 125; Hall, "Recollections."

37. Schmitt to Hanson, Jan. 14, 1942, and "Progress Report," July 28, 1942, both in RF group 1.1, series 224, box 4, folder 36.

38. Schmitt, Search, 131–40; Hall, "Recollections." On Beadle, see Kay, "Selling Pure Science." On personal contacts in obtaining OSRD contracts, see Dupree, Pursell.

39. Hanson to Weaver, July 1, 1942; Weaver to Hanson, July 3, 1942; Schmitt to Hanson, July 28, 1942; Hanson diary entries on interviews with Schmitt at Woods Hole, Aug. 25–31, 1942, and Aug. 17–Sept. 1, 1943; all in RF group 1.1, series 224, box 4, folder 36. Cf. Schmitt, Search, 141, where Schmitt recalls that he actually finagled a prototype EMU machine rather than a model B.

40. Jakus interview. One instance suggesting the degree of wartime opportunity for women to enter biological electron microscopy is Stuart Mudd's advice to a laboratory director, about to obtain a microscope and searching for a bacteriologist competent to use it, that he scour the local universities and women's colleges for an advanced female student; Mudd to Donald Abel, Sept. 26, 1944, MMB carton 15, folder A 1944–45. Cf. Newberry, EMSA and Its People, 62–67.

41. Jakus interview; Schmitt to Miller, Apr. 6, 1946, RF group 1.1, series 224, box 4, folder 36. Much of the paramecium research is reported in Jakus.

42. Jakus has recalled that during trial and error with various metal solutions, she recalled phosphotungstate and its recipe by consulting her notes from a conventional cytology course (Jakus interview).

43. Schmitt et al., "Structure of Collagen"; Hall et al., "Certain Muscle Fibrils"; Schmitt et al., "Electron Microscope and X-ray Diffraction."

44. "Resolution RF47039," Apr. 2, 1947, RF group 1.2, series 224, box 1, folder 1.

45. Marton to Wilbur, Dec. 15, 1942, SUA-RLW box 122, "Electron Microscope" folder. Marton to Wilbur, Nov. 14, 1943, SUA-RLW box 127, "Rockefeller" folder.

46. Marton, "Alice." See offprints in SUA-RLW box 122, "Electron Microscope" folder, and box 127, "Electron Optics" folder.

47. Marton, "Alice," 247.   48. Ibid., 252.

49. Mudd and Lackman.   50. Marton, "Alice," 252.

51. Kay, "Stanley's Crystallization."   52. Marton, "Alice," 252.

53. Ibid., 254.

54. McBain to Wilbur, July 1, 1943, SUA-RLW box 122, "Electron Microscope" folder. McBain here says that for months Marton has been claiming that the microscope is nearly ready, but that it finally does seem complete. The microscope is featured in the *Stanford Alumni Review* 44, no. 8 (May 1943): 7–9. Cf. "Minutes 3 November 1943, Committee on the Electron Microscope," SUA-RLW box 127, "Electron Microscope" folder, confirming the microscope's operational status in the autumn of 1943.

55. Marton to Hanna, June 4, 1943, and Marton to Tate, June 4, 1943, both in SUA-RLW box 122, "Electron Microscope" folder.

56. O'Shaughnessy to Wilbur, June 21, 1943, and McBain to Wilbur, July 1, 1943, both in SUA-RLW box 122, "Electron Microscope" folder.

57. H. L. Snyder to Kirkpatrick, Dec. 9, 1945, SUA-DT box 5, folder 8. Quotation is from Weaver to "FEB" (identity unknown), Sept. 24, 1945, RF group 1.1, series 205, box 10, folder 139.

58. McBain to Wilbur, Jan. 10, 1941, SUA-RLW box 114, "Rockefeller" folder.

59. The cell is featured in the *Stanford Alumni Review* 45, no. 1 (Jan. 1944): 19–20. Miller memo of interview with Marton, June 19, 1945, RF group 1.1, series 205, box 10, folder 139. Protecting specimens from vacuum is mentioned in Marton's second draft proposal of early 1941, and the cell idea is explicitly included in the official grant proposal of May 1941.

60. Marton to Tresidder, Mar. 22, 1944, SUA-DT box 127, "Electron Microscope" folder. See Loring et al.; and E. W. Schultz et al.

61. Carlton Schwerdt, telephone interview with author, Oct. 1991.

Schwerdt and his advisor considered the results on polio particle size inconclusive, as did most other polio researchers.

62. Marton to Eurich, May 6, 1946, and Eurich to Marton, May 7, 1946, both in SUA-DT, box 5, folder 8. The plan involved an expected polio research grant from the National Foundation for Infantile Paralysis. Cf. Miller memo of telephone call with Marton, Dec. 16, 1946, RF group 1.1, series 205, box 10, folder 139. On the Stanford physicists' interest in electron optics, see Hevley.

63. Galison, *Image and Logic*, chap. 9.

64. Bourdieu, *Homo Academicus*, 128, 136; idem, "Specificity."

65. Hillier interview; see T. Smith.

66. Paul Green, personal communication. Anderson, "Personal Memories." There are numerous letters in Anderson's papers about his job search.

67. Stanley to Mudd, Sept. 14, 1940, WMS carton 16, "RCA Fellowship" folder. Here Stanley says that the final decision on who should be the RCA fellow was Zworykin's.

68. Mudd to H. Morton, July 21, 1940, MMB carton 14, "Reappointment of RCA Fellow" folder.

69. V. Zworykin to Griggs, May 14, 1941, Mudd to V. Zworykin, May 16, 1941, and Griggs to Committee, May 28, 1941; all in MMB, carton 14, "Reappointment of RCA Fellow" folder. Mudd and Metz were heads of academic departments at Penn; Demerec was at that time passing from assistant director to director of the Carnegie Institution Laboratory for Quantitative Biology at Cold Spring Harbor and was also the new director of the adjacent labs of the Long Island Biological Association, where he ran a lively summer research colony along the lines of Woods Hole's Marine Biological Laboratory; Haskins, who would become president of the Carnegie Institution in 1956, served in key offices of the OSRD throughout the war. See *American Men of Science*, 9th ed.

70. *Homo Academicus*, 105, 147–51.

71. Mudd to R. F. Griggs, Sept. 10, 1940, WMS, carton 16, "RCA Fellowship" folder.

72. "Report of the Committee on the RCA Fellowship on the Biological Applications of the Electron Microscope, NRC Division of Biology and Agriculture" [meeting minutes], Sept. 25, 1940, MMB, carton 14, "Reappointment of RCA Fellow" folder.

73. [Anderson], "Progress Report—Biological Applications of Electron Microscope, November 1940, December 1940," undated, and [Anderson], "Distribution of Efforts and Results to May 12, 1940," undated; both in TFA carton 3.

74. "Report," Sept. 25, 1940. See "Proceedings of Local Branches."

75. "Report," Sept. 25, 1940.

76. Tension between sales and Anderson's work was clearly a chronic problem. In June 1941, for example, filmmakers from Paramount Pictures took up a week of Anderson's time shooting in the microscope lab and asking him to

mount specimens for the piece they were producing (see Anderson, "Notebook 10/7/36–6/26/41," TFA carton 16, 57–58). The Committee itself also had conflicts with sales. The model B's product manager, Theodore Smith, grudgingly submitted biological micrographs to Mudd before publishing a sales brochure with a reminder that Anderson was using the demonstration instrument, even though Zworykin had not originally cleared that arrangement with marketing. Mudd replied that Smith should worry that sales might suffer if inferior work became associated with the microscope. Smith to Mudd, May 29, 1941, and Mudd to Smith, May 31, 1941, both in MMB carton 14, "Reappointment of RCA Fellow" folder.

77. "Notebook 10/7/36–6/26/41," 51–57. Anderson used Marton's model A well into the spring of 1941, at least occasionally, because the model B often was either under repair or otherwise engaged. Thus the model A was not immediately "scrapped" (Reisner, "Early History"), explaining how Anderson was able to continue working when the prototype model B was taken apart for delivery to American Cyanamid in late Nov. 1940.

78. Hillier, personal communication. Princeton chemist (and V. Zworykin's friend) John Turkevitch later recounted taking Manhattan samples to RCA; Sterling Newberry, personal communication.

79. "Proceedings of Local Branches." Katherine Polevitzky, "Pictures of Bacterial Forms Taken with the Electron Microscope" abstract, Oct. 30, 1940; Stuart Mudd and David Lackman, "Structural Differentiation Within the Bacterial Cell" abstract, Nov. 30, 1940; T. F. Anderson, "Report for 1940–1941–RCA Fellow of the National Research Council," Oct. 7, 1941; all in TFA, carton 3.

80. For instance, the cytogeneticists seem to have grown disgruntled with what they perceived as Anderson's lack of enthusiasm and effort on behalf of their chromosomes; Metz to Mudd, May 7, 1941, MMB carton 14, folder "Reappointment of RCA Fellow May 1941." Anderson, "Notebook 10/7/36–6/26/41," 51, entry dated May 15, 1941.

81. Anderson, "Distribution of Efforts."

82. Anderson, "Notebook 10/7/36–6/26/41," 57–58. Here one finds Anderson pleased that the schedule will prevent "certain people" (i.e., Mudd) from making him produce micrographs out of turn as "special favors." Anderson's summer and autumn schedule was determined from Anderson, "Progress Report for September, 1941–Biological Applications," Oct. 1, 1941; idem, "Report for 1940–1941–RCA Fellow of the National Research Council," Oct. 7, 1941; and idem, handwritten loose schedule for June, July, and Aug. 1941, undated; all in TFA, carton 3.

83. E.g., Spencer; "Electron Microscope" editorial.

84. Heidelberger was added in Jan. 1941. Former NRC Medical Sciences chairman E. V. Cowdry, from Washington University, wrote Mudd and NRC officials to complain that the RCA committee was too exclusive, and that Scott,

who had done earlier work incinerating tissue in an emission type microscope, deserved to be included. See Griggs to Mudd, Jan. 18, 1941, Cowdry to Mudd, Apr. 19, 1941, and V. Zworykin to Griggs, May 14, 1941, all in MMB carton 14, "Reappointment of RCA Fellow" folder. Scott had actually just built a transmission microscope for biological work in St. Louis with Sterling Newberry, but this work was unpublished; see Reisner, "Early History."

85. "Transcript of Meeting of the Committee on the Biological Applications of the Electron Microscope," Oct. 8, 1941, 1–2, TFA carton 3. War Production Board files on the priority decision could not be located at the United States National Archives; possibly they remained classified after the war because of the atomic bomb connection.

86. Seven of the sixteen microscopes delivered in the model B's first ten months seem to have gone to the chemical industry, and three to other industrial labs; all of these industrial users may have had war production contracts. In contrast, no more than a dozen of the 58 RCA model Bs sold before the microscope was superseded in 1944 were delivered to labs where they saw much use for biological work. For a list of model B microscopes and order of delivery, see Reisner, "Reflections."

87. Publicity took on an importance in World War II far beyond that in any previous conflict; indeed, Paul Fussell (chap. 11) has suggested that the "War of Publicity" would be as apt a designation as the (more traditional) "War of Mass Production."

88. Anderson, handwritten loose schedule of Feb.–May 1942, undated; idem, "Progress Report for February, 1942," Mar. 2, 1942; idem, "Progress Report for March, 1942," Apr. 4, 1942; idem, "Progress Report for April, 1942," May 4, 1942; idem, "Progress Report for the Month of May, 1942," undated; all in TFA carton 3.

89. "Transcript of Meeting of the Committee on the Biological Applications of the Electron Microscope," Oct. 8, 1941, 6–7, 12–14, TFA carton 3.

90. Anderson, "Progress Report for October, 1941," undated; idem, "Progress Report for November, 1941," Dec. 4, 1941; idem, "Progress Report for December, 1941," Jan. 2, 1942; idem, "Progress Report for January, 1942," Feb. 3, 1942; all in TFA carton 3. Scott looked at mammalian nerves and membranes, fixed in formalin, chromic acid, and osmic acid. Heidelberger was scheduled for January, and was probably the unnamed researcher who brought in highly purified hemoglobin molecules, which were successfully resolved for the first time, according to the reports.

91. Anderson, handwritten loose schedule of Feb.–May 1942; idem, "Progress Report for February, 1942," "Progress Report for March, 1942," "Progress Report for April, 1942," and "Progress Report for the Month of May, 1942," all in TFA carton 3.

92. Woods Hole did not view RCA's behavior as altruism. The firm not

only paid for transport and set-up of the model B, but also $200 rent for laboratory space there; V. Zworykin to Mudd, Aug. 28, 1942, MMB carton 14, "Reappointment of RCA Fellow" folder.

93. Anderson, "Personal Memories." At Woods Hole, Anderson attended various lectures and used the electron microscope on bacteria, antibodies, invertebrate sperm, more phage from Luria and Max Delbrück, and no doubt other unrecorded specimens; see red and black notebook labeled "Record," 30–35, TFA carton 16. In the third week of June 1942, Anderson's microscope had been working for several days at Wood's Hole, and Heidelberger was already doing experiments with it; Richards to Mudd, June 24, 1942, MMB carton 14, "Reappointment of RCA Fellow" folder.

94. Chambers to Anderson, Sept. 10, 1942; Anderson to Bronk, Sept. 21, 1942; Anderson to Harrison, Oct. 13, 1942; all in TFA carton 1. It is unclear who was behind Anderson's special NRC funding, because almost every Committee member might have had the necessary connections in Washington. See R. G. Harrison to M. H. Jacobs, Sept. 15, 1942, Merkel Jacobs papers, box G, folder 4, Archives of the Marine Biological Laboratories, Woods Hole. For Anderson's later wartime work, see "Record" notebook, 36–170.

95. Mudd to V. Zworykin, May 16, 1941, MMB carton 14, "Reappointment of RCA Fellow" folder. For papers, see listings in volumes of *Index Medicus*, covering 1940–44. Beyond publications counted here, there were also papers in unlisted chemistry journals, and possibly others not bearing Anderson's name but on which he worked. Anderson has placed the grand total of papers resulting from his RCA fellowship at 31 (Anderson, "Personal Memories," 8).

96. Mudd to Stern, May 31, 1941, Mudd to V. Zworykin, May 31, 1941, and Stern to Mudd, June 3, 1941, all in MMB carton 14, "Reappointment of RCA Fellow" folder. Cf. "Transcript of Meeting."

97. *Homo Academicus*, 97. On the related topic of shared experimental resources in the building of the genetics community, see Kohler, *Lords of the Fly*.

98. Seymour to Mudd, July 1, 1941, MMB carton 14, "Reappointment of RCA Fellow" folder. The paper was Seymour and Benmosche.

99. "Notebook 10/7/36–6/26/41," 57, entry dated June 26, 1941.

100. Mudd to R. L. Dickinson, Feb. 12, 1942, Mudd to Seymour, Mar. 26, 1942, both in MMB carton 14, "Reappointment of RCA Fellow" folder.

101. As Adele Clarke has pointed out, some American sex researchers in the early part of this century maintained high scientific status by portraying their work as fundamental endocrinology or physiology. Seymour, as someone actively practicing artificial insemination, was unable to escape the moral opprobrium surrounding the sex field by these means.

102. On the centrality of the science-resort Woods Hole Marine Biological Laboratory in the establishment of American biological disciplines on a truly

national level, see Pauly, "Summer Resort." Cold Spring Harbor's importance as social focus of the nascent molecular biology of the 1940s will be familiar to readers of the reminiscences in Cairns et al. Essentially all the biologists who collaborated with Anderson during his RCA-NRC fellowship had social links to Committee members (many though Demerec, head of the Cold Spring Harbor summer program) or did summer research at Woods Hole.

103. Bourdieu, *Homo Academicus*, 56–59; idem, *Distinction*. Also see Gerson.

104. Anderson, "Electron Microscopy of Phages." The controversial view of phage, consistent with cyclogenic theories of bacteria that were becoming unfashionable in the 1940s, was that of Hadley and other proponents of the idea that phages were a normal part of bacterial life cycles. On cyclogeny, see Amsterdamska; Summers; also see Chapter 2 on the sperm-phage resemblance.

105. Mudd to Stanley, Feb. 26, 1941, MMB carton 14, "Reappointment of RCA Fellow" folder. Anderson, "Electron Microscopy of Phages."

106. Mudd to Smith, May 31, 1941, MMB carton 14, "Reappointment of RCA Fellow" folder.

107. Mudd to V. Zworykin, May 16, 1941, MMB carton 14, "Reappointment of RCA Fellow" folder.

108. Harvey and Anderson.

109. Anderson, "Personal Memories."

110. Mudd and Anderson, "Selective 'Staining' for Electron Micrography"; Richards et al., "Microtome Sectioning"; Stanley and Anderson.

111. See Rasmussen, "Facts."

112. Shapin brings a related, and considerably more elaborate, analysis to the earliest foundations of experimental science.

113. See Chambers et al.; Green et al.; Luria et al., "Electron Microscope Studies."

114. Chambers et al; Luria et al., "Electron Microscope Studies."

115. See Bracegirdle, 63–64.

116. Richards et al., "Squid"; for an example of similar reasoning, see Mudd and Anderson, "Selective 'Staining' for Electron Micrography." The cognitive strategies and tactics by which artifacts are identified in electron microscopy has attracted considerable recent interest in the philosophy of science; see Introduction, and Rasmussen, "Facts," and works cited therein.

117. Chambers et al.; Richards et al., "Squid."

118. Luria et al., "Electron Microscope Studies."

119. Anderson and Richards.

120. [Anderson], "Transcript of Meeting of the Committee on the Biological Applications of the Electron Microscope," Oct. 8, 1941, 10; [Anderson], "Report for 1940–1941, RCA Fellow of the National Research Council," Oct. 7, 1941; both in TFA carton 3.

121. Anderson and Richards.

122. A. Franklin, *Neglect*, chap. 6; Hacking, *Representing*, 186–209; Kosso, "Dimensions"; and idem, "Objectivity."

123. These new members were R. Bowling Barnes (V. Zworykin's friend at American Cyanamid), Princeton physical chemist Henry Eyring, and Kodak physicist Lloyd Jones (whose presence might have quelled fears that RCA's commercial interests were too closely served). Mudd to Griggs, Jan. 28, 1942, and Mudd to Griggs, Mar. 14, 1942; both in MMB carton 14, "Reappointment of RCA Fellow" folder.

124. Mudd to Harrison, Mar. 13, 1942, and Harrison to Mudd, Mar. 20, 1942; both in MMB carton 14, "Reappointment of RCA Fellow" folder. The announcement can be found in *Science* 95 (1942): 348–49.

125. H. A. Barton to Anderson, June 3, 1942, TFA carton 1. Spangenberg to A. N. Goldsmith, June 19, 1942; Goldsmith to Spangenberg (cc Mudd), June 26, 1942; both in MMB carton 14, "Reappointment of RCA Fellow" folder. Spangenberg taught several courses with Marton before he was transferred to Sperry Gyroscopes in 1943.

126. Anderson to Barton, "Draft," undated, TFA carton 1.

127. Though my use of *discipline* here is in conformity with the ordinary dictionary definition of the verb, viz., to train people so as to make them behave in particular ways, I do also mean to evoke some connotations with which the word has recently been freighted, especially Foucault's idea (see *Discipline and Punish*) that the specifically modern form of social power depends essentially on the standardization of behavior through education and other constructive training. I also wish to bring to mind some arguments in recent history of science that build on Foucault to suggest that scientific disciplines are founded on standardization of practices, which is to say, on discipline. See Lenoir.

128. Mudd intervened with editors when stories presented an insufficiently serious image or made exaggerations so great that he feared "discredit" might result. See Mudd to Watson Davis, Oct. 4, 1940, MMB carton 14, "Reappointment of RCA Fellow" folder.

129. Clark to Anderson, Nov. 2, 1942, TFA carton 1 (quotation). Newberry, *EMSA and Its People*, 38–41, and "Yearbook 1943" appendix, 10–11 (pp. 53–54 lists the 76 registrants); see also Reisner, "Reflections." A version of the keynote address was published: V. Zworykin, "Electron Microscopy in Chemistry."

130. Newberry, *EMSA and Its People*, "Yearbook 1943" appendix, 58–60.

131. Ibid., 9–52.

132. Richards et al., "Microtome Sectioning."

133. Newberry, *EMSA and Its People*, "Yearbook 1943" appendix.

134. Newberry, *EMSA and Its People*, 62–64; "Proceedings of the Electron

Microscope Society of America" (1945); "Proceedings of the Electron Microscope Society of America" (1946).

135. Newberry, *EMSA and Its People*, 43.

136. V. Zworykin et al.

137. Griggs to Committee, Apr. 26, 1944, WMS carton 8, folder "Griggs, Robert F."

138. Penick et al.; Sapolosky; Strickland, *Politics*; and idem, *Story*. On the EMU, see Reisner, "Reflections."

139. MacDonald; Reisner gives 290 as the total of RCA microscopes shipped through 1950 (Reisner, "Early History"). EMSA membership in 1950 obtained by counting list of members, dated July 1950, MMB carton 20, folder "Electron Microscope Society." The membership of EMSA in 1952, at the end of the group's first decade, had grown from an original 153 to 433; see Newberry, *EMSA and Its People*, 35.

140. Sobel, 122–67; Abramson, 108–225, 248–72. Cf. Federal Communications Commission, *Sixth Annual Report*, 70–72.

141. Federal Communications Commission, *Thirteenth Annual Report*, 22–26; Sobel, 143–67.

142. "Return to Physics."

143. For a rather more literal use of this evolutionary concept in history, see Kohler, "Drosophila."

144. Lenoir and Lécuyer describe another case of a new instrument's domestication to science under the aegis of industry. For a comparison of electron microscopy to the practices surrounding other novel scientific instruments, see Rasmussen "Making a Machine Instrumental." Bourdieu, *Homo Academicus*, 84–90, notes other relevant ways in which practice is unconsciously disciplined in academic fields, such as concentration of resources among the eminent, reward of those who conform to accepted exemplars of practice, and punishment of those who move too quickly or beyond their proper place.

145. See Theresa's autobiography; on the politics of Theresa and her confessors, see Bilinkoff, chap. 5.

146. De Solla Price.

147. James, 31.

148. Thompson.

149. Bourdieu, *Logic of Practice*, book 1, chap. 3; Fleck; Kuhn. The Kuhnian "paradigm" cannot easily explain conservative yet continuous gradual change in either practice or theory if we read the notion in its original, strongly discontinuous sense. The *habitus* would seem to explain the conservation of practice and authority structure best, while the force of the *Denkkollectiv* would seem to explain the conservation of facts and theories about the world. Wittgensteinian "meaning webs" are less capable of dealing with practice the more the terrain

shifts from the linguistic to the material. A single theory capable of capturing both aspects of conservation is lacking.

150. Williams to L. Hadley, June 29, 1956, RW uncataloged carton 1, folder "H."

Chapter 2

1. Serres, 1–3.
2. George Chapman, interview with James Strick and author, Dec. 1993.
3. Silverstein, chap. 4.
4. Pauling and Campbell; also see Kay, "Pauling's Immunochemistry."
5. [Mudd], May 15, 1944, "Project X-187. Structural Changes in *Eberthella typhosa* After Irradiation with UV Light as Viewed with the Electron Microscope," MMB carton 15, folder "N, 1945–46."
6. Freund to Mudd, Aug. 24, 1944; Mudd to Freund, Nov. 30, 1944; Mudd to Freund, Dec. 7, 1944; Mudd to Joseph Felsen, Apr. 23, 1945; Freund to Mudd, May 7, 1945; all in MMB carton 15, folder "F, 1944–45." Mudd to Karl Meyer, Dec. 22, 1944, MMB carton 15, folder "M, 1944–45." Also see Rasmussen, "Freund's Adjuvant."
7. [Mudd], Oct. 23, 1944, "Research Project X-165, Report No. 1. A Laboratory Study of Bacteriophages for the Control of Bacillary Dysentery," MMB carton 15, folder "N, 1945–46." Demerec to Mudd, May 18, 1945; Mudd to Demerec, June 13, 1945; both in MMB carton 15, folder "D, 1944–1945."
8. Frances Young, [Nov. 1944?], "Technique Employed for Electron Microscopic Studies of Bacteriophages at Naval Medical Research Institute in 1944"; idem, Aug. 1945, "Project X-445. Electron Microscopic Studies on Dysentery Bacteriophages. Summary of Work Accomplished to 22 August 1945"; idem, Sept. 11, 1945, "Plans for Electron Microscopic Studies of Bacteria"; Mudd to E. G. Hakansson, Sept. 27, 1945; all in MMB carton 15, folder "N, 1945–46."
9. Smith to Mudd, Aug. 27, 1944, MMB carton 15, folder "S, 1944–45." Cf. W. E. Smith and S. A. Mudd, "An Unfamiliar Pattern of Bacteria Morphology," in "Proceedings of the Electron Microscope Society of America" (1945), 265.
10. Mudd to Richard Henry, Sept. 16, 1948, MMB carton 18, folder "H, 1948–49."
11. Robinow to Mudd, Aug. 30, 1947; Mudd to Robinow, Apr. 27, 1948; Robinow to Mudd, Aug. 28, 1948; all in MMB carton 17, folder "R, 1947–48." Mudd to Rene Dubos, Apr. 24, 1948, MMB carton 17, folder "D, 1947–48."
12. Mudd to Mitchell, Jan. 9, 1951; [Mudd?], n.d., "1950–51" departmental roster; both in MMB carton 19, folder "Mitchell, Dean . . . 1950–51."
13. Mitchell to Mudd, Jan. 5, 1952, MMB carton 21, folder "Mitchell, Dean . . . 1951–52."
14. By 1954 Morton's salary and budget were a black box to Mudd because

of Big Ben's classified status. Only $4,000, no more than half of Morton's salary, came from Department of Microbiology funds. See Mudd to Dean's Office, Oct. 22, 1954, MMB carton 26, folder "Mitchell, Dean . . . 1954–55."

15. Mudd to Mitchell, Sept. 29, 1950, MMB carton 19, folder "Mitchell, Dean . . . 1950–51."

16. Mudd to A. N. Richards, May 16, 1945, MMB carton 15, folder "R, 1944–45." Mudd to J. Mitchell, June 5, 1958, MMB carton 28, folder "Mitchell, Dean . . . 1957–58."

17. Mudd to Hartman, July 2, 1956, MMB carton 27, folder "H, 1956–57."

18. Schmitt, "A Program of Training for Post-Doctoral Medical Fellows at the Massachusetts Institute of Technology," Dec. 1950, FOS carton 1, folder 49. See Chapter 4.

19. Mudd to Mitchell, May 22, 1957, MMB carton 27, folder "Mitchell, Dean . . . 1956–57."

20. [Mudd?], draft NIH grant proposal, Dec. 15, 1947, MMB carton 25, folder "USPHS, Correspondence . . . 1947–48–49–50 RG 760." [Mudd?], Feb. 1949, "Scientific Progress Report, USPHS # RG 760," MMB carton 19, folder "Nutrition Foundation Inc, 1949–." Mudd to Mitchell, Dec. 17, 1952; Mudd to Mitchell, Mar. 17, 1953; both in MMB carton 22, folder "Mitchell, Dean . . . 1952–53." See Rasmussen, "Biophysics Bubble."

21. Mudd to Demerec, Jan. 16, 1953, MMB carton 22, folder "D, 1952–53." Gots to Mudd, Feb. 15, 1955, MMB carton 26, folder "Mitchell, Dean . . . 1954–55."

22. Admiral Stephenson to Mudd, [Oct. 1948?], MMB carton 18, Folder "S, 1948–49." Mudd to DeLamater, Oct. 12, 1950, MMB carton 19, folder "D, 1950–51."

23. Mudd to Shields Warren, July 30, 1951, MMB carton 23, folder "Atomic Energy Commission, 1951–1955."

24. Mudd to Harold Plough, Feb. 18, 1951; Warren to Mudd, Aug. 13, 1951; Plough to Mudd, Feb. 13, 1952; Plough to Mudd, Feb. 25, 1952; all in MMB carton 23, folder "Atomic Energy Commission, 1951–1955."

25. Robert Shannon to Mudd, Nov. 2, 1955; Jean Hasselberg memo to Mudd, Apr. 2, 1958; Mudd handwritten summary of AEC funding, [Mar. 1959?]; all in MMB carton 27, folder "AEC, 1955, 56, 57, 58."

26. For instance, Mudd had to urge student Phil Hartman to publish quickly on lysogenic phage because "Seymour Cohen is working full blast with several technicians" on the same topic; Mudd to Hartman, Dec. 23, 1954, MMB carton 26, folder "H, 1954–55." The work in question was Hartman et al., and Payne et al.

27. Mudd to Britton Chance, Sept. 27, 1949; Mudd to Chance, Oct. 1, 1949; both in MMB carton 18, folder "C, 1949–50." Mudd to William Siefriz, Apr. 23, 1952, MMB carton 21, folder "S, 1951–52" (quotation).

28. Hillier to Mudd, Apr. 22, 1946; Mudd to Hillier, June 12, 1946; both in MMB carton 15, folder "H, 1945–46." Mudd to Knaysi, May 9, 1946; Mudd to Knaysi, May 19, 1946; both in MMB carton 15, folder "I, J, K, 1945–46." Cf. Hillier to Mudd, July 22, 1946; Jane Henry to Mudd, July 22, 1946; both in MMB carton 16, folder "H, 1946–47."

29. Mudd to V. Zworykin, Nov. 5, 1948, MMB carton 18, folder "X, Y, Z, 1948–49." Mudd to Hillier, Apr. 8, 1949, MMB carton 18, folder "H, 1948–49" (quotation).

30. Mudd to Harry Eagle, Nov. 10, 1948, MMB carton 18, folder "E, 1948–49." Mudd to Hillier, Dec. 9, 1948, MMB carton 18, folder "H, 1948–49." Mudd to Richard Baker and Daniel Pease, June 6, 1949, MMB carton 17, folder "B, 1948–49."

31. For instance, in late 1951 this administration concern for operating expenses nearly prevented Mudd's AEC grant for purchase of an ultracentrifuge; Mudd to Mitchell, Dec. 3, 1951, MMB carton 23, folder "Atomic Energy Commission, 1951–1955."

32. Mudd to Hillier, June 1, 1949, MMB carton 18, folder "H, 1948–49." Hillier to Mudd, Mar. 27, 1951, MMB carton 19, folder "H, 1950–51." Mudd to Hillier, July 13, 1951, MMB carton 20, folder "H, 1951–52."

33. Mudd to William Siefriz, Apr. 23, 1952, MMB carton 21, folder "S 1951–52" (quotation).

34. DeLamater to Thomas Fitz-Hugh, Jr., May 28, 1954, MMB carton 22, folder "D, 1953–54."

35. Mudd to Wyckoff, Nov. 1, 1957, MMB carton 28, folder "W, 1957–58." Mudd to Moselio Schaechter, Nov. 15, 1957, MMB carton 28, folder "S, 1957–58."

36. Williams and Wyckoff, "Shadowed Electron Micrographs of Bacteria."

37. Porter et al.

38. Claude and Fullam, "Preparation of Sections"; Pease and Baker, "Sectioning Techniques."

39. Claude, "Method of Replicas"; Palay and Claude.

40. Mudd and Anderson, "Pathogenic Bacteria I"; and Mudd, "Pathogenic Bacteria II."

41. Knaysi and Mudd. Mudd seemed certain that his high-voltage studies had definitively revealed the bacterial nucleus at the first meeting of what became EMSA; see Newberry, *EMSA and Its People*, appendix p. 23. On the 1941 high-voltage microscope, see Reisner, "Early History."

42. See Robinow and Kellenberger.

43. Mudd and Anderson, "Pathogenic Bacteria I"; and Mudd, "Pathogenic Bacteria II."

44. Mudd to Detlev Bronk, Dec. 8, 1944, MMB carton 15, folder "B, 1944–45." Mudd to J. F. Whelen, Nov. 13, 1944, MMB carton 15, folder "WXYZ,

1944–45." Mudd to Frank Fremont-Smith, July 27, 1945; Mudd to Fremont-Smith, Feb. 14, 1946; both in MMB carton 15, folder "Macy Foundation, 1945–46."

45. Knaysi and Baker. Knaysi worked in Mudd's lab, according to Mudd to Hillier, June 12, 1946; MMB carton 15, folder "H, 1945–46," and also according to the acknowledgments in Knaysi and Baker.

46. These findings were alluded to in Mudd's 1947 Sweden talk, and ultimately published as Hillier et al., "Internal Structure."

47. Mudd, "Submicroscopical Structure."

48. Spath.

49. Robinow to Mudd, Aug. 30, 1947, MMB carton 17, folder "R, 1947–48." The work was first published as Robinow and Coslett. Cf. Coslett.

50. Mudd to Robinow, Feb. 18, 1948, MMB carton 17, folder "R, 1947–48." Hillier to Mudd, Apr. 15, 1948, MMB carton 17, folder "H, 1947–48." For van Iterson's work at Delft, see van Iterson, "Some Electron-Microscopical Observations." Also see Strick.

51. Pijper.

52. Strick.

53. Hillier to Mudd, May 21, 1946; Mudd to Hillier, May 27, 1946; both in MMB carton 15, folder "H, 1945–46."

54. Hillier et al., "Internal Structure." Cf. Robinow and Kellenberger.

55. Hillier et al., "New Preparation." Mudd was also using Hillier's experimental objective lens, the testing of which provided the occasion for Mudd's access to the RCA microscopes. See Hillier, "Some Remarks."

56. Hillier et al., "Internal Structure." Cf. Robinow and Kellenberger.

57. Robinow seminar plan, "Protonuclei in Bacteria," Mar. 18, 1949, MMB carton 18, folder "Robinow, C. F."

58. Mudd and Smith, "Bacterial Nuclei II."

59. Mudd and Smith, "Bacterial Nuclei I."

60. Mudd and DeLamater paper summary, "Nuclear Cytology of the Vegetative Cell of *Bacillus megatherium*," Aug. 1950, MMB carton 19, folder "Rio Conference." See DeLamater and Hunter.

61. DeLamater to Mudd, July 19, 1950; Mudd to DeLamater, July 31, 1950, MMB carton 19, folder "D, 1950–51."

62. See, e.g., DeLamater and Mudd; DeLamater, "Preliminary Observations"; DeLamater and Woodburn; DeLamater, "Evidence."

63. DeLamater et al.

64. Bisset's interpretation of bacterial division was consistent with the older cyclogeny model that Olga Amsterdamska has recently characterized as obsolete and fundamentally opposed to the atomistic approach to bacteria that dominated the postwar period. The hostility of postwar bacteriology to cyclogeny is perhaps exaggerated, particularly with regard to those who, like Mudd, were open to the

close relation between bacteria and higher forms. Mudd, at least, was initially not altogether hostile to Bisset, and felt that Bisset's book "persuasively" homologized "the phases of a bacterial culture with the alternating sporophyte-gametophyte generations of higher plants." Mudd to DeLamater, July 31, 1950, MMB carton 19, folder "D, 1950–51."

65. Bisset, "Genetical Implications."

66. Bisset to Mudd, Mar. 14, 1953, MMB carton 22, folder "B, 1952–53."

67. Bisset, *Cytology*, chaps. 5–7.

68. DeLamater, "New Cytological Basis" (with following discussion). Bisset quote on p. 410.

69. Bisset, "Spurious Mitotic Spindles," and DeLamater reply following.

70. Mudd to Bisset, Mar. 2, 1953, MMB carton 22, folder "B, 1952–53" (quotation). Mudd to Bisset, Dec. 15, 1953; MMB carton 22, folder "B, 1953–54."

71. Mudd to Bisset, Feb. 3, 1953; MMB carton 22, folder "B, 1952–53" (quotation). Cf. Robinow; Robinow and Kellenberger.

72. Mudd to Bisset, Mar. 2, 1953, MMB carton 22, folder "B, 1952–53" (quotation). Mudd to Mitchell, Oct. 11, 1951; Mitchell to Mudd, Jan. 5, 1952; both in MMB carton 21, folder "Mitchell, Dean . . . 1951–52."

73. The first sign of this appears in 1952, when DeLamater apologetically insisted on splitting a joint review they were supposed to write into two single-authored papers. DeLamater to Mudd (telegram), July 24, 1952, MMB carton 22, folder "D, 1952–53."

74. Mudd to Bisset, Feb. 11, 1953, MMB carton 22, folder "B, 1952–53" (quotation). As to Mudd's liberties, for instance Ralph Wyckoff was loath to send Mudd any unpublished micrographs on the grounds that he was a "tricky person to deal with"; Wyckoff to Williams, Feb. 5, 1946, RW carton 3, folder "Society of American Bacteriologists." Similarly, Robinow objected violently when Mudd arranged for publication of some micrographs taken by himself and van Iterson, without permission and according to Robinow "over our heads," in a friend's textbook. Robinow to Mudd, Nov. 30, 1948, MMB carton 18, folder "Robinow, C. F."

75. Mudd to Mitchell, May 21, 1954, MMB carton 21, folder "Mitchell, Dean . . . 1953–54."

76. Mudd paper text, "The Nuclear Vesicles of Young Gram-Negative Rods: Fixation of the Specimen by Irradiation in the Electron Microscope," Aug. 1950, MMB carton 19, folder "Rio Conference."

77. Ibid. Mudd attributed this theory to R. R. Mellon in 1927.

78. Cf. Spath.

79. Mudd et al., "Evidence Suggesting That the Granules of Mycobacteria."

80. Mudd et al., "Further Evidence"; J. Davis et al.

81. DeLamater et al.; "Proceedings of the Electron Microscope Society of America" (1952), 162.

82. Bisset, "Interpretation."

83. Mudd to Bisset, Feb. 3, 1953, MMB carton 22, folder "B, 1952–53." See Strick.

84. Mudd to Bisset, Feb. 11, 1953, MMB carton 22, folder "H, 1952–53."

85. Mudd to Bisset [not mailed], Apr. 22, 1953, MMB carton 22, folder "H, 1952–53." DeLamater to John MacLennan, May 26, 1953, MMB carton 22, folder "D, 1952–53." See Maclean et al.

86. Mudd and Winterscheid.

87. Mudd to Bisset, Dec. 15, 1953; Bisset to Mudd, Jan. 22, 1954; Mudd to Bisset, Feb. 4, 1954; Bisset to Mudd, Feb. 16, 1954; all in MMB carton 22, folder "B, 1953–54."

88. Mudd to Novikoff, May 14, 1952, MMB carton 20, folder "Histochemical Society."

89. Mudd to Hillier, Feb. 11, 1953, MMB carton 22, folder "H, 1952–53."

90. Ibid. See Palade, "An Electron Microscope Study," and "Discussion" following, 265–75.

91. Mudd to Hillier, Feb. 11, 1953; Hillier to Mudd, Mar. 2, 1953; both in MMB carton 22, folder "H, 1952–53."

92. Mudd, "Mitochondria of Bacteria."

93. Anderson, "Personal Memories"; and idem, "Electron Microscopy of Phages."

94. Gabor, 86–87.

95. Anderson, "Personal Memories"; and idem, "Electron Microscopy of Phages." See Anderson, "Techniques for the Preservation."

96. Chapman interview.

97. Chapman; Chapman and Hillier.

98. Chapman and Hillier.

99. Chapman interview.

100. Mudd to Andrew Smith, Oct. 30, 1953,MMB carton 22, folder "S, 1952–53."

101. Mudd, "Cytology of Bacteria I."

102. Mudd to Mitchell, 17 Sept. 1954, MMB carton 26, folder "Mitchell, Dean . . . 1954–55." By the time of his retirement in 1960, Mudd had nine Japanese researchers in his department, up from zero in 1954. Mudd to Mitchell, Nov. 12, 1954, MMB carton 26, folder "Mitchell, Dean . . . 1954–55." Cf. Mudd to Mitchell, Feb. 3, 1960 memo, "Foreign Personnel, Department of Microbiology," MMB carton 29, folder "Mitchell, Dean . . . 1959–1960."

103. Chapman interview. See K. Zworykin and Chapman; Chapman and K. Zworykin.

104. Mudd to Chapman, Sept. 27, 1954 (quotation); Chapman to Mudd, Oct. 6, 1954; [Mudd?], [1954], experiment outline, "The Structure of Bacterial Cytoplasmic Granules and Membranes as Revealed by Various Instrumentalities"; all in MMB carton 26, folder "C, 1954–55."

105. Mudd et al., "Electron-Scattering Granules."

106. P. Hartman to Mudd, Apr. 14, 1957, MMB carton 28, folder "H, 1957–58." For the loosening definition of bacterial mitochondria, compare Mudd, "Cytology of Bacteria I," with idem, "Cellular Organisation."

107. See reviews of early material in Alexander; Mitchell and Moyle.

108. Stanier.

109. See, e.g., Storck and Wachsman; Weibull et al.

110. Cota-Robles et al.

111. Chapman et al.; Glauert and Hopwood; Shinohara et al.

112. Mudd et al., "Plasma Membranes."

113. Hopwood and Glauert; van Iterson, "Some Features"; Fitz-James.

114. Rasmussen, "Fact."

115. Bracegirdle.

116. See Mudd to Hillier, Mar. 15, 1953; Hillier to Mudd, Apr. 3, 1953; both in MMB carton 22, folder "H, 1952–53."

117. Robinow to Mudd, Apr. 4, 1948, MMB carton 17, folder "R, 1947–48."

118. Mudd and Smith, "Bacterial Nuclei II."

119. Hillier et al., "Internal Structure."

120. Cf. Rasmussen, "Facts"; Strick.

Chapter 3

1. Dewey, 102–37.

2. Kuhn, especially chaps. 2–3. On normative versus cognitive consensus, see, e.g., Law.

3. See Rasmussen, "Facts."

4. Mary Bonneville interview with author, Sept. 1993.

5. Corner; A. Harvey, chap 5; Lewis.

6. Claude, "Coming of Age." Cf. Loewy; Rheinberger, "From Microsomes to Ribosomes."

7. Porter to Murphy, June 15, 1939; Porter to F. Hisaw, June 26, 1939; both in KPA carton 1, earliest box. Cf. Porter in Mary Bonneville interviews with Palade, Porter, and students, tape KC2B.

8. For an analysis of social forces contributing to the demise of the plasmagene theory, see Sapp.

9. Porter was at Ray Brook Sanatorium in upstate New York for several months in late 1942 and 1943, for example; see Murphy to Porter, Nov. 24, 1942; Murphy to Porter, Dec. 4, 1942; and Murphy to Porter, May 11, 1943; all in KPA carton 1, earliest box. Porter's wife also required treatment for tuberculosis during this period.

10. Porter on Bonneville tape KC1B.

11. Reisner and Fullam. Also Porter on Bonneville tape GEPRU8.

12. Claude and Fullam, "Isolated Mitochondria." See Murphy, "Reports to the Board of Scientific Directors of the Rockefeller Institute for Medical Research . . . April 21, 1945," "Report of Dr. Murphy," RF RG 439.

13. Porter et al.

14. See Claude and Fullam, "Preparation of Sections"; also see Gessler and Fullam, for which Claude provided tissue specimens. Cf. Murphy, [Apr. 1946], "Reports to the Corporation and to the Board of Scientific Directors of the Rockefeller Institute for Medical Research . . . 1945–1946," "Report of Doctor Murphy," RF RG 439.

15. Claude, "Studies on Cells." Cf. Loewy.

16. Palade, Bonneville tape GEPAS2.

17. Murphy, "Report of Doctor Murphy," 1946; Murphy, "Reports to the Board of Scientific Directors of the Rockefeller Institute for Medical Research . . . April 19, 1947," "Report of Dr. Murphy," RF RG 439.

18. Rheinberger, "From Microsomes to Ribosomes"; Rasmussen, "Mitochondrial Structure."

19. Porter to Pickels, June 3, 1946, KPA carton 1, second box. See Corner, 401–5.

20. Claude et al.; Porter and Thompson, "Particulate Body."

21. Porter and Thompson, "Some Morphological Features."

22. In 1953–54 Porter, in preliminary thin-sectioning studies on actively dividing cells, thought he identified the "growth granules" associated with the "centrosphere" of interphase cells only. This is the latest reference to them that I can find. See Porter, [Apr.?] 1954, "Report of Dr. Porter (with Drs. Fawcett, Palade, Palay, Pappas, Robinow, Rudzinska, and Sedar)," semifinal draft, KPN unmarked carton, folder "Annual Report 1954."

23. Porter and Kallman, "Significance of Cell Particulates." The 30 other talks at the Nov. 1951 conference on viruses and cancer that appear in the same number as this piece attest to what a popular notion the cancer-virus link had already become.

24. Rous to Porter, May 9, 1952; Porter to Rous, May 13, 1952; Rous to Porter, May 15, 1952; all in KPA carton 5, box "1952–1955, P–W."

25. "Cancer Research Marches Ahead."

26. Corner, 401–5; Porter on Bonneville tape KC3P (quotation).

27. Porter to Rivers, Jan. 25, 1955; Porter to Rivers, Mar. 8, 1955; both in KPA carton 5, box "1952–1955, P–W."

28. Porter, [1950?], "Personnel" (handwritten budget), KPA, carton 5, unlabeled box.

29. Newman et al., "Ultra-Microtomy"; idem, "A New Sectioning Technique." Palade to Porter, Oct. 28, 1949, KPA carton 2, unlabeled box.

30. Porter, [Apr.?] 1949, "Annual Report—1949"; and Porter, Palade, and

Parker Vanamee, [Apr. ?] 1949, untitled progress report; both in KPN unlabeled carton, folder "Report to Dr. Murphy, 1949." Porter, [Apr. ?] 1950, "Annual Report—1950," KPN unlabeled carton, folder "Report to Dr. Murphy, 1950." Porter, [Apr.?] 1951, "Report of Dr. Porter (with Drs. Palade and Vanamee)," KPN unlabeled carton, folder "Annual Report—1951." Porter and Gasser, [Apr.?] 1952, "Reports to the Corporation of the Rockefeller Institute for Medical Research by the Board of Scientific Directors . . . 1951–1952," "Report of Dr. Porter (with Drs. Palade, Vanamee, and Wolken)," RF RG 439.

31. Porter on Bonneville tape GEPRU7.

32. See 1949–52 annual reports cited above; also cf. Porter and Kallman, "Properties and Effects."

33. Palade, "Study of Fixation."

34. Porter and Blum.

35. See Pease and Porter.

36. Jerome Gross, interview with Susan Mehrtens, FSOH box 1, folder "J.G."

37. Porter on Bonneville tape COLO4.

38. Irene Manton (quotation), cited in Pease and Porter.

39. Porter to Rivers, Jan. 25, 1955; Porter to Rivers, Mar. 8, 1955; Palade to Porter, Sept. 6, 1955; all in KPA carton 5, box "1952–1955, P–W." I have not been able to ascertain precisely when the second EMU-2 was acquired, but it seems to have been in place by mid 1956.

40. Porter et al.

41. Porter to Pickels, June 3, 1946, KPA carton 1, second box.

42. Rheinberger, "From Microsomes to Ribosomes."

43. Murphy, "Reports to the Board . . . April 19, 1947," 81.

44. Ibid., 80.

45. Porter, [Apr. ?] 1948, "Annual Report—1948," KPN unlabeled carton, folder "Report to Dr. Murphy 1948," 4.

46. Porter, [Apr.?] 1951, "Report of Dr. Porter," 4.

47. Porter to Gasser, Mar. 8, 1951, and attached "Statement for Descriptive Pamphlet," KPA carton 5, unlabeled box.

48. Porter, [Apr.?] 1951, "Report of Dr. Porter."

49. As Porter himself acknowledged; Porter and Kallman, "Properties and Effects." There is no reason to regard this situation as a scandal of epistemological inconsistency, once it is grasped that experimentalists, at least in biology, may be pragmatists and not correspondence theory "realists."

50. Porter and Gasser, [Apr.?] 1952, "Report of Dr. Porter . . . 1951–1952."

51. See Rasmussen, "Facts."

52. Porter, "Submicroscopic Structure."

53. Cf. Rheinberger, "From Microsomes to Ribosomes." I use the term

without adopting the supposed autonomy of the representational systems as Rheinberger conceives them.

54. Porter, "Submicroscopic Basophilic Component."

55. Palade and Porter.

56. Porter, "Electron Microscopy of Basophilic Components."

57. Ibid.

58. Palay and Palade.

59. See Loewy; Rheinberger, "From Microsomes to Ribosomes."

60. Claude, "Studies on Cells"; De Duve and Beaufay.

61. Porter and Gasser, [Apr.?] 1952, "Report of Dr. Porter."

62. Barnum and Huseby; Petermann et al.

63. Porter, "Electron Microscopy of Basophilic Components"; Palade, "Small Particulate Component."

64. Bonneville interview (quotation). Porter, [Apr.?] 1955, "Report of Dr. Porter (with Drs. Moses, Palade, Pappas, Rudzinska, Scherer, Sedar, Siekevitz, Sotelo, Teng and Watson)," semifinal draft, KPN unlabeled carton, folder "Annual Report 1955."

65. Siekevitz and Zamecnik. For early examples of the "integrated" approach of the Rockefeller group, see Palade and Siekevitz, "Liver Microsomes"; and idem, "Pancreatic Microsomes." Also see Rheinberger, "Experiment, Difference, and Writing"; and idem, "From Microsomes to Ribosomes."

66. Porter and Palade, [Apr.?] 1957, "Cytology" annual report, KPN unlabeled carton, folder "Annual Report 1956–1957."

67. Porter to Charles Huttrer, Nov. 5, 1954, KPA carton 5, box "1952–1955, H–O." See Proceedings of a Conference.

68. Porter and Bennett. On eagerness to establish the fractionation-friendly nature of the journal, see Porter to Alfred Marshak, Dec. 29, 1954, KPA carton 5, box labeled "1952–1955, H–O."

69. Porter on Bonneville tape UCOLO6.

70. Porter, "Submicroscopic Structure."

71. Porter on Bonneville tape UCOLO5.

72. Sjöstrand, "Ultrastructure of Cells."

73. Porter to Gross, June 2, 1955, KPA carton 4, box "1952–1955, A–G."

74. My use of "bundle" here is compatible with common usage in this age of mass-marketed computer systems and, similarly, with Fujimura's notion of "package."

75. Altmann; Michaelis.

76. Bensley and Hoerr.

77. Palade, Bonneville tape GEPRU7; Murphy, "Reports to the Board . . . April 19, 1947," 72–75. Hogeboom et al., "Isolation of Morphologically Intact Mitochondria"; and idem, "Cytochemical Studies."

78. Lazarow and Barron; Claude, "Studies on Cells."

79. De Duve and Beaufay.

80. Fritiof Sjöstrand, interview with author, Apr. 1993.

81. Sjöstrand, "Fixation and Preparation"; and idem, "Electron-Microscopic Examination."

82. Schmitt, Apr. 1954, "A Review of the Program for Training in Medical Research at the Department of Biology, MIT," "Appendix D," FOS carton 1, folder 51. Cf. Sjöstrand interview.

83. For an early use of the term, see Bear et al.

84. Sjöstrand interview; Sjöstrand, "Electron Microscopic Demonstration of Membrane Structure"; and idem, "Method for Making Ultra-Thin Tissue Sections." See "Program of the E. F. Burton Memorial Meeting."

85. Sjöstrand to Porter, Sept. 29, 1949, and Sjöstrand to Porter, Jan. 14, 1950; both in KPA carton 2, first unlabeled box.

86. Sjöstrand interview.

87. Pease and Porter.

88. Sjöstrand stated that he learned Palade's fixation protocol by "personal communication," in his first mitochondrion publication of Jan. 1953 (Sjöstrand, "Electron Microscopy of Mitochondria"). Moreover, he had visited the Rockefeller labs in 1949 to discuss microtome design with Porter and to learn Porter's method for preparing tissue culture cells as specimens; Sjöstrand interview.

89. See Sjöstrand abstract in "Proceedings of the Electron Microscope Society of America" (1953), 116. Sjöstrand interview.

90. Palade, "Fine Structure of Mitochondria."

91. Schneider and Hogeboom.

92. Sjöstrand, "Electron Microscopy of Mitochondria."

93. Mudd to Hillier, Feb. 11, 1953; Hillier to Mudd, Mar. 2, 1953; both in MMB carton 22, folder "H 1952–53."

94. Palade, "Electron Microscope Study of Mitochondrial Structure."

95. Plate 3.15 is a fine example of what Michael Lynch (in "Externalized Retina") has called a "split-screen" representation meant to focus attention selectively on certain elements of the micrograph.

96. Palade, "Electron Microscope Study of Mitochondrial Structure."

97. See comments by Ross McArdle, Earl Newcomer, George Palade, and Emma Shelton in "Discussion."

98. Sjöstrand and Rhodin.

99. Rhodin.

100. Sjöstrand and Hanzon.

101. Sjöstrand, "Recent Advances."

102. Porter on Bonneville tape GEPRU8.

103. Ibid.

104. Sjöstrand, "Recent Advances."

105. Pease and Porter.

106. Sjöstrand interview.

107. Sjöstrand, "Recent Advances."

108. Palade, "Fixation of Tissues."

109. Ibid.

110. For a discussion of the importance of aesthetic considerations in the epistemology of electron microscopists, see Rasmussen, "Facts."

111. Palade to Porter, Aug. 30, 1954, KPA carton 5, unlabeled box.

112. Porter, preface, *Proceedings of a Conference*. Porter to Sjöstrand, Dec. 7, 1955, KPA carton 5, box "1952–1955, P–W."

113. See *Proceedings of a Conference*.

114. Sjöstrand, preface, *Stockholm Conference*, 1.

115. Schulz et al.

116. Eckholm and Sjöstrand, "Ultrastructure of the Thyroid."

117. For instance, in a 1958 review article written by two of the most important European biochemists studying mitochondria, both the Palade notion of folds and the Sjöstrand notion of septa are presented as live possibilities of mitochondrial structure, the truth about which is of great concern to biochemists. See Ernster and Lindberg.

118. Palade, "Mitochondria and Other Cytoplasmic Structures."

119. Sedar and Porter; Sedar and Rudzinska.

120. As suggested by, for example, Ernster and Lindberg.

121. Palade, "Mitochondria and Other Cytoplasmic Structures."

122. Ibid.

123. Pease; Bargmann et al.

124. Eckholm and Sjöstrand, "Ultrastructural Organization."

125. In this paper, the principle is manifest in the assumption that all the preparative processes of fixation, dehydration, embedding, and polymerization result in "disorientation."

126. Sjöstrand, "Ultrastructure of Cells."

127. Ibid.

128. Eckholm and Sjöstrand, "Ultrastructural Organization"; idem, "Ultrastructure of the Thyroid"; Anderssen-Cedergren.

129. See Lehninger, 24.

130. On changing biochemical theories of energy production, see Ernster and Schatz; Slater; Robinson; Allchin.

131. Sjöstrand, "Molecular Structure of Mitochondrial Cristae."

132. See Ernster and Schatz.

133. Cf. Law.

134. On Fawcett's work and that of some other visitors, see Porter and Gasser, [Apr.?] 1953, "Reports to the Corporation of the Rockefeller Institute for Medical Research by the Board of Scientific Directors . . . 1952–1953,"

"Report of Dr. Porter (with Drs. Fawcett, Palade, Pappas, Robinow, Rudzinska, and Wolken)," RF RG 439.

135. Porter on Bonneville tape KC4B.

136. L. Peachy, P. Satir, and M. Bonneville, Bonneville tape PHILA1.

137. Indeed, Sjöstrand recalls more than one occasion on which infuriated biochemists have disruptively stormed out of the room in the course of his seminar presentations; Sjöstrand interview.

138. See Sjöstrand, "Critical Evaluation," and references therein.

139. Shapere.

140. Finean et al.

141. Sjöstrand was able to convince the exceptional biochemist of his doubts about fractionation, judging from his recollection that when Hogeboom saw Sjöstrand's mitochondria pictures, he decided that the fractionation procedure needed to be reconstructed completely, and invited Sjöstrand to collaborate on this project in Bethesda. Sjöstrand interview.

142. This is essentially the position of sociologists who maintain that objects of scientific inquiry are nothing but the socially processed tokens exchanged among scientists, such as cell lines, micrographs, or DNA sequences.

143. Indeed, for Dewey, the embedding of a finding about the world in a systematic history, which then finds further use, is what makes it "real."

144. See Gold.

145. Pierre Bourdieu has suggested that heterogeneity of both practice and the standards by which it is judged is especially great in the academic domain; see *Homo Academicus*, 11–14, 114, 128. Peter Galison has described how cloud chamber users and Geiger counter users represent different experimental traditions in particle physics that have coexisted, sometimes in tension, since the early twentieth century (see *How Experiments End*). Scientists employing an unusual set of technical or epistemological practices often pay a price in relevance of work to others, when deviant procedures mean that results cannot readily be compared, or used at all except by a small minority.

Chapter 4

1. Husserl, 40–41. On mathematization and the electron microscope in particular, see Lynch, "Externalized Retina."

2. Rasmussen, "Biophysics Bubble."

3. On the physics community, see Kevles, *Physicists*; and A. Smith. On the much-discussed role of physicists in founding molecular biology, see Abir-Am, "Discourse of Physical Power"; idem, "Beyond Deterministic Sociology"; Fleming; Kay, "Conceptual Models"; Keller, "Physics and the Emergence"; idem, *Refiguring Life*, chap. 3; Olby, *Path*; idem, "Schrödinger's Problem"; Yoxen, "Where Does Schroedinger's 'What Is Life' Belong."

4. Schmitt, "Freshman Convocation" speech, Feb. 13, 1947 (quotation p. 7); and idem, "Molecular Cytology" talk, Dec. 30, 1946 (quotation p. 5); both in FOS carton 11, folder 3. See Kettering.

5. Schmitt, "Structural Basis of Life," 232.

6. J. T. Bonner, interview with J. Strick, Princeton, New Jersey, Nov. 1993.

7. Marie Jakus, telephone interview with author, Feb. 1994. See Schmitt, *Search*.

8. Bud, 86–87; Kohler, *Partners*, 318–21. Cf. Schmitt, *Search*, 119–20.

9. Betty Geren Uzman, interview with Susan Mehrtens; FSOH box 2, folder "Uzman."

10. Schmitt, *Search*, chap 9; David Crockett, interview with Susan Mehrtens, FSOH box 1, folder "Crockett." Cf. Schmitt speech at MGH, dated Oct. 15, 1946; FOS carton 11, folder 3.

11. See the *MIT Catalog* for years 1944–50; "Dorrance Building"; Proctor; Schmitt, *Search*, 126.

12. G. R. Harrison to Killian, Nov. 23, 1946, CK box 31, folder 5; Weaver to Schmitt, Apr. 14, 1947, FOS carton 2, folder 24. For comparison with Cal Tech, see Kay, "Cooperative Individualism"; Rasmussen, "Biophysics Bubble."

13. F. M. Rhind to Killian, Dec. 5, 1952; FOS carton 2, folder 27.

14. "Proposed Biology and Food Technology Building," undated; also untitled mimeograph with same content, Feb. 14, 1946; both in CK box 31, folder 5. Compton, "Memorandum for Executive Committee," Dec. 6, 1947 CK box 31, folder 7. Schmitt to Stanley, June 2, 1953, with attached program, "Dedication of the John Dorrance Building" (dated June 25, 1953), WMS carton 13, folder "Schmitt, Francis." See also "Dorrance Building"; Killian, 206–11.

15. Anonymous, "History of Biology at MIT," dated "autumn 1952," CK box 31, folder 7, p. 16. "Dorrance Building"; Buchanan, "Biochemistry at MIT"; plans MG 16 A08.00, MG 16 A07.00, MG 16 A06.00, MG 16 A05.00, MIT Buildings and Grounds records collection.

16. Schmitt, "Memorandum to the Staff," Apr. 21, 1943, FSOH box 2, folder "Talalay." Robert Bud's dismissal of Schmitt as a biologist "who had other ideas" inconsistent with biotechnology (Bud, 87) seems a bit unfair, given his interest in developing the first principals by which biological molecules could be engineered.

17. See *Opportunities in Chemical Biology and Physical Biology at MIT* (available at MIT Archives and Special Collections).

18. Pauly, *Controlling Life*.

19. "Society of Arts Lecture," Mar. 1946, FOS carton 11, folder 3.

20. In 1950–51, the number of postdocs was put unofficially at sixteen, though only twelve "research associates" are listed officially in the MIT catalog, whereas in 1948–49, eighteen "research associates" are listed. Cf. *Catalog*; also "Proposal for a Grant from the Commonwealth Fund," Dec. 1950; and "A

Program of Training for Post-Doctoral Medical Fellows at the Massachusetts Institute of Technology," Dec. 1950; both in FOS carton 1, folder 49.

21. Uzman interview.

22. [Schmitt], "A Review of the Program for Training in Medical Research at the Department of Biology, MIT," "Appendix D," Apr. 1954, FOS carton 1, folder 51.

23. "Memorandum to the Staff."

24. Alan Hodge, interview with Susan Mehrtens, FSOH box 1, folder "Hodge"; Uzman interview.

25. Schmitt, Search, 146. James Ebert, interview with Susan Mehrtens, FSOH box 1, folder "Ebert"; Uzman interview.

26. Jerome Gross, interview with Susan Mehrtens, FSOH box 1, folder "Gross."

27. Pauly, "General Physiology."

28. Schmitt, Search, chap 8.

29. Schmitt to "Director ONR, attn. Mr. Barry," Mar. 28, 1952, FOS carton 2, folder 14. "Draft of Proposal," Jan. 9, 1951, FOS carton 2, folder 17.

30. Schmitt, Search, chap 10. See H. Smith (in FOS carton 3, folder 16).

31. "Progress Report, NR-119-100, Investigations on the Ultrastructure and Chemistry of Nerve," Nov. 15, 1957, FOS carton 2, folder 15.

32. See Pauly, "General Physiology"; and idem, Controlling Life, chap 8.

33. Schmitt, Search, chap. 10; Gross interview; David Robertson, interview with Susan Mehrtens, FSOH box 2, folder "Robertson."

34. Uzman interview.

35. Anon., "The World Do Move, Yes," Feb. 1947 (flyer advertising speech by Schmitt for the MIT Alumni Association of Northern New Jersey), FOS carton 11, folder 4.

36. Schmitt, "Structural Basis of Life." The organism-as-factory metaphor goes back at least to Wilhelm Roux, who likened his defect experiments in embryology to throwing a bomb into the works.

37. "Protoplasmic Ultrastructure and Its Physiological Implications," the Lowell Lecture Series, Oct. 17–Nov. 11, 1947, FOS carton 11, folder 1 (quotations pp. 8, 15, 64). Opportunities in Chemical Biology and Physical Biology at MIT. On the equipment available, also see "A Program of Training for Post-Doctoral Medical Fellows at the Massachusetts Institute of Technology," Dec. 1950, FOS carton 1, folder 49.

38. Schmitt to H. M. Miller, Apr. 6, 1946, RF RG 1.1, ser. 224, box 4, folder 36.

39. For a popular summary of collagen work in the Schmitt lab in the period under discussion, see Schmitt, "Giant Molecules."

40. Schmitt, "Morphology in Muscle and Nerve Physiology." On neurophysiology in the early twentieth century, see Frank.

41. Richards et al., "Squid."

42. De Robertis and Schmitt, "Certain Nerve Axon Constituents" (quotation p. 1).

43. Ibid; also see "Appendix D," Apr. 1954.

44. Schmitt to H. van Riper, Sept. 23, 1947, FOS carton 31, folder 24. See Schmitt, "A Program of Biophysical and Biochemical Research on Nerve," Jan. 8, 1948, FOS, carton 31, folder 24.

45. De Robertis and Schmitt, "Study of Nerves"; De Robertis.

46. Schmitt to Harold Faber, Sept. 15, 1948, FOS carton 31, folder 24. "Polio at Work."

47. Rasmussen, "Facts."

48. Sanford Palay, interview with Susan Mehrtens, FSOH box 2, folder "Palay."

49. "Appendix D," Apr. 1954.

50. Schmitt, "Morphology in Muscle and Nerve Physiology."

51. Schmitt and Geren; Schmitt, "Structure of the Axon Filaments."

52. Schmitt, Search, 168. For the status of Schmitt's neurofilament research at five years, see Schmitt, "Fibrous Protein of the Nerve Axon"; also "Progress Report," Nov. 15, 1957.

53. Geren and McCulloch.

54. Geren and Raskind; Geren and Schmitt.

55. Geren and Schmitt.

56. Schmitt, Search, 149–50; Uzman interview.

57. Geren.

58. See Meyerhoff, chap. 4, for a summary of early twentieth-century muscle work.

59. A. Huxley, Reflections; Mommaerts.

60. Schmitt to H. M. Miller, Aug. 8, 1945, FOS carton 2, folder 25.

61. The Szeged results were made widely available in English in Szent-Györgyi, The Chemistry of Muscular Contraction. Astbury's findings were known by the Sixth International Conference of Experimental Cytology, held in Stockholm in July 1947; see Astbury. Also see Hall et al., "Investigation of Cross Striation."

62. Jakus and Hall, "Studies of Actin and Myosin" (which was submitted Nov. 1946); Hall et al., "Investigation of Cross Striation." The doubt about actin is reflected in comments on the work of A. Sandow in Schmitt to Miller, Aug. 8, 1945, FOS carton 2, folder 25; and about Astbury's conclusion in Schmitt, "Morphology in Muscle and Nerve Physiology."

63. Lowell lecture, 33–34; Schmitt et al., "Electron Microscope and X-Ray Diffraction."

64. Rosza et al., "Electron Microscopy of F-Actin"; and idem, "Fine Structure of Myofibrils."

65. Pease and Baker, "Fine Structure of Skeletal Muscle."

66. Draper and Hodge.

67. Ashley et al.

68. Schmitt, *Search*, 148; Gross interview. See Latta and Hartman; Geren and McCulloch; Hodge et al., "Simple New Microtome."

69. Farrant and Mercer.

70. Bennett and Porter.

71. See Szent-Györgyi, "Lost." Though Szent-Györgyi's listing in *American Men of Science* gives 1947 as the start of his Muscle Institute directorship, the actual move took place in 1949, judging by the masthead indicating both Muscle Institute and NIH affiliation, on Szent-Györgyi, "Free Energy."

72. On the purchase of the EMU, see Philpott to A. C. Maher, July 13, 1947, box "Philpott, Delbert," archives of the Marine Biological Laboratories, Woods Hole.

73. Philpott and Szent-Györgyi.

74. Schmitt, *Search*, 154; Randall. Randall recalls that Huxley went in Sept. 1952 and Hanson in Feb. 1953, whereas Schmitt's records (cf. "Appendix D") list them on the lab roster from Nov. 1952 and July 1953 respectively, which may simply date their official MIT registration. Both left in the spring of 1954.

75. Hanson.

76. Randall. H. Huxley, "X-Ray Analysis."

77. Schmitt, "Appendix D." H. Huxley, "Electron Microscopic Studies"; Hodge et al., "Studies on Ultrathin Sections of Muscle." Both of these papers were submitted in Aug. 1953. See Randall on chronology of Huxley's joint work with Hanson.

78. Randall.

79. Hanson and Huxley, "Structural Basis of the Cross-Striations in Muscle."

80. See Randall, esp. 323.

81. Huxley and Hanson, "Changes in the Cross-Striations."

82. Muscle as prepared for electron microscopy was, of course, no longer in a state capable of contraction, so a given specimen could only be observed at one stage of contraction; moreover, different preparations even of initially identical biological material could easily undergo different degrees of shrinkage during preparation. X-ray crystallography of muscle only gave information on cross-sectional structure.

83. Quoted in Randall, 325.

84. [Schmitt?], "NAS Symposium on Muscle—Record of Conference of Oct. 27 with Proposed Topic Assignments," Oct. 1953; Schmitt to Szent-Györgyi, Nov. 2, 1953; both in FOS carton 31, folder 14.

85. Schmitt, quoted by Randall, 325.

86. Gross, Robertson interviews. Jakus interview.

87. A. Huxley and Niedergerke.

88. Randall. A. Huxley, "Looking Back on Muscle"; here Huxley suggests that the sliding filament model might have been proposed anytime after 1900, a claim that clearly puts little weight on the value of microstructural evidence.

89. Hodge, "Studies of Insect Flight Muscle."

90. Hodge, "Studies on the Structure of Muscle III."

91. Hanson and Huxley, "Structural Changes."

92. Hanson and Huxley, "Structural Basis of Contraction."

93. See "Discussion" following Hanson and Huxley, "Structural Changes" (580–82).

94. Huxley and Hanson, "Preliminary Observations."

95. Sjöstrand and Andersson, "Ultrastructure of Skeletal Muscle Myofilaments."

96. Sjöstrand and Andersson, "Ultrastructure of Skeletal Muscle Myofilaments at Various Stages."

97. Huxley, "Double Array of Filaments."

98. Bennett; Hodge, "Fibrous Proteins"; Morales.

99. A. Huxley, *Reflections*, 38.

100. Gross interview.

101. Pauling; see Kay's (*Molecular Vision*) astute analysis.

102. Schmitt, "Chairman's Prefatory Remarks."

103. Killian, memorandum of understanding, June 27, 1952, CK box 31, folder 6.

104. Killian to Schmitt, Mar. 4, 1954, CK box 31, folder 6.

105. Uzman, Crockett interviews.

106. Schmitt, *Search*, 186–87. Schmitt to Denues, June 19, 1956, FOS carton 21, folder 11.

107. [Schmitt], "Investigations in Molecular Biology" grant application, Dec. 1955, FOS carton 2, folder 2. Hodge interview.

108. The Philips was, when purchased, the top-of-the-line $20,000 model; see Harrison to Killian, Dec. 14, 1950, CK box 31, folder 6. Cf. "Progress Report," Nov. 15, 1957.

109. "Investigations in Molecular Biology" grant application. See Schmitt, "Chromosomes, Genes, and Macromolecular Systems"; and idem, "Patterns of Interaction."

110. Rasmussen, "Biophysics Bubble."

111. Ibid.; Schmitt, *Search*, chap. 11; Weart, chap. 8. On the NIH, see Strickland, *Politics*; and idem, *Story*.

112. Ernest Allen to Schmitt, Nov. 9, 1954, FOS carton 20, folder 37; Schmitt, *Search*, chap. 12. See Oncley, 1.

113. "Biophysics and Biophysical Chemistry Study Section, Minutes of Meeting," Jan. 21, 1955, FOS carton 20, folder 37. Schmitt recalls (*Search*, 190) that physical chemist John Kirkwood was particularly insistent on the name

change, but if so, it must have been at later meetings, because the first meeting's minutes do not show him as present.

114. [Bolt?], Jan. 1958, "Development of NIH Extramural Programs for Training and Research in Physical Biology," FOS carton 20, folder 38. The study section members came to include physicists Jesse Beams (ultracentrifuge developer and "general physicist"), Philip Abelson (radioisotope specialist), Gordon Sutherland (spectroscopist), and Robley Williams (electron microscopist); medical biochemists and physiologists from government medical research labs Irving Gray, Murray Rosenberg, David Goldman, and James Hardy, as well as Fuhr and Sober; John Ferry, who, like Oncley and Neurath, was an academic biochemist working on the physical chemistry of proteins; protein chemist John Kirkwood; and nerve electrophysiologist Haldan Hartline.

115. "Conference on the Status of Biophysics, Ann Arbor, Michigan," Sept. 16, 1955, and attached seating plan, FOS carton 21, folder 12. Attending were Fuhr, Schmitt, Oncley, Neurath, Sutherland, Williams, Sober, and Gray of the study section, George Bishop of Washington University, William Bloom and Ray Zirkle of the University of Chicago, George Bowen and Robert Sinsheimer of Iowa State College, Alan Burton of the University of Western Ontario, Robert Emerson and William Fry of the University of Illinois in Urbana, Joseph Foster and Lorin Mullins of Purdue University, Ralph Gerard and Victor Guillemin Jr. of the University of Illinois in Chicago, Paul Kaesberg of the University of Wisconsin, Wilfried Mommaerts of Case Western Reserve University, Irvine Page of the Cleveland Clinic Foundation, Max Lauffer of the University of Pittsburgh, and F. O. Schmitt's brother, Otto, of the University of Minnesota.

116. "Conference on the Status of Biophysics, Berkeley, California," Dec. 1, 1955, FOS carton 21, folder 12. Attending were Fuhr, Schmitt, Neurath, Oncley, Sutherland, and Williams, all of the study section, James Bartlett of the University of Illinois in Urbana, H. Stanley Bennett, Robert Rushmer, and Allan Young of the University of Washington, Sidney Benson of the University of Southern California, Lawrence Blinks and Paul Kirkpatrick of Stanford University, Pauling, Dan Campbell, Jerome Vinograd, and Max Delbrück of the California Institute of Technology, Herbert Jehle of the University of Nebraska, Dan Mazia, Howard Schachman, Cornelius Tobias, and Wendell Stanley of the University of California in Berkeley (Schachman, like Williams, was a member of Stanley's virus laboratory), and W. A. Selle of the University of California in Los Angeles. Cf. Kohler, *From Medical Chemistry*.

117. "Biophysics and Biophysical Chemistry Study Section, Minutes of Meeting," Jan. 14–15, 1956, FOS carton 21, folder 18.

118. Neurath to Fuhr, Jan. 3, 1956; Schmitt memo, "Impressions of Tangible Results," Dec. 27, 1955; Oncley memo, "Comments—JL Oncley," undated; Williams to Fuhr, "Conferences Held by the Biophysics and Biophysical Chemis-

try Study Section," undated; Sutherland, "Interim Conclusions of a Member (GBBM Sutherland) of the Study Group," undated; all in FOS carton 21, folder 6.

119. Schmitt memo to study section, Mar. 23, 1956; FOS carton 21, folder 37. "Biophysics and Biophysical Chemistry Study Section, Minutes of Meeting," Apr. 26–27, 1956, FOS carton 21, folder 18. Schmitt to Sinsheimer, May 10, 1956, FOS carton 21, folder 11.

120. Schmitt, *Search*, chap 12. Fuhr, "Progress Report to the National Advisory Health Council," Oct. 23, 1956, FOS carton 21, folder 4. Allen to Schmitt, Nov. 1, 1956, FOS carton 21, folder 9. On Bolt's ultrasound work, see Blume, chap 3.

121. [Schmitt], handwritten comment on Williams to Schmitt, Dec. 16, 1958, FOS carton 21, folder 3.

122. Anonymous [Bolt?], "Special Programming in Biophysical Science: A Status Report on Catalytic Activities," June 1957, FOS carton 21, folder 10. On the republication of Hill's paper, see A. G. Lewey to Schmitt, Nov. 1, 1956, FOS carton 21, folder 6; Schmitt to Hill, Feb. 12, 1957; both in FOS carton 21, folder 8. On objections to subsidizing a professional society, see Neurath to Bolt, Mar. 25, 1957, FOS carton 21, folder 8.

123. Bolt and M. D. Rosenberg to Study Section, Sept. 11, 1958, FOS carton 21, folder 3. Cf. Schmitt, *Search*, chap 12. On Schmitt's effort to bring about consensus among the "heavy brass" on the Boulder program, see Hodge interview.

124. [Bolt?], "Development of National Institutes of Health Extramural Programs for Training and Research in Physical Biology," Jan. 1958, FOS carton 20, folder 38.

125. Schmitt, "Molecular Biology," 5.

126. Ibid., 6.

127. This would seem to lend historical validity to Richard Burian's contention that molecular biology is a general style of life science, whereas molecular genetics is (today) a discipline characterized by that style.

128. This distinction between methodological and metaphysical reductionism by such a leading character as Schmitt calls into question the *kind* of reductionism supposed to be entailed in molecular biology by philosophers; see, e.g., Fuerst.

129. On Bolt's presentation before the Surgeon General on Mar. 6, 1959, see David Price, "RMD Executive Memorandum," Mar. 11, 1959, FOS carton 21, folder 4. [Fuhr?], "Biophysical Science and Its Relation to Medical Research: A Report to Congress," Apr. 1959, FOS carton 31, folder 26. Cf. Sen. Lister Hill, "Report of the Senate Committee on Appropriations," June 16, 1958, Calendar 1753, Report 1719 (Washington: General Printing Office, 1958).

130. K. S. Cole, E. C. Pollard, O. H. Schmitt, and S. A. Talbot, "National

Biophysics Conference. March 4, 1957. Report of Planning Committee"; "Objectives of a Biophysical Society," Mar. 1957 (quotation); "Minutes of First Business Session. National Biophysics Conference" and "Minutes of Second Business Session. National Biophysics Conference," Mar. 5, 1957; all in FOS carton 20, folder 32. Registrant list of Ohio meeting, FOS carton 21, folder 6.

131. I adopt an eclectic and intuitive approach to "discipline," justifiable given theoretical confusion surrounding the notion.

132. "Conference on the Status of Biophysics, Ann Arbor, Michigan."

133. Perhaps ironically, it seems that physical chemistry itself achieved disciplinary status at the turn of the century because structural chemists were slow to take interest in, and thus to absorb within the discipline of chemistry, the field's central problems and methods; see Dolby.

134. The postwar expansion of biochemistry is impressive even in the context of the general expansion of life science, and deserves close study. Kohler (*From Medical Chemistry*) shows how biochemistry only slowly grew as a basic science from a role as a medical school service department. Arguably, even today's biochemistry is medically oriented to the extent that the life processes of higher vertebrates are its predominant topic.

135. That Schmitt used much of the same discipline-organizing tactics here seems implicit in the structure of the resultant book; cf. Quarton et al.

Chapter 5

1. Boyer, chaps. 4–6; Kevles, *Physicists*, chaps. 20–21; Rasmussen, "Biophysics Bubble"; A. Smith, esp. chap 2; Strickland, *Scientists in Politics*; Yavendetti; Weart, esp. chap 6.

2. Delbrück; Warren, "Radioactivity." On Delbrück, see Kay, "Conceptual Models"; also Fischer and Lipson.

3. Warren, "Atomic Energy and Medicine." For further examples, see Rasmussen, "Biophysics Bubble."

4. See footnote in *Scientific Monthly* 64 (1947): 213.

5. See group portrait in Schmitt, *Search*, 197. Francis Schmitt later recalled a "lack of focus" in the discussion at this meeting, which no doubt revolved largely around Schrödinger's speculations in his bestselling *What Is Life*. See Schmitt memo to Biophysics and Biophysical Chemistry Study Section, Mar. 23, 1956, FOS carton 20, folder 37. On *What Is Life* and the reactions among physicists, see Keller, "Physics and the Emergence"; Olby, "Schrödinger's Problem"; Yoxen, "Where Does Schrödinger's 'What Is Life?' Belong."

6. Carlson; Kay, "Stanley's Crystallization"; Keller, "Physics and the Emergence"; Muller; Ravin.

7. Rasmussen, "Biophysics Bubble." On Cal Tech, see Kay, *Molecular Vision*.

8. On the "end-in-view," see Hickman, chap 3. sec. 6.

9. Stanley to Hugh Wolfe, July 12, 1949, WMS carton 16, folder "Federation of American Societies of Experimental Biology." On the Federation of American Scientists and related groups, see Strickland, *Scientists in Politics*; Hodes. On the loyalty oaths, see Gardner.

10. Almási, 169–70.

11. Loewy argues very cogently that such false "straightenings" of history are typical of participants' accounts.

12. Wyckoff, "Ultraviolet Microscopy."

13. Alfred Weissler, "Biophysics at the National Institutes of Health," [Sept. 1955], in "Conference on Status of Biophysics, Ann Arbor, Michigan," Sept. 16, 1955, FOS carton 21, folder 12; [Williams], "Three Dimensional Electron Microscopy," undated [probably early 1945], attached to Harold Cole to Williams, Apr. 3, 1945, black folder "C," RW carton 1; Wyckoff, "Reminiscences." See Williams and Wyckoff, "Thickness." Williams to V. M. Slipher, Sept. 30, 1945, RW carton 1, black folder "S" (quotation).

14. Williams to Lindsay Black, Sept. 24, 1946, RW carton 1, red folder "B." Philip Owen to Williams, Apr. 10, 1946; Williams to Fritz Linke, Apr. 18, 1946; Williams to Linke, May 29, 1947; all in RW carton 3, folder "Cancer Research." Williams to Hayden Nicholson, June 4, 1947, RW carton 1, red folder "H" (quotation).

15. Williams, "Progress Report" PHS Grant C444, June 8, 1948, RW carton 3, folder "Lymphomatosis." Williams to Cole, May 3, 1947, RW carton 1, red folder "Cole."

16. Press release, n.d., "The Phoenix Project: A Statement of Aims and Policies," in RW carton 1, red folder "P."

17. Williams, "Application for Grant," Oct. 4, 1948; Williams to Sawyer, Oct. 4, 1948; both in RW carton 3, folder "Phoenix." Sawyer, "Atomic Research in the Physical and Biological Sciences," June 12, 1949; [Williams], speech outline "Applications of Atomic Research," Feb. 21, 1950; both in RW carton 3, folder "Phoenix Project Speech." [Williams], index card speech outline "Alumni, Canton Ohio," Mar. 22, 1950, RW carton 3, folder "Phoenix." *Michigan, the Atom and Peace*, General Bulletin 48, Bureau of Alumni Relations, University of Michigan, Dec. 1949 (quotation). Minutes of General Committee of the Division of Biological Sciences, Mar. 4, 1950, RW carton 3, folder "Biological Research and Division." See Rasmussen, "Biophysics Bubble."

18. Williams and Wyckoff, "Shadow-Micrography of Virus Particles"; idem, "Shadow-Micrography of Tobacco Mosaic"; idem, "Shadowed Electron Micrographs of Bacteria"; idem, "Electron Shadow-Micrographs of Hemocyanin"; Williams et al.; Price et al.; Taylor and Williams.

19. Williams, "Electron Microscope in Biology," 212.

20. Gerald Oster to Williams, May 8, 1948, RW carton 1, folder "O" (red). Williams to Price, Feb. 10, 1949, and May 24, 1949, RW carton 1, folder "P" (red). Rhian et al.

21. Steere and Williams, "Simplified Method of Purifying"; Williams and Steere; Steere and Williams, "Observations of the Unit of Length."

22. Wyckoff, "Developing Bacteriophage."

23. See Kellenberger; here thin sections appear with a few X-ray style micrographs, but no shadowed preparations.

24. Backus and Williams, "Small Spherical Particles"; Gerould; Backus and Williams, "Use of Spraying."

25. Williams and Backus; Williams et al., "Macromolecular Weights"; Luria et al., "Counts of Bacteriophage Particles."

26. Carlton Schwerdt, interview with author, Palo Alto, California, Jan. 30, 1994. See Cohen.

27. Sproul to Stanley, Feb. 19, 1940; Stanley to Sproul, Mar. 1, 1940; Stanley to Sproul, Nov. 4, 1940; Sproul to Stanley, Mar. 18, 1946; Stanley to Sproul, June 25, 1946; Stanley to Sproul, Oct. 2, 1946; Stanley to Sproul, Nov. 15, 1946; Sproul to Stanley, Jan. 10, 1947; Sproul to Stanley, Jan. 20, 1947; Stanley to Sproul, Jan. 27, 1947; all in WMS carton 13, folder "Sproul, Robert." Max Lauffer, "Biophysics at the University of Pittsburgh," [Sept. 1955], in "Conference on Status of Biophysics, Ann Arbor, Michigan," Sept. 16, 1955, FOS carton 21, folder 12. See Cohen; Corner; Craeger; Rasmussen, "Biophysics Bubble."

28. Stanley to Sproul, Oct. 2, 1946; Stanley to Sproul, Mar. 4, 1947; Stanley to Sproul, July 11, 1947; Sproul to Stanley, Feb. 6, 1948; Sproul to Stanley, Mar. 3, 1948; all in WMS carton 13, folder "Sproul, Robert." See Craeger; Westwick.

29. Stanley to Sproul, Dec. 10, 1947; Sproul to Stanley, Dec. 29, 1947; WMS carton 13, folder "Sproul, Robert." Weaver to Stanley, Dec. 1, 1947; Stanley to Weaver, Jan. 8, 1948; both in WMS carton 14, folder "Weaver, Warren." Stanley to Harry Weaver, June 17, 1948; WMS carton 14, folder "Weaver, Harry." [Stanley], "A Brief Summary of the Grants-in-Aid and Some Accomplishments," Dec. 1951, WMS carton 24, folder "Written Reports." See Craeger; Rasmussen, "Biophysics Bubble."

30. Stanley to Hillier, May 31, 1949, and Stanley to Hillier, July 12, 1949; WMS carton 2, folder "May 1949." "Departmental Budget, 1949–50, Virus Laboratory—National Foundation for Infantile Paralysis Grant 39042," June 1949, WMS carton 2, folder "June, 1949." "Key Individuals—$37,900 Charged Against Five Year Grant of $239,500," Apr. 1, 1952, WMS carton 27, untitled folder. "1952–53 Salary Budget," "Virus Laboratory," "Biochemistry," and "National Foundation for Infantile Paralysis," Oct. 17, 1952; all in WMS carton 27, folder "MBVL—Reports and Other Information." [Stanley], "Explanatory Statement," May 31, 1954, WMS carton 27, folder "Opening Ceremonies,

News, Information." Cf. *American Men of Science*, 9th ed.; Rasmussen, "Biophysics Bubble."

31. "Memorandum on the Relations Existing Between the Department of Biochemistry of the College of Arts and Sciences, the Department of Plant Biochemistry of the College of Agriculture, and the Division of Biochemistry of the School of Medicine," Apr. 1952, WMS carton 27, folder "MBVL—Reports and Other Information." "Biochemistry and Virus Laboratory Open House," Oct. 1952, WMS carton 27, folder "Opening Ceremonies, News, Information." Untitled list of contractor costs, Dec. 26, 1951, WMS carton 27, folder "Virus Laboratory, Administrative."

32. "Biophysics at the University of California," [Dec. 1955], in "Conference on Status of Biophysics, Berkeley, California," Dec. 1, 1955, FOS carton 21, folder 16. [Stanley], "Explanatory Statement," May 31, 1954, WMS carton 27, folder "Opening Ceremonies, News, Information." "Biochemistry and Virus Laboratory Open House"; "Rockefeller Foundation Grant, January 1949–January 1952," Aug. 15, 1951; both in WMS carton 27, folder "Virus Laboratory—Grants."

33. [Stanley], "Semi-Annual Report of Progress" to NFIP, Jan. 11, 1950, WMS carton 2, folder "January 1950." [Stanley], "Progress Report" to NFIP, July 1952, WMS carton 24, folder "Written Reports." [Stanley], untitled NFIP progress report, Dec. 16, 1952, WMS carton 27, folder "Virus Laboratory—Grants." [Knight?], "Statement in Layman's Language of Research," Feb. 1954, WMS carton 27, untitled folder (quotation). Fraenkel-Conrat, "Protein Chemists."

34. [Stanley], "Explanatory Statement." Fraser to A. D. Hershey, Jan. 27, 1949 (quotation), and Fraser to Hayden Nicholson, Feb. 28, 1949; both in WMS carton 5, folder "Virus Laboratory Outgoing Correspondence January–June 1949."

35. See Stanley, "Abstract of Final Report to the Committee on Medical Research, OSRD," [1945], WMS carton 24, folder "Annual Research Reports." Judging by correspondence with Hillier, the Princeton Rockefeller labs did not yet have their microscope in early 1945, contrary to George Corner's dating of the purchase to 1944 (Corner, 404); see RCA micrographs dated Jan. 10, 1945, WMS carton 24, folder "CMR4 Vaccine," and Hillier to Stanley, Jan. 20, 1945, WMS carton 9, folder "Hillier, James." Electron microscopy first makes its reappearance in Stanley's annual report for 1946; "Report of Dr. Stanley for the Board of Scientific Directors," Feb. 1947, WMS carton 24, "Annual Research Reports." The Princeton microscope was evidently the cheaper RCA console model released in 1944, the lens specially modified by Hillier (at least according to Keith Porter, who had never seen it himself); Porter to R. E. Shank, Aug. 25, 1949, KPA carton 2, first box.

36. Stanley, "Application for Grant," Feb. 13, 1949, WMS carton 27, folder "Virus Laboratory—Grants."

37. Williams, "High-Resolution."

38. Williams, "Method of Freeze-Drying"; and idem, "Shapes and Sizes."

39. [Williams], NIH grant extension application "C-2245(S)," Oct. 22, 1954, WMS carton 27, untitled folder.

40. Williams to Gordon Ellis, Nov. 7, 1955, RW uncataloged carton 1, folder "E." Kallman to Porter and Palade, Aug. 30, 1954, KPA carton 5 [?], box "1952–1955, H–O." The lack of BVL contribution to the cell biology of viral infection during the 1950s is clear in the citations one finds even in the article on the topic in Stanley's own textbook on viruses; see Bang.

41. Bachrach and Schwerdt. Also see Schachman and Williams, 301.

42. Schwerdt et al.; Fenton. On the new culture methods, see Kevles, "Renato Dulbecco."

43. E.g., Kay, *Molecular Vision*, 269–82.

44. Dekker and Schachman. Delbrück to Stanley, May 5, 1953, WMS carton 7, folder "Delbrück, Max" (quotation). Fraser and Williams.

45. Tsugita et al.

46. Kuhn and Lakatos are, of course, associated with contemporary forms of the view that theories determine research programs. The leading role of theory in experimental fields has come under fire in recent years from many quarters. For an example of conserved method in immunology despite theoretical change, see Rasmussen, "Freund's Adjuvant."

47. See Commoner et al.; Rice et al.; R. Franklin; Schramm et al.

48. Fraenkel-Conrat,"Chemical Basis."

49. Fraenkel-Conrat and Williams; "Change from Mushrooms."

50. Fraenkel-Conrat, "Role of Nucleic Acid"; Fraenkel-Conrat et al., "Virus Reconstitution" (quotation); Fraenkel-Conrat et al., "Infectivity."

51. See Fraenkel-Conrat, "Chemical Basis," 452–53.

52. Williams, "Replication of Nucleic Acids," 240. The importance of the pneumococcal transforming factor and of phage for the development of molecular genetics is documented in general histories such as Judson, and Olby.

53. Schwerdt et al.

54. Stanley and Williams to Chancellor E. W. Strong, Nov. 7, 1963, WMS carton 15, folder "Biochemistry and Virus Laboratory Building." For other examples, see Rasmussen, "Biophysics Bubble."

55. Craeger.

56. [Stanley], provisional program "Second National Cancer Conference," [Apr. 1952?], WMS carton 18, folder "2nd National Cancer Conference." Stanley to Harry Weaver, Nov. 30, 1955, WMS carton 14, folder "Weaver, Harry." For the increase in cancer applications, see, e.g., [Stanley], "Application for Institutional Research Grant," May 11, 1955, WMS carton 27, folder "Virus Laboratory—Grants." This last application asks for an additional $64,000 supplement on renewal of the previous year's $90,000 grant from the American Cancer

Society, "Studies on the Interaction of Viruses and Cells in Tissue Cultures with Special Reference to Viruses Possessing Tumor Inciting or Tumor Regressing Properties."

57. Untitled song, WMS carton 29, folder "MBVL—Administrative."

58. [Stanley], "The Virus Etiology of Cancer," June 1956, WMS carton 24, folder "WMS—Manuscripts for Talks."

59. Kevles, "Renato Dulbecco."

60. For instance, see tables in Fraenkel-Conrat et al., "Infectivity."

Chapter 6

1. Merleau-Ponty, 159–90.

2. Rotman.

3. Lakoff; Johnson; also see E. Keller, "Paradox."

4. As will soon become evident, in all of this I am indebted to Patrick Heelans's study, *Space Perception*.

5. On "the Body" of the perceiver, see Heelan, "After Experiment."

6. "Console Model Microscope"; Spencer.

7. W. Davis.

8. Spencer; "Speaking of Pictures" 1942. Also see, e.g., "Smaller and Smaller"; G. Hawley, 11–13.

9. G. Hawley, 5; Hillier, "Story."

10. Spencer; "Speaking of Pictures" 1942 (here the hair is "an enormous cable 30 feet thick"); "Can You Identify These Objects?"

11. "Speaking of Pictures" 1945.

12. Editorial: "Electron Microscope."

13. Marton, "Alice."

14. "Speaking of Pictures" 1942; "Into the Invisible World."

15. "Into the Invisible World"; "Smaller and Smaller"; "Speaking of Pictures" 1942; "Electron Microscope Magnifies Invisible World."

16. W. Davis.

17. Ibid.

18. Spencer.

19. "Console Model Microscope."

20. It was typical of hygiene films in both the First and Second World Wars to cast women as the source of venereal infection; see Brandt; Colwell.

21. Spencer.

22. "Into the Invisible World."

23. Anonymous [Watson Davis?], "Weekly Science Page" Aug. 30, 1942, of the Science Service, Washington D.C.; MMB, box 14. "Console Model Microscope."

24. "Into the Invisible World." Marton mentioned the Siegfried line in

many writings around 1940, for instance in his first informal grant applications for Rockefeller Foundation support (see Chap. 1).

25. As noted in Chap. 1, publicity took on previously unheard of importance in the Second World War.

26. "Speaking of Pictures" 1942; "Console Model Microscope."

27. Habermas's "positive prejudices" count under this head, if the notion can be applied to apprehension of natural as well as social reality. On the later Husserl's cognate notion of the pre-given (*Vorgegeben*) aspect of the lifeworld, see Carr. For Heidegger's late concern with technology as (material) culture deeply prestructuring the world, see Heidegger; also Zimmerman; Margolis.

28. Ihde, *Technology and the Lifeworld*; cf. Suchman, 53–54. See Polanyi, chap. 1.

29. O. S. Duffendack to Clarence Yoakum, Feb. 10, 1943, RW carton 3, "Electron Microscope" folder.

30. Of the characters discussed in this work, Porter, Palade, Hall, and Williams by the mid-1950s all had "personal" electron microscopes that the other people in their labs were not to touch without special permission.

31. Fritiof Sjöstrand, interview with author, Apr. 1993.

32. Reisner, "Early History."

33. George Chapman, interview with J. Strick and author, Dec. 1993.

34. See Haraway; Stone.

35. Palade, Bonneville tape KC4B

36. Robinow to Mudd, Feb. 18, 1948, MMB carton 17, folder "R 1947–48."

37. [Williams], "Three Dimensional Electron Microscopy," [early 1945?], attached to Harold Cole to Williams, Apr. 3, 1945, RW carton 1, black folder "C."

38. Williams and Wyckoff, "Shadow-Micrography of Virus Particles."

39. Williams to Rubin Borasky, Nov. 30, 1959, NMAH, Borasky Papers, carton 3.

40. Bastide.

41. Recognized by Barnes et al. For a nice example of early insensitivity to this effect, see Jakus and Hall, "Electron Microscope Studies," in which the authors oriented consecutive figures (all on two facing pages), with apparent illumination from below right, above left, and above right.

42. For some thoughts on the conventions of cartography, see Wood.

43. Newberry, *EMSA and Its People*, "Yearbook Appendix," pp.42–43.

44. Hillier to Mudd, Nov. 18, 1948, MMB carton 18, folder "H, 1948–49." Williams to Fraser, Aug. 16, 1957, RW uncataloged carton 1, folder "F." See Benjamin.

45. Undated letter, Charles Gerould [of Dow Chemical] to "electron microscopists"; Williams to Gerould, Nov. 4, 1948; both in RW carton 1, red folder "E." Williams to H. L. Nixon, July 18, 1956, RW uncataloged carton 2, folder "N." Cf. Backus and Williams, "Small Spherical Particles"; "Smallest Measuring Sticks."

46. Newberry, *EMSA and Its People*, "Yearbook Appendix," 45.

47. Ibid., 45–46.

48. Cited in Bourdieu, *Logic of Practice*, 34.

49. I would maintain that orientation is the primary purpose of the magnification series and not, as Bastide has suggested, "restriction of interpretation," unless by this she simply means interpretation of the scale and location of the objects in the micrograph.

50. Porter, undated lecture, "The Unit of Life," KPN unlabeled carton, folder "Univ. of Penna—Institute of Humanistic Studies."

51. On human sight, see Jonas.

52. Heelan, *Space Perception*.

53. Karl Popper is, of course, famous for making this argument. However, this view is by no means peculiar to Popper or his specific followers, but underlies essentially all claims to universality for scientific knowledge—or even an improvement in validity of scientific knowledge over time in any absolute sense (i.e., validity not only for present *Homo sapiens*).

54. Hacking, *Representing and Intervening*.

55. At least according to Piaget; see *Construction*, chaps. 2–3, and *Origins*, chaps. 4–5. However, still more would be attributed to innate cognitive structures by rival schools such as Chomsky's, making the case for the biological basis of these epistemological maneuvers stronger, if anything.

56. Piaget and Garcia.

57. For some recent work in evolutionary epistemology, see Callebaut and Pinxten; Hahlweg and Hooker; and Rescher. Of course, evolutionary epistemology is dependent on a scientific theory itself.

58. Lakoff; M. Johnson.

59. Richards et al., "Squid."

60. Bonneville tape UCOLO5.

61. See Schmitt, "Festvortrag," and references therein.

62. Chapman and Hillier.

63. I describe the pluralistic epistemology of the electron microscopist elsewhere (Rasmussen, "Facts") in greater detail.

64. Schmitt, "Festvortrag," 6.

65. Schaefer and Harker; Keller and Geisler; Heidenreich and Peck.

66. Newberry, *EMSA and Its People*, "Yearbook Appendix," 35–40. See Heidenreich and Matheson.

67. Newberry, *EMSA and Its People*, 44, and "Yearbook Appendix," 42–43. See C. Burton et al.

68. Heidenreich.   69. Schaefer and Harker.

70. Heidenreich and Peck.   71. Hillier and Baker.

72. See, e.g., Heidenreich.

73. Claude, "Method of Replicas"; Palay and Claude.

74. Margolis. For further thoughts on the relation between pragmatism and

phenomenology, see Rorty, *Consequences of Pragmatism*, viii; also Ihde, *Consequences*, chap. 9.

75. Foucault, *Birth of the Clinic*, 169–70.

76. Rous, "Copy: X/15/51 Comment on Dr. Porter's Pictures," May 1952, KPA carton 5, box "1952–1955, P–W."

77. Blume, chap. 4.

78. Edge and Mulkay.

79. See Chap. 3. Also Rasmussen, "Facts."

80. Nicholas Jardine (*Scenes*, 98) suggested this same sense of déjà vu regarding the notion of scientific knowledge as based on common sense. Common sense is important in the present view as a manifestation of pre-understandings, both learned and innate, that can also be experienced as intuition.

81. Crease distinguishes performance from reception hermeneutics, as we might also do here. But this would serve no function for the present argument.

82. Ricoeur, 145.

Works Cited

The following abbreviations are used throughout the Works Cited:

BBA	*Biochimica et Biophysica Acta*
CSH	Cold Spring Harbor
ECR	*Experimental Cell Research*
HSPS	*Historical Studies in the Physical and Biological Sciences*
JAP	*Journal of Applied Physics*
JB	*Journal of Bacteriology*
JCB	*Journal of Cell Biology* (and before 1962 *Journal of Biochemical and Biophysical Cytology*)
JEM	*Journal of Experimental Medicine*
JHB	*Journal of the History of Biology*
OSRD	Office of Scientific Research and Development
PNAS	*Proceedings of the National Academy of Sciences, U.S.A.*
PSEBM	*Proceedings of the Society for Experimental Biology and Medicine*
SHPS	*Studies in the History and Philosophy of Science*
SUNY	State University of New York

Abir-Am, P. "Beyond Deterministic Sociology and Apologetic History: Reassessing the Impact of Research Policy upon New Scientific Disciplines." *Social Studies of Science* 14 (1984): 252–63.

——. "The Discourse of Physical Power and Biological Knowledge in the 1930s: A Reappraisal of the Rockefeller Foundation's 'Policy' in Molecular Biology." *Social Studies of Science* 12 (1982): 341–82.

Abramson, A. *The History of Television, 1880–1941*. London: McFarland, 1987.

Alexander, M. "Localisation of Enzymes in the Microbial Cell." *Bacteriological Reviews* 20 (1956): 67–93.

Allchin, D. "Resolving Disagreement in Science: The Ox-Phos Controversy, 1961–1977." Ph.D. diss., University of Chicago, 1991.

Almási, M. *The Philosophy of Appearances*. Trans. A. Vitányi. Dordrecht: Kluwer, 1989.

Altmann, R. *Die Elementarorganismen und ihre Bezeihungen zu den Zellen*. Leipzig: Veit, 1890.

Amann, K., and K. Knorr-Cetina. "The Fixation of Visual Evidence." In Lynch and Woolgar, 85–121.

Amsterdamska, O. "Stabilizing Instability: The Controversy over Cyclogenic Theories of Heredity During the Interwar Period." *JHB* 24 (1991): 191–222.

Anderson, T. F. "Electron Microscopy of Phages." In Cairns et al., 63–78.

———. "Some Personal Memories of Research." *Annual Review of Microbiology* 29 (1975): 1–18.

———. "Techniques for the Preservation of Three-Dimensional Structure in Pre-paring Specimens for the Electron Microscope." *Transactions of the New York Academy of Sciences* 2, no. 13 (1951): 130–34.

Anderson, T. F., and A. G. Richards. "An Electron Microscope Study of Some Structural Colors of Insects." *JAP* 13 (1942): 748–58.

Anderssen-Cedergren, E. "Ultrastructure of Motor Endplate and Sarcoplasmic Components of Mouse Skeletal Muscle Fiber." *Journal of Ultrastructure Research* supp. 1 (1959): 1–19.

Ashley, C., K. Porter, D. Philpott, and G. Hass. "Observations by Electron Microscopy on Contraction of Skeletal Myofibrils Induced with Adenosine-triphosphate." *JEM* 94 (1951): 9–19.

Astbury, W. "X-Ray and Electron Microscope Studies, and Their Cytological Significance, of the Recently-Discovered Muscle Proteins, Tropomyosin and Actin." *ECR* supp. 1 (1949): 234–46.

Bachelard, G. *The New Scientific Spirit*. Trans. A. Goldhammer. Boston: Beacon, 1984.

Bachrach, H., and C. Schwerdt. "Purification Studies on Lansing Poliomyelitis Virus: II. Analytical Electron Microscopic Identification of the Infectious Particle in Preparations of High Specific Infectivity." *Journal of Immunology* 72 (1954): 30–38.

Backus, R., and R. C. Williams. "Small Spherical Particles of Exceptionally Uniform Size." *JAP* 20 (1949): 224–25.

———. "The Use of Spraying Methods and of Volatile Suspending Media in the Preparation of Specimens for Electron Microscopy." *JAP* 21 (1950): 11–15.

Bang, F. B. "The Morphological Approach." In Burnet and Stanley, 3:63–110.

Bargmann, W., E. Knoop, and T. Schiebler. "Histologische, cytochemische, und elektronen-mikroskopische Untersuchungen am Nephron." *Zeitschrift für Zellforschung und mikroskopische Anatomie* 42 (1955): 386–422.

Barnes, R. B., C. J. Burton, and R. G. Scott. "Electron Microscopical Replica Techniques for the Study of Organic Surfaces." *JAP* 16 (1945): 730–39.

Barnouw, E. *Tube of Plenty: The Evolution of American Television.* New York: Oxford University Press, 1975.

Barnum, C. P., and R. A. Huseby. "Some Quantitative Analyses of the Particulate Fractions from Mouse Liver Cell Cytoplasm." *Archives of Biochemistry* 19 (1948): 17–23.

Bastide, F. "The Iconography of Scientific Texts: Principles of Analysis." In Lynch and Woolgar, 187–229.

Bear, R. A., F. O. Schmitt, and J. Z. Young. "The Ultrastructure of Nerve Axoplasm." *Proceedings of the Royal Society of London* B 123 (1937): 505–19.

Benjamin, W. "The Work of Art in the Age of Mechanical Reproduction." Trans. H. Zohn. In *Illuminations*, ed. H. Arendt, 219–53. New York: Harcourt, 1968.

Bennett, H. S. "Structure of Muscle Cells." In Oncley, 394–401.

Bennett, H. S., and K. R. Porter. "An Electron Microscope Study of Sectioned Breast Muscle of the Domestic Fowl." *American Journal of Anatomy* 93 (1953): 61–106.

Bensley, R., and N. Hoerr. "The Preparation and Properties of Mitochondria." *Anatomical Record* 60 (1934): 449–55.

Bilinkoff, J. *The Avila of Saint Theresa: Religious Reform in a Sixteenth-Century City.* Ithaca: Cornell University Press, 1989.

Bisset, K. A. *Cytology and Life-History of Bacteria.* Edinburgh: Livingstone, 1950.

——. "The Genetical Implications of Bacterial Cytology." *Cold Spring Harbor Symposia for Quantitative Biology* 16 (1951): 373–79.

——. "The Interpretation of Appearances in the Cytological Staining of Bacteria." *ECR* 3 (1952): 681–88.

——. "Spurious Mitotic Spindles and Fusion Tubes in Bacteria." *Nature* 169 (1952): 247–48.

Blume, S. *Insight and Industry.* Cambridge, Mass.: MIT Press, 1992.

Bonneville, M. A. "Profile: Keith Roberts Porter." *Ultrastructural Pathology* 4 (1983): 401–8.

Borges, J. L. *A Universal History of Infamy.* Trans. N. di Giovanni. New York: Dutton, 1972.

Bourdieu, P. *Distinction: A Social Critique of the Judgment of Taste.* Trans. R. Nice. Cambridge, Mass.: Harvard University Press, 1984.

——. *Homo Academicus.* Trans. P. Collier. Palo Alto, Calif.: Stanford University Press, 1988.

——. *The Logic of Practice.* Trans. R. Nice. Palo Alto, Calif.: Stanford University Press, 1990.

——. "The Specificity of the Scientific Field and the Social Conditions for the Progress of Reason." *Social Science Information* 6 (1975): 19–47.

Boyer, P. *By the Bomb's Early Light: American Thought and Culture at the Dawn of the Atomic Age.* New York: Pantheon, 1985.

Bracegirdle, B. *A History of Microtechnique.* 2d ed. Lincolnwood, Ill.: Science Heritage, 1986.

Brandt, A. *No Magic Bullet: A Social History of Venereal Disease in the United States Since 1880.* New York: Oxford University Press, 1985.

Bridgman, P. *The Logic of Modern Physics.* New York: Macmillan, 1927.

Buchanan, J. "Biochemistry at MIT." *Technology Review* 56 (Dec. 1953): 87–88.

Bud, R. *The Uses of Life.* Cambridge, Eng.: Cambridge University Press, 1993.

Bugos, G. "Managing Cooperative Research and Borderland Science in the National Research Council, 1922–1942." *HSPS* 20 (1989): 1–32.

Burian, R. "Technique, Task Definition, and the Transition from Genetics to Molecular Genetics: Aspects of the Work on Protein Synthesis in the Laboratories of J. Monod and P. Zamecnik." *JHB* 26 (1993): 387–407.

Burnet, F. M., and W. M. Stanley, eds. *The Viruses.* New York: Academic Press, 1959.

Burton, C. J., R. B. Barnes, and T. G. Rochow. "The Electron Microscope: Calibration and Use at Low Magnification." *Industrial and Engineering Chemistry* 34 (1942): 1429–36.

Burton, E. F., and W. H. Kohl. *The Electron Microscope.* New York: Reinhold, 1942.

Bush, V. *Science: The Endless Frontier.* Washington: OSRD, 1945.

Cairns, J., G. Stent, and J. Watson, eds. *Phage and the Origins of Molecular Biology.* Cold Spring Harbor, N.Y.: CSH Laboratory of Quantitative Biology, 1966.

Callebaut, W., and R. Pinxten, eds. *Evolutionary Epistemology: A Multiparadigm Program With a Complete Evolutionary Epistemology Bibliography.* Dordrecht: Reidel, 1987.

"Can You Identify These Objects?" *Time,* July 4, 1949, 73.

"Cancer Research Marches Ahead." *JB* 74 (1957) [no page number].

Carlson, E. A. *Genes, Radiation, and Society: The Life and Work of H. J. Muller.* Ithaca, N.Y.: Cornell University Press, 1981.

Carr, D. "The Life-World Revisited." In *Husserl's Phenomenology: A Textbook,* ed. J. N. Mohanty and W. McKenna, 291–308. Washington: University Press of America, 1989.

Cartwright, N. *Nature's Capacities and Their Measurement.* Oxford: Clarendon, 1989.

Chalmers, A. *Science and Its Fabrication.* Minneapolis: University of Minnesota Press, 1990.

———. *What Is This Thing Called Science?* 2d ed. St. Lucia, Qld.: University of Queensland Press, 1982.

Chambers, L. A., W. Henle, M. A. Lauffer, and T. F. Anderson. "Studies on the Nature of the Virus of Influenza: II. The Size of the Infectious Unit in Influenza A." *JEM* 77 (1943): 265–76.

Chang, H. "Circularity and Reliability in Measurement." *Perspectives on Science* 3 (1995): 153–72.

"Change from Mushrooms." *Punch*, Nov. 9, 1955, 533.

Chapman, G. "Electron Microscopy of Ultra-Thin Sections of Bacteria." Ph.D. diss., Princeton University, 1953.

Chapman, G., J. H. Hanks, and J. H. Wallace. "An Electron Microscope Study of the Disposition and Fine Structure of *Mycobacterium lepraemurium* in Mouse Spleen." *JB* 77 (1959): 205–11.

Chapman, G., and J. Hillier. "Electron Microscopy of Ultra-Thin Sections of Bacteria: I. Cellular Division in *Bacillus cereus*." *JB* 66 (1953): 362–73.

Chapman, G., and K. P. Zworykin. "Study of Germinating *Bacillus cereus* Spores Employing Television Microscopy of Living Cells and Electron Microscopy of Ultrathin Sections." *JB* 74 (1957): 126–32.

Clarke, A. "Embryology and the Rise of American Reproductive Sciences, Circa 1910–1940." In *The Expansion of American Biology*, ed. K. Benson, J. Maienschein, and R. Rainger, 107–32. New Brunswick, N.J.: Rutgers University Press, 1991.

Claude, A. "The Coming of Age of the Cell." *Science* 189 (1975): 433.

——. "Electron Microscope Studies of Cells by the Method of Replicas." *JEM* 89 (1949): 425–30.

——. "Studies on Cells: Morphology, Chemical Constitution, and Distribution of Biochemical Function." *Harvey Lectures* 43 (1950): 121–64.

Claude, A., and E. Fullam. "An Electron Microscope Study of Isolated Mitochondria." *JEM* 81 (1945): 51–62.

——. "The Preparation of Sections of Guinea Pig Liver for Electron Microscopy." *JEM* 83 (1946): 499–503.

Claude, A., K. R. Porter, and E. Pickels. "Electron Microscope Study of Chicken Tumor Cells." *Cancer Research* 7 (1947): 421–30.

Cohen, S. S. "Some Contributions of the Princeton Laboratory of the Rockefeller Institute on Proteins, Viruses, and Nucleic Acids." *Annals of the New York Academy of Sciences* 325 (1979): 303–6.

Collins, H. *Changing Order: Replication and Induction in Scientific Practice*. London: Sage, 1985.

——. "A Strong Confirmation of the Experimenter's Regress." *Studies in the History and Philosophy of Science* 25 (1994): 493–503.

Colwell, S. "The End of the Road: Gender, the Dissemination of Knowledge, and the American Campaign Against Venereal Disease During World War I." *Camera Obscura* 29 (1994): 90–129.

Commoner, B., M. Yamada, S. D. Rosenberg, T.-Y. Wang, and E. Basler Jr. "The Proteins Synthesized in Tissue Infected with Tobacco Mosaic Virus." *Science* 118 (1953): 529–34.

"Console Model Microscope." *Newsweek*, May 15, 1944, 74–76.

Corner, G. *A History of the Rockefeller Institute, 1901–1953.* New York: Rockefeller Institute Press, 1964.

Coslett, V. E. "Random Recollections of the Early Days." In Hawkes, 23–61.

Cota-Robles, E., A. G. Marr, and E. H. Nilson. "Submicroscopic Particles in Extracts of *Azotobacter agilis.*" *JB* 75 (1958): 243–52.

Craeger, A. "Stanley's Dream of a Freestanding Biochemistry Department." *JHB* 29 (1996): 331–60.

Crease, R. *The Play of Nature: Experimentation as Performance.* Bloomington: Indiana University Press, 1993.

Culp, S. "Defending Robustness: The Bacterial Mesosome as a Test Case." *Proceedings of the Philosophy of Science Association* 1 (1994): 46–57.

Davis, J., L. Winterscheid, P. E. Hartman, and S. A. Mudd. "A Cytological Investigation of the Mitochondria of Three Strains of Salmonella Typhosa." *Journal of Histochemistry and Cytochemistry* 1 (1953): 123–37.

Davis, W. "30,000 Times Life Size!" *Reader's Digest* 37 (1940): 13–16.

De Duve, C., and H. Beaufay. "A Short History of Tissue Fractionation." *JCB* 91 (1981): 293s–99s.

Dekker, C. A., and H. K. Schachman. "On the Macromolecular Structure of Deoxyribonucleic Acid: An Interrupted Two-Strand Model." *PNAS* 40 (1954): 894–909.

DeLamater, E. "Evidence for the Occurrence of True Mitosis in Bacteria." *Science* 113 (1951): 477.

———. "A New Cytological Basis for Bacterial Genetics." *Cold Spring Harbor Symposia for Quantitative Biology* 16 (1951): 381–412.

———. "Preliminary Observations on the Occurrence of Mitosis in *Caryophanon latum.*" *Mycologia* 44 (1952): 203–6.

DeLamater, E., and M. E. Hunter. "Preliminary Report of True Mitosis in the Vegetative Cell of *Bacillus megatherium.*" *American Journal of Botany* 38 (1951): 659–62.

DeLamater, E., M. E. Hunter, and S. A. Mudd. "Current Status of the Bacterial Nucleus." *ECR* supp. 2 (1952): 319–43.

DeLamater, E., and S. A. Mudd. "The Occurrence of Mitosis in the Vegetative Phase of *Bacillus megatherium.*" *ECR* 2 (1951): 499–512.

DeLamater, E., and M. Woodburn. "Evidence for the Occurrence of Mitosis in the Micrococci." *JB* 64 (1952): 793–803.

Delbrück, M. "Experiments with Bacterial Viruses." *Harvey Lectures* 41 (1946): 161–87.

De Robertis, E. "An Electron Microscope Analysis of Nerves Infected with the B Virus." *JEM* 90 (1949): 291–96.

De Robertis, E., and F. O. Schmitt. "An Electron Microscope Analysis of Certain Nerve Axon Constituents." *Journal of Cellular and Comparative Physiology* 31 (1948): 1–23.

———. "An Electron Microscope Study of Nerves Infected with Human Polio-myelitis Virus." *JEM* 90 (1949): 283–90.

De Solla Price, D. J. "Of Sealing Wax and String: A Philosophy of the Experimenter's Craft and Its Role in the Genesis of High Technology." In *Little Science, Big Science . . . and Beyond*, 237–53. New York: Columbia University Press, 1986.

Dewey, J. *Experience and Nature*. 2d ed. Chicago: Open Court, 1929.

"Discussion." *Journal of Histochemistry and Cytochemistry* 1 (1953): 265–75.

Dolby, R. G. "The Case of Physical Chemistry." In *Perspectives on the Emergence of Scientific Disciplines*, ed. G. Lemaine, R. Macleod, M. Mulkay, and P. Weingart, 63–73. The Hague: Mouton, 1976.

"The Dorrance Building." *Technology Review* 56 (Dec. 1953): 81–84.

Draper, M., and A. Hodge. "Sub-Microscopic Localization of Minerals in Skeletal Muscle by Internal 'Micro-Incineration' Within the Electron Microscope." *Nature* 163 (1949): 576–77.

Dupré, J. *The Disorder of Things: Metaphysical Foundations of the Disunity of Science*. Cambridge, Mass.: Harvard University Press, 1993.

Dupree, A. H. "The Great Instauration of 1940: The Organization of Scientific Research for War." In *The Twentieth-Century Sciences: Studies in the Biography of Ideas*, ed. Gerald Holton, 443–67. New York: Norton, 1972.

Eckholm, R., and F. Sjöstrand. "The Ultrastructural Organization of the Mouse Thyroid Gland." *Journal of Ultrastructure Research* 1 (1957): 178–99.

———. "The Ultrastructure of the Thyroid Gland of the Mouse." In *Stockholm Conference*, 171–73.

Edge, D. O., and Mulkay, M. *Astronomy Transformed: The Emergence of Radio Astronomy in Britain*. New York: Wiley, 1976.

Editorial: "Electron Microscope." *New York Times*, Apr. 13, 1942, sec. 4, 8.

"Electron Microscope Magnifies Invisible World 25,000 Times." *Life*, Apr. 29, 1940, 54.

Ernster, L., and O. Lindberg. "Animal Mitochondria." *Annual Review of Physiology* 20 (1958): 13–42.

Ernster, L., and G. Schatz. "Mitochondria: A Historical Review." *JCB* 91 (1981): 227s–55s.

Farrant, J., and E. Mercer. "Studies on the Structure of Muscle: II. Arthropod Muscle." *ECR* 3 (1952): 553–63.

Federal Communications Commission. *Sixth Annual Report*. Washington: Government Printing Office, 1940.

———. *Thirteenth Annual Report*. Washington: Government Printing Office, 1947.

Fenton, J. "Polio Virus Definitely Identified and Photographed for First Time." *New York Times*, Nov. 12, 1953, sec. 1, 1.

Ferguson, T. "Industrial Conflict and the Coming of the New Deal: The Triumph of Multinational Liberalism in America." In *The Rise and Fall of the*

New Deal Order, 1930–1980, ed. S. Fraser and G. Gerstle, 3–31. Princeton, N.J.: Princeton University Press, 1989.

Feyerabend, P. K. *Against Method*. Rev. ed. London: Verso, 1988.

Finean, J. B., F. Sjöstrand, and E. Steinmann. "Submicroscopic Organisation of Some Layered Lipoprotein Structures (Nerve Myelin, Retinal Rods, and Chloroplasts)." *ECR* 5 (1953): 557–59.

Fischer, E. P., and C. Lipson. *Thinking About Science: Max Delbrück and the Origins of Molecular Biology*. New York: Norton, 1988.

Fitz-James, P. "Participation of the Cytoplasmic Membrane in the Growth and Spore Formation of Bacilli." *JCB* 8 (1960): 507–28.

Fleck, L. *Genesis and Development of a Scientific Fact*. Trans. F. Bradley and T. J. Trenn. Chicago: University of Chicago Press, 1979.

Fleming, D. "Émigré Physicists and the Biological Revolution." *Perspectives in American History* 2 (1968): 152–89.

Foreman, P. "Inventing the Maser in Postwar America." In Van Helden and Hankins, 105–34.

Foucault, M. *The Birth of the Clinic: An Archaeology of Medical Perception*. Trans. A. Sheridan. New York: Vintage, 1975.

———. *Discipline and Punish: The Birth of the Prison*. Trans. A. Sheridan. New York: Pantheon, 1978.

Fraenkel-Conrat, H. "The Chemical Basis of the Infectivity of Tobacco Mosaic Virus and Other Plant Viruses." In Burnet and Stanley, 1:429–57.

———. "Protein Chemists Encounter Viruses." *Annals of the New York Academy of Sciences* 325 (1979): 309–18.

———. "The Role of Nucleic Acid in the Reconstitution of Active Tobacco Mosaic Virus." *Journal of the American Chemical Society* 78 (1956): 882–92.

Fraenkel-Conrat, H., B. Singer, and R. C. Williams. "The Infectivity of Viral Nucleic Acid." *BBA* 25 (1957): 87–96.

———. "Virus Reconstitution: II. Combination of Protein and Nucleic Acid from Different Strains." *BBA* 24 (1957): 540–48.

Fraenkel-Conrat, H., and R. C. Williams. "Reconstitution of Active Tobacco Mosaic Virus from Its Inactive Protein and Nucleic Acid Components." *PNAS* 41 (1955): 690–98.

Frank, R. "Instruments, Nerve Action, and the All-or-None Principle." *Osiris* 9 (1994): 208–35.

Franklin, A. *Experiment, Right or Wrong*. Cambridge, Eng.: Cambridge University Press, 1990.

———. "How to Avoid the Experimenter's Regress." *Studies in the History and Philosophy of Science* 25 (1994): 463–91.

———. *The Neglect of Experiment*. Cambridge, Eng.: Cambridge University Press, 1986.

Franklin, R. "Structure of Tobacco Mosaic Virus." *Nature* 175 (1955): 379–81.

Fraser, D., and R. C. Williams. "Electron Microscopy of the Nucleic Acid Released from Individual Bacteriophage Particles." *PNAS* 39 (1953): 750–52.

Fuerst, J. "The Role of Reductionism in the Development of Molecular Biology: Peripheral or Central?" *Social Studies of Science* 12 (1982): 241–78.

Fujimura, J. "Constructing 'Do-able' Problems in Cancer Research: Articulating Alignment." *Social Studies of Science* 17 (1987): 257–93.

Fussell, P. *Wartime.* New York: Oxford University Press, 1989.

Gabor, D. *The Electron Microscope.* New York: Chemical Publishing, 1948.

Galison, P. "History, Philosophy, and the Central Metaphor." *Science in Context* 2 (1988): 197–212.

———. *How Experiments End.* Chicago: University of Chicago Press, 1987.

———. *Image and Logic: A Material Culture of Microphysics.* Chicago: University of Chicago Press, 1997.

Gardner, D. P. *The California Oath Controversy.* Berkeley: University of California Press, 1967.

Geison, G., and F. L. Holmes, eds. *Research Schools. Osiris* 8 (1993).

Geren, B. "The Formation from the Schwann Cell Surface of Myelin in the Peripheral Nerves of Chick Embryos." *ECR* 7 (1954): 558–62.

Geren, B., and D. McCulloch. "Development and Use of the Minot Rotary Microtome for Thin Sectioning." *ECR* 2 (1951): 97–102.

Geren, B., and J. Raskind. "Development of the Fine Structure of the Myelin Sheath in Sciatic Nerves of Chick Embryos." *PNAS* 39 (1953): 880–84.

Geren, B., and F. O. Schmitt. "The Structure of the Schwann Cell and Its Relation to the Axon in Certain Invertebrate Nerve Fibers." *PNAS* 40 (1954): 863–71.

Gerould, C. "Comments on the Use of Latex Spheres as Size Standards in Electron Microscopy." *JAP* 21 (1950): 183–84.

Gerson, E. "Scientific Work and Social Worlds." *Knowledge: Creation, Diffusion, Utilization* 4 (1983): 357–77.

Gessler, A., and E. Fullam. "Sectioning for the Electron Microscope Accomplished by the High Speed Microtome." *American Journal of Anatomy* 78 (1946): 245–80.

Giere, R. N. *Explaining Science: A Cognitive Approach.* Chicago: University of Chicago Press, 1988.

Glauert, A. M., and D. A. Hopwood. "A Membranous Component of the Cytoplasm in *Streptomyces coelicolor.*" *JCB* 6 (1959): 515–16.

Gold, M. *A Conspiracy of Cells.* Albany: SUNY Press, 1986.

Green, R. H., T. F. Anderson, and J. E. Smadel. "Morphological Structure of the Virus of Vaccinia." *JEM* 75 (1942): 651–57.

Grubb, D., and A. Keller. "Beam-Induced Radiation Damage in Polymers and Its Effect on the Image Formed in the Electron Microscope." *Proceedings of the Fifth European Conference of Electron Microscopy* (1972): 554–60.

Habermas, J. *The Theory of Communicative Action.* Vol. 1. Boston: Beacon Press, 1983.

Hacking, I. *Representing and Intervening.* Cambridge, Eng.: Cambridge University Press, 1983.

———. "The Self-Vindication of the Laboratory Sciences." In *Science as Culture and Practice*, ed. A. Pickering, 29–64. Chicago: University of Chicago Press, 1992.

Hadley, P. "The Twort-D'Herelle Phenomenon: A Critical Review and Presentation of a New Conception (Homogamic Theory) of Bacteriophage Action." *Journal of Infectious Diseases* 42 (1928): 263–434.

Hahlweg, K., and C. A. Hooker, eds. *Issues in Evolutionary Epistemology.* Albany: SUNY Press, 1989.

Hall, C. *Introduction to Electron Microscopy.* New York: McGraw-Hill, 1953.

———. "Recollections from the Early Years: Canada-USA." In Hawkes, 275–96.

Hall, C., M. Jakus, and F. O. Schmitt. "An Investigation of Cross Striation and Myosin Filaments in Muscle." *Biological Bulletin* 90 (1946): 32–50.

———. "The Structure of Certain Muscle Fibrils as Revealed by the Use of Electron Stains." *JAP* 16 (1945): 459–65.

Hanson, E. J. "Changes in the Cross-Striation of Myofibrils During Contraction Induced by Adenosine Triphosphate." *Nature* 169 (1952): 530–33.

Hanson, E. J., and H. Huxley. "The Structural Basis of Contraction in Striated Muscle." *Symposia of the Society for Experimental Biology and Medicine* 9 (1955): 228–63.

———. "Structural Basis of the Cross-Striations in Muscle." *Nature* 172 (1953): 530–32.

———. "Structural Changes in Striated Muscle During Contraction and Stretch." In *Proceedings of the Third International Conference*, 576–80.

Haraway, D. J. *Simians, Cyborgs, and Women: The Reinvention of Nature.* New York: Routledge, 1991.

Hartman, P. E., J. I. Payne, and S. A. Mudd. "Cytological Analysis of Ultraviolet Irradiated *Escherichia coli*: I. Cytology of Lysogenic *E. coli* and a Non-Lysogenic Derivative." *JB* 70 (1955): 531–39.

Harvey, A. *Science at the Bedside: Clinical Research in American Medicine, 1905–1945.* Baltimore: Johns Hopkins University Press, 1981.

Harvey, E. B., and T. F. Anderson. "The Spermatozoon and Fertilization Membrane of *Arbacia punctulata* as Shown by the Electron Microscope." *Biological Bulletin* 85 (1943): 151–56.

Hawkes, P. W., ed. *Beginnings of Electron Microscopy. Advances in Electronics and Electron Physics* supp. 16 (1985).

Hawley, E. *The Great War and the Search for a Modern Order: A History of the American People and Their Institutions, 1917–1933.* New York: Saint Martin's Press, 1979.

Hawley, G. *Seeing the Invisible: The Story of the Electron Microscope.* New York: Knopf, 1946.

Heelan, P. "After Experiment: Realism and Research." *American Philosophical Quarterly* 26 (1989): 297–307.

——. *Space Perception and the Philosophy of Science.* Berkeley: University of California Press, 1983.

Heidegger, M. "The Question Concerning Technology." In *The Question Concerning Technology and Other Essays,* ed. and trans. W. Lovitt, 3–35. New York: Garland, 1977.

Heidenreich, R. D. "Interpretation of Electron Micrographs of Silica Surface Replicas." *JAP* 14 (1943): 312–20.

Heidenreich, R. D., and L. A. Matheson. "Electron Microscopic Determination of Surface Elevations and Orientations." *JAP* 15 (1944): 423–35.

Heidenreich, R. D., and V. G. Peck. "Fine Structure of Metallic Surfaces with the Electron Microscope." *JAP* 14 (1943): 23–29.

Heilbron, J., and R. W. Seidel. *Lawrence and His Laboratory: A History of the Lawrence Berkeley Laboratory.* Vol. 1. Berkeley: University of California Press, 1989.

Hessen, B. "The Social and Economic Roots of Newton's Principia." In *Science at the Crossroads,* ed. H. Dingle. London: Kniga, 1933.

Hevley, B. "Stanford's Supervoltage X-ray Tube." *Osiris* 9 (1994): 85–100.

Hickman, L. *John Dewey's Pragmatic Technology.* Bloomington: Indiana University Press, 1990.

Hillier, J. "Some Remarks on the Image Contrast in Electron Microscopy and the Two-Component Objective." *JB* 57 (1949): 313–17.

——. "The Story of the Electron Microscope." In Weaver, ed., 78–81.

Hillier, J., and R. F. Baker. "The Mounting of Bacteria for Electron Microscope Examination." *JB* 52 (1946): 411–16.

Hillier, J., G. Knaysi, and R. F. Baker, "New Preparation Techniques for the Electron Microscopy of Bacteria." *JB* 56 (1948): 569–76.

Hillier, J., S. A. Mudd, and A. G. Smith. "Internal Structure and Nuclei in Cells of *Escherichia coli* as Shown by Improved Electron Microscopic Technique." *JB* 57 (1949): 319–38.

Hillman, H. "Artefacts in Electron Microscopy and the Consequences for Biological and Medical Research." *Medical Hypotheses* 6 (1980): 233 44.

——. "Towards a Classification of Evidence." *Acta Biotheoretica* 25 (1976): 153–62.

Hillman, H., and P. Sartory. *The Living Cell.* Chichester: Packard, 1980.

Hodes, E. "Precedents for Social Responsibility Among Scientists: The American Association of Scientific Workers and the Federation of American Scientists, 1938–1948." Ph.D. diss., University of California, Santa Barbara, 1982.

Hodge, A. "Electron Microscopic Studies of Insect Flight Muscle." In *Proceedings of the Third International Conference,* 572–76.

——. "Fibrous Proteins of Muscle." In Oncley, 409–25.

——. "Studies on the Structure of Muscle: III. Phase Contrast and Electron Microscopy of Dipteran Flight Muscle." *JCB* 1 (1955): 361–80.

Hodge, A., H. E. Huxley, and D. Spiro. "Electron Microscope Studies on Ultrathin Sections of Muscle." *JEM* 99 (1954): 201–6.

——. "A Simple New Microtome for Ultrathin Sectioning." *Journal of Histochemistry and Cytochemistry* 2 (1954): 54–61.

Hogeboom, G., W. Schneider, and G. E. Palade. "Cytochemical Studies of Mammalian Tissues: I. Isolation of Intact Mitochondria from Rat Liver." *Journal of Biological Chemistry* 172 (1948): 619–35.

——. "The Isolation of Morphologically Intact Mitochondria from Rat Liver." *PSEBM* 65 (1947): 320–21.

Hook, S. *John Dewey: An Intellectual Portrait*. Amherst, N.Y.: Prometheus, 1995.

Hopwood, D., and A. Glauert. "A Fine Structure of *Streptomyces coelicolor*." *JCB* 8 (1960): 267–78.

Hughes, T. "Technological Momentum." In Smith and Marx, 101–13.

Husserl, E. *The Crisis of European Sciences and Transcendental Phenomenology*. Trans. D. Carr. Evanston, Ill.: Northwestern University Press, 1970.

Huxley, A. "Looking Back on Muscle." In *The Pursuit of Nature*, ed. A. L. Hodgkin, 23–64. Cambridge, Eng.: Cambridge University Press, 1977.

——. *Reflections on Muscle*. Princeton, N.J.: Princeton University Press, 1980.

Huxley, A., and R. Niedergerke. "Structural Changes in Muscle During Contraction." *Nature* 173 (1954): 971–73.

Huxley, H. "The Double Array of Filaments in Cross-Striated Muscle." *JCB* 3 (1957): 631–46.

——. "Electron Microscopic Studies of the Organization of the Filaments in Striated Muscle." *BBA* 12 (1953): 387–94.

——. "X-Ray Analysis and the Problem of Muscle." *Proceedings of the Royal Society* B 141 (1953): 59–62.

Huxley H., and E. J. Hanson. "Changes in the Cross-Striations of Muscle During Contraction and Stretch and Their Structural Interpretation." *Nature* 173 (1954): 973–76.

——. "Preliminary Observations on the Structure of Insect Flight Muscle." In *Stockholm Conference*, 202–4.

Ihde, D. *Consequences of Phenomenology*. Albany: SUNY Press, 1986.

——. *Technology and the Lifeworld*. Bloomington: Indiana University Press, 1990.

"Into the Invisible World." *Newsweek*, Feb. 12, 1945, 81–82.

Jakus, M. A. "The Structure and Properties of the Trichocysts of Paramecium." *Journal of Experimental Zoology* 100 (1945): 457–80.

Jakus, M. A., and C. E. Hall. "Electron Microscope Studies on Muscle." *ECR* supp. 1 (1949): 262–66.

——. "Studies of Actin and Myosin." *Journal of Biological Chemistry* 167 (1947): 705–14.

James, W. *Pragmatism*. Buffalo: Prometheus, 1991.

Jardine, N. *The Fortunes of Inquiry*. Oxford: Clarendon Press, 1986.

———. *The Scenes of Inquiry*. Oxford: Clarendon Press, 1991.

Johnson, D., and M. Cantino. "Artifacts of Analysis in Biological Electron Microscopy." In *Artifacts in Biological Electron Microscopy*, ed. R. Crang and K. Klomparens, 219–27. New York: Plenum, 1988.

Johnson, M. *The Body in the Mind: The Bodily Basis of Meaning, Imagination, and Reason*. Chicago: University of Chicago Press, 1987.

Jonas, H. "The Nobility of Sight: A Study in the Phenomenology of the Senses." In *The Phenomenon of Life*, 135–56. Chicago: University of Chicago Press, 1982.

Judson, H. *The Eighth Day of Creation*. New York: Simon & Schuster, 1979.

Kargon, R. "Temple to Science: Cooperative Research and the Birth of the California Institute of Technology." *HSPS* 8 (1977): 3–31.

Kay, L. E. "Conceptual Models and Analytical Tools: The Biology of Physicist Max Delbrück." *JHB* 18 (1985): 207–46.

———. "Cooperative Individualism and the Growth of Molecular Biology at Caltech 1928–1953." Ph.D. diss., Johns Hopkins University, 1987.

———. "Molecular Biology and Pauling's Immunochemistry: A Neglected Dimension." *History and Philosophy of Life Science* 11 (1989): 211–19.

———. *The Molecular Vision of Life*. New York: Oxford University Press, 1993.

———. "Selling Pure Science in Wartime: The Biochemical Genetics of G. W. Beadle." *JHB* 22 (1989): 73–101.

———. "W. M. Stanley's Crystallization of the Tobacco Mosaic Virus, 1930–1940." *Isis* 77 (1986): 450–72.

Kellenberger, E. "Vegetative Bacteriophage and the Maturation of the Virus Particle." *Advances in Virus Research* 8 (1961): 2–61.

Keller, E. F. "The Paradox of Scientific Subjectivity." *Annals of Scholarship* 9 (1992): 135–53.

———. "Physics and the Emergence of Molecular Biology: A History of Cognitive and Political Synergy." *JHB* 23 (1990): 389–409.

———. *Refiguring Life: Metaphors of Twentieth-Century Biology*. New York: Columbia University Press, 1995.

Keller, F., and A. H. Geisler. "Extending Microscopic Examination of Metals." *JAP* 15 (1944): 696–704.

Kettering, C. F. "The Future of Science." *Science* 104 (1946): 609–14.

Kevles, D. "George Ellery Hale, the First World War, and the Advancement of Science in America." *Isis* 59 (1968): 427–37.

———. *The Physicists*. New York: Knopf, 1978.

———. "Renato Dulbecco and the New Animal Virology: Medicine, Methods, and Molecules." *JHB* 26 (1993): 409–42.

Killian, J., Jr. *The Education of a College President*. Cambridge, Mass.: MIT Press, 1985.

Knaysi, G., and R. F. Baker. "Demonstration, with the Electron Microscope, of a Nucleus in *Bacillus mycoides* Grown in a Nitrogen-Free Medium." *JB* 53 (1947): 539–53.

Knaysi, G., and S. A. Mudd. "The Internal Structure of Certain Bacteria as Revealed by the Electron Microscope: A Contribution to the Study of the Bacterial Nucleus." *JB* 45 (1943): 349–59.

Knorr-Cetina, K., and M. Mulkay, eds. *Science Observed: Perspectives on the Social Study of Science*. London: Sage, 1983.

Kohler, R. "Drosophila: A Life in the Laboratory." *JHB* 26 (1993): 281–310.

——. *From Medical Chemistry to Biochemistry*. Cambridge, Eng.: Cambridge University Press, 1982.

——. *Lords of the Fly*. Chicago: University of Chicago Press, 1994.

——. *Partners in Science: Foundations and Natural Scientists, 1900–1945*. Chicago: University of Chicago Press, 1991.

Kosso, P. "Dimensions of Observability." *British Journal of the Philosophy of Science* 39 (1988): 449–67.

——. *Observability and Observation in Physical Science*. Dordrecht: Kluwer, 1989.

——. "Science and Objectivity." *Journal of Philosophy* 86 (1989): 245–57.

Kuhn, T. *The Structure of Scientific Revolutions*. Chicago: University of Chicago Press, 1962.

Kunkle, G. "Technology in the Seamless Web: 'Success' and 'Failure' in the History of the Electron Microscope." *Technology and Culture* 36 (1994): 80–103.

Lakatos, I. "History of Science and Its Rational Reconstructions." In *Method and Appraisal in the Physical Sciences*, ed. C. Howson, 1–39. Cambridge, Eng.: Cambridge University Press, 1976.

Lakoff, G. *Women, Fire, and Dangerous Things: What Our Categories Reveal About the Mind*. Chicago: University of Chicago Press, 1987.

Latour, B. "Drawing Things Together." In Lynch and Woolgar, 19–68.

——. *Science in Action*. Cambridge, Mass.: Harvard University Press, 1987.

Latour, B., and S. Woolgar. *Laboratory Life: The Social Construction of Scientific Facts*. London: Sage, 1979.

Latta, H., and J. Hartman. "Use of a Glass Edge in Thin Sectioning for Electron Microscopy." *PSEBM* 74 (1950): 436–39.

Laudan, L. "The Pseudo-Science of Science." *Philosophy of the Social Sciences* 11 (1981): 173–98.

Law, J. "Theories and Methods in the Sociology of Science: An Interpretative Approach." In *Perspectives on the Emergence of Scientific Disciplines*, ed. G. Lemaine, R. Macleod, M. Mulkay, and P. Weingart, 221–31. The Hague: Mouton, 1976.

Lazarow, A., and E. G. Barron. "The Oxygen Uptake of Mitochondria and Other Cell Fragments." *Anatomical Record* 79, supp. 2 (1941): 41–42.

Lehninger, A. *The Mitochondrion*. New York: Benjamin, 1965.

Lenoir, T. "The Discipline of Nature and the Nature of Disciplines." In *Knowledges: Historical and Critical Studies in Disciplinarity*, ed. E. Messer-Davidow, D. Sylvan, and D. Shumway, 70–102. Charlottesville: University Press of Virginia, 1993.

Lenoir, T., and C. Lécuyer. "Instrument Makers and Discipline Builders: The Case of Nuclear Magnetic Resonance." *Perspectives on Science* 3 (1995): 276–345.

Lewis, S. *Arrowsmith*. New York: Grosset & Dunlap, 1925.

Loewy, I. "Variances in Meaning in Discovery Accounts: The Case of Contemporary Biology." *HSPS* 21 (1990): 87–121.

Loring, H. S., L. Marton, and C. E. Schwerdt. "Electron Microscopy of a Purified Lansing Virus." *PSEBM* 62 (1946): 291–92.

Lukács, G. *History and Class Consciousness: Studies in Marxist Dialectics*. Trans. R. Livingstone. Cambridge, Mass.: MIT Press, 1971.

Luria, S. E., M. Delbrück, and T. F. Anderson. "Electron Microscope Studies of Bacterial Viruses," *JB* 46 (1943): 57–77.

Luria, S. E., R. C. Williams, and R. Backus. "Electron Micrographic Counts of Bacteriophage Particles." *JB* 61 (1951): 179–88.

Lynch, M. *Art and Artifact in a Laboratory Science*. London: Routledge, 1985.

——. "The Externalized Retina: Selection and Mathematization in the Visual Documentation of Objects in the Life Sciences." In Lynch and Woolgar, 153–86.

Lynch, M., and D. Edgerton. "Aesthetics and Digital Image Processing: Representational Craft in Contemporary Astronomy." In *Picturing Power*, ed. G. Fyfe and J. Law, 185–220. New York: Routledge, 1988.

Lynch, M., and S. Woolgar, eds. *Representation in Scientific Practice*. Cambridge, Mass.: MIT Press, 1990.

MacDonald, W. W. "Electron Microscopy in the United States." *Electronics* 23, no. 2 (1950): 66–69.

Maclean, R. A. , W. L. Sulzbacher, and S. A. Mudd. "*Micrococcus cryophilus*, Spec. Nov.: A Large Coccus Especially Suitable for Cytologic Study." *JB* 62 (1951): 723–28.

Margolis, J. "Pragmatism, Praxis, and the Technological." In *The Philosophy of Technology*, ed. P. T. Turbin, 113–30. Dordrecht: Kluwer, 1989.

Marton, L. "Alice in Electronland." *American Scientist* 31 (1943): 247–54.

——. *The Early History of the Electron Microscope*. San Francisco: San Francisco Press, 1968.

——. "The Electron Microscope: A New Tool for Bacteriological Research." *JB* 41 (1941): 397–413.

——. "Electron Microscopy of Biological Objects." *Nature* 133 (1934): 911.

——. "A New Electron Microscope." *Physical Review* 58 (1940): 57–60.

Medewar, P. "Is the Scientific Paper a Fraud?" *The Listener* 12 (1963): 377–78.

Merleau-Ponty, M. "Eye and Mind." Trans. C. Dallerey. In *The Primacy of Perception*, ed. J. Edie, 159–90. Evanston, Ill.: Northwestern University Press, 1964.

Meyerhoff, O. *Chemical Dynamics of Living Phenomena*. Philadelphia: J. B. Lippincott, 1924.

Michaelis, L. "Die vitale Farbung, eine Darstellungsmethode der Zellgranula." *Archiv für mikroskopische Anatomie* 55 (1899): 558–75.

Misa, T. "Retrieving Sociotechnological Change from Technological Determinism." In Smith and Marx, 115–41.

Misak, C. "Pragmatism in Focus." *SHPS* 25 (1994): 123–29.

Mitchell, P., and J. Moyle. "Osmotic Function and Structure in Bacteria." *Symposia of the Society for General Microbiology* 6 (1956): 150–80.

Mommaerts, W. *Muscular Contraction: A Topic in Molecular Physiology*. New York: Interscience, 1950.

Morales, M. "Mechanisms of Muscle Contraction." In Oncley, 426–32.

Mudd, S. A. "Cellular Organisation in Relation to Function." *Bacteriological Reviews* 20 (1956): 268–71.

———. "Cytology of Bacteria: I. The Bacterial Cell." *Annual Review of Microbiology* 8 (1954): 1–22.

———. "The Mitochondria of Bacteria." *Journal of Histochemistry and Cytochemistry* 1 (1953): 248–53.

———. "Pathogenic Bacteria, Rickettsias and Viruses as Shown by the Electron Microscope: II. Relationships to Immunity." *Journal of the American Medical Association* 126 (1944): 632–39.

———. "Submicroscopical Structure of the Bacterial Cell, as Shown by the Electron Microscope." *Nature* 161 (1948): 302–3.

Mudd, S. A., and T. F. Anderson. "Pathogenic Bacteria, Rickettsias and Viruses as Shown by the Electron Microscope: I. Morphology." *Journal of the American Medical Association* 126 (1944): 561–71.

———. "Selective 'Staining' for Electron Micrography." *JEM* 76 (1942): 103–8.

Mudd, S. A., A. Brodie, L. Winterscheid, P. Hartman, E. H. Beutner, and R. A. Maclean. "Further Evidence of the Existence of Mitochondria in Bacteria." *JB* 62 (1952): 729–39.

Mudd, S. A., T. Kawata, J. I. Payne, T. Sall, and A. Takagi. "Plasma Membranes and Mitochondrial Equivalents as Functionally Coordinated Structures." *Nature* 189 (1961): 79–80.

Mudd, S. A., and D. Lackman. "Bacterial Morphology as Shown by the Electron Microscope: I. Structural Differentiation Within the Streptococcal Cell." *JB* 41 (1941): 415–20.

Mudd, S. A., and A. Smith. "Electron and Light Microscopic Studies of Bacterial Nuclei: I. Adaptation of Cytological Processing to Electron Microscopy; Bacterial Nuclei as Vesicular Structures." *JB* 59 (1950): 561–73.

——. "Electron and Light Microscopic Studies of Bacterial Nuclei: II. An Improved Staining Technique for the Nuclear Chromatin of Bacterial Cells." *JB* 59 (1950): 575–87.

Mudd, S. A., K. Takeya, and H. J. Henderson. "Electron-Scattering Granules and Reducing Sites in Mycobacteria." *JB* 72 (1956): 767–83.

Mudd, S. A., and L. Winterscheid. "A Note Concerning Bacterial Cell Walls, Mitochondria, and Nuclei." *ECR* 5 (1953): 251–54.

Mudd, S. A., L. Winterscheid, E. DeLamater, and H. J. Henderson. "Evidence Suggesting That the Granules of Mycobacteria Are Mitochondria." *JB* 62 (1952): 459–75.

Muller, H. J. "The Production of Mutations." In *Nobel Lectures in Physiology or Medicine, 1942–1962*, 154–71. New York: Elsevier, 1967.

Newberry, S. "Electron Microscopy, the Early Years: Part 1." *Bulletin of the Electron Microscope Society of America* 15, no. 1 (1985): 39.

——. *EMSA and Its People: The First Fifty Years.* Woods Hole, N.Y.: Electron Microscopy Society of America, 1992.

Newman, S., E. Borysko, and M. Swerdlow. "A New Sectioning Technique for Light and Electron Microscopy." *Science* 110 (1949): 66–68.

——. "Ultra-Microtomy by a New Method." *Journal of Research of the National Bureau Standards* 43 (1949): 183–99.

Olby, R. *The Path to the Double Helix.* Seattle: University of Washington Press, 1974.

——. "Schrödinger's Problem: What Is Life?" *JHB* 4 (1971): 119–48.

Oncley, J. L., ed. *Biophysical Science—A Study Program.* New York: John Wiley, 1959.

Opportunities in Chemical Biology and Physical Biology at MIT. Cambridge, Mass.: Massachusetts Institute of Technology, 1949.

Palade, G. E. "An Electron Microscope Study of Mitochondrial Structure." *Journal of Histochemistry and Cytochemistry* 1 (1953): 188–211.

——. "Electron Microscopy of Mitochondria and Other Cytoplasmic Structures." In *Enzymes: Units of Biological Structure and Function*, ed. Oliver Gaebler, 185–215. New York: Academic Press, 1956.

——. "The Fine Structure of Mitochondria." *Anatomical Record* 114 (1952): 427–51.

——. "The Fixation of Tissues for Electron Microscopy." In *Proceedings of the Third International Conference*, 129–42.

——. "A Small Particulate Component of the Cytoplasm." *JCB* 1 (1955): 59–68.

——. "A Study of Fixation for Electron Microscopy." *JEM* 95 (1952): 285–97.

Palade, G. E., and K. R. Porter. "Studies on the Endoplasmic Reticulum: I. Its Identification in Cells *in Situ*." *JEM* 100 (1954): 641–56.

Palade, G. E., and P. Siekevitz. "Liver Microsomes: An Integrated Morphological and Biochemical Study." *JCB* 2 (1956): 171–200.

——. "Pancreatic Microsomes: An Integrated Morphological and Biochemical Study." *JCB* 2 (1956): 671–90.

Palay, S., and A. Claude. "An Electron Microscope Study of Salivary Gland Chromosomes by the Replica Method." *JEM* 89 (1949): 431–38.

Palay, S., and G. E. Palade. "The Fine Structure of Neurons." *JCB* 1 (1955): 69–88.

Pauling, L. "Molecular Architecture and Medical Progress." In Weaver, ed., 110–14.

Pauling, L., and D. Campbell. "The Artificial Production of Antibodies in Vitro." *JEM* 76 (1942): 211–20.

Pauly, P. *Controlling Life: Jacques Loeb and the Engineering Ideal in Biology.* New York: Oxford University Press, 1987.

——. "General Physiology and the Discipline of Physiology, 1890–1935." In *Physiology in the American Context 1850–1940,* ed. G. Geison, 195–207. Baltimore: American Physiological Society, 1987.

——. "Summer Resort and Scientific Discipline: Woods Hole and the Structure of American Biology 1882–1925." In *The American Development of Biology,* ed. R. Rainger, K. Benson, and J. Maienschein, 121–50. Philadelphia: University of Pennsylvania Press, 1988.

Payne, J. I., P. E. Hartman, S. A. Mudd, and C. Liu. "Cytological Analysis of Ultraviolet Irradiated *Escherichia coli:* II. Ultraviolet Induction of Lysogenic *E. coli.*" *JB* 70 (1955): 540–46.

Pease, D. C. "Electron Microscopy of the Tubular Cells of the Kidney Cortex." *Anatomical Record* 121 (1955): 723–43.

Pease, D. C., and R. F. Baker. "The Fine Structure of Skeletal Muscle." *American Journal of Anatomy* 84 (1949): 175–200.

——. "Sectioning Techniques for Electron Microscopy Using a Conventional Microtome." *PSEBM* 67 (1948): 470–74.

Pease, D. C., and K. R. Porter. "Electron Microscopy and Ultramicrotomy." *JCB* 91 (1981): 287s–92s.

Penick, J., C. Purcell, M. Sherwood, and D. Swain. *The Politics of American Science.* Chicago: Rand McNally, 1965.

Petermann, M., N. A. Mizen, and M. G. Hamilton. "The Macromolecular Particles of Normal and Regenerating Rat Liver." *Cancer Research* 13 (1953): 372–75.

Philpott, D., and A. Szent-Györgyi. "The Series Elastic Component in Muscle." *BBA* 12 (1953): 128–33.

Piaget, J. *The Construction of Reality in the Child.* Trans. M. Cook. New York: Basic Books, 1954.

——. *The Origins of Intelligence in Children.* Trans. M. Cook. New York: International University Press, 1952.

Piaget, J., and R. Garcia. *Psychogenesis and the History of Science.* Trans. H. Feider. New York: Columbia University Press, 1989.

Pijper, A. "Shape and Motility of Bacteria." *Journal of Pathology and Bacteriology* 58 (1946): 325–42.

Pinch, T. *Confronting Nature: The Sociology of Solar Neutrino Detection*. Dordrecht: Reidel, 1986.

Polanyi, M. *The Tacit Dimension*. London: Routledge, 1967.

"Polio at Work." *Time*, Sep. 20, 1948, 46.

Popper, K. *The Logic of Scientific Discovery*. New York: Basic Books, 1959.

Porter, K. R. "Electron Microscopy of Basophilic Components of Cytoplasm." *Journal of Histochemistry and Cytochemistry* 2 (1954): 346–75.

——. "Observations on a Submicroscopic Basophilic Component of Cytoplasm." *JEM* 97 (1953): 727–50.

——. "The Submicroscopic Structure of Protoplasm." *Harvey Lectures* 51 (1956): 175–28.

Porter, K. R., and H. S. Bennett. "Introduction: Recollections on the Beginnings of the Journal of Cell Biology." *JCB* 91 (1981): IXs–XIs.

Porter, K. R., and J. Blum. "A Study in Microtomy for Electron Microscopy." *Anatomical Record* 117 (1953): 685–710.

Porter, K. R., A. Claude, and E. Fullam. "A Study of Tissue Culture Cells by Electron Microscopy." *JEM* 81 (1945): 233–46.

Porter, K. R., and F. Kallman. "The Properties and Effects of Osmium Tetroxide as a Tissue Fixative with Special Reference to Its Use for Electron Microscopy." *ECR* 4 (1953): 127–41.

——. "Significance of Cell Particulates as Seen by Electron Microscopy." *Annals of the New York Academy of Sciences* 54 (1952): 882–91.

Porter, K. R., and H. Thompson. "A Particulate Body Associated with Epithelial Cells Cultured from Mammary Carcinomas of a Milk Factor Strain." *JEM* 88 (1948): 15–24.

——. "Some Morphological Features of Cultured Rat Sarcoma Cells as Revealed by the Electron Microscope." *Cancer Research* 7 (1947): 431–38.

Price, W. C., R. C. Williams, and R. Wyckoff. "Electron Micrographs of Crystalline Plant Viruses." *Archives of Biochemistry* 9 (1946): 175–85.

Proceedings of a Conference on Tissue Fine Structure. JCB supp. 2 (1956).

"Proceedings of the Electron Microscope Society of America." *JAP* 16 (1945): 263–66.

"Proceedings of the Electron Microscope Society of America." *JAP* 17 (1946): 66–68.

"Proceedings of the Electron Microscope Society of America." *JAP* 23 (1952): 156–64.

"Proceedings of the Electron Microscope Society of America." *JAP* 24 (1953): 111–18.

"Proceedings of Local Branches of the Society of American Bacteriologists: Eastern Pennsylvania Chapter." *JB* 41 (1941): 259–60.

Proceedings of the Third International Conference on Electron Microscopy, 1954. London: Royal Microscopical Society, 1956.

Proctor, B. "Food Technology at MIT." *Technology Review* 56 (Dec. 1953): 89–92.

"Program of the E. F. Burton Memorial Meeting of the Electron Microscope Society of America." *JAP* 19 (1948): 1186–92.

Pursell, C., Jr. "Science Agencies in World War II: The OSRD and Its Challengers." In *The Sciences in the American Context: New Perspectives*, ed. Nathan Reingold, 359–78. Washington: Smithsonian Institution Press, 1979.

Qing, L. *Zur Frühgeschichte des Elektronenmikroskops.* Stuttgart: GNT-Verlag, 1995.

Quarton, G., T. Melnechuk, and F. Schmitt, eds. *The Neurosciences: A Study Program.* New York: Rockefeller University Press, 1967.

Rajewsky, B., and M. Schon, eds. *Biophysics: FIAT Review of German Science, 1939–1946.* Vol. 22. Wiesbaden: Office of Military Government for Germany, Field Information Agencies—Technical, 1948.

Randall, J. T. "Emmeline Jean Hanson." *Biographical Memoirs of the Fellows of the Royal Society* 21 (1975): 313–44.

Rasmussen, N. "Facts, Artifacts, and Mesosomes: Practicing Epistemology with the Electron Microscope." *SHPS* 24 (1993): 227–65.

——. "Freund's Adjuvant and the Realization of Questions in Postwar Immunology." *HSPS* 23 (1993): 337–66.

——. "Making a Machine Instrumental: RCA and the Wartime Beginnings of Biological Electron Microscopy." *SHPS* 27 (1996): 311–49.

——. "The Midcentury Biophysics Bubble: Hiroshima and the Biological Revolution in America, Revisited." *History of Science* 35 (1997): 245–99.

——. "Mitochondrial Structure and the Practice of Cell Biology in the 1950s." *JHB* 28 (1995): 1–49.

Ravin, A. "The Gene as Catalyst; The Gene as Organism." *Studies in the History of Biology* 1 (1977): 1–45.

Reingold, N. *Science, American Style.* New Brunswick, N.J.: Rutgers University Press, 1991.

——. "Science and Government in the United States Since 1945." *History of Science* 32 (1994): 361–86.

Reisner, J. "An Early History of the Electron Microscope in the United States." *Advances in Electronics and Electron Physics* 73 (1989): 134–233.

——. "Reflections." *Bulletin of the Electron Microscope Society of America* 20, no. 2 (1990): 49–53.

Reisner, J., and E. Fullam. "Reflections." *Bulletin of the Electron Microscope Society of America* 16 (1986): 35–43.

Rescher, N., ed. *Evolution, Cognition, and Realism: Studies in Evolutionary Epistemology.* Lanham, Maryland: University Press of America, 1990.

"Return to Physics." *Newsweek*, Oct. 26, 1953, 109–10.

Rheinberger, H.-J. "Experiment, Difference, and Writing: I. Tracing Protein Synthesis." *SHPS* 23 (1992): 305–31.

——. *Experiment, Differenz, Schrift*. Marburg: Basilisken Presse, 1992.

——. "From Microsomes to Ribosomes: 'Strategies' of 'Representation' 1930–1955." *JHB* 28 (1995): 49–89.

Rhian, M., S. Lensen, and R. C. Williams. "An Electron Microscopic Study of Material from Tissue of the Central Nervous System of Poliomyelitic and Normal Mice and Cotton Rats." *Journal of Immunology* 62 (1949): 487–504.

Rhodin, J. "Correlation of Ultra-Structural Organization and Function in Normal and Experimentally Changed Proximal Convoluted Tubules of the Mouse Kidney." Ph.D. diss., Karolinska Institute, 1954.

Rice, R. V., P. Kaesberg, and M. A. Stahmann. "The Breaking of Tobacco Mosaic Virus Using a New Freeze Drying Method." *BBA* 11 (1953): 337–43.

Richards, A. G., T. F. Anderson, and R. T. Hance. "Microtome Sectioning Technique for Electron Microscopy Illustrated with Sections of Striated Muscle." *PSEBM* 51 (1942): 148–52.

Richards, A. G., H. B. Steinbach, and T. F. Anderson. "Electron Microscope Studies of Squid Giant Nerve Axoplasm." *Journal of Cellular and Comparative Physiology* 21 (1943): 129–43.

Ricoeur, P. "Metaphor and the Main Problem of Hermeneutics." In *The Philosophy of Paul Ricoeur*, ed. C. Reagan and D. Stewart, 134–48. Boston: Beacon, 1978.

Robinow, C. F. "The Chromatin Bodies of Bacteria." *Bacteriological Reviews* 20 (1956): 207–42.

Robinow, C. F., and V. E. Coslett. "Nuclei and Other Structures of Bacteria." *JAP* 19 (1948): 124.

Robinow, C. F., and E. Kellenberger. "The Bacterial Nucleoid Revisited." *Microbiological Reviews* 58 (1994): 211–32.

Robinson, J. "The Chemiosmotic Hypothesis of Energy Coupling and the Path of Scientific Opportunity." *Perspectives in Biology and Medicine* 27 (1984): 367–83.

Rockefeller Foundation Annual Report. New York: Rockefeller Foundation, 1933.

Rorty, R. "An Anti-Representationalist View." In *Realism and Representation*, ed. G. Levine, 125–33. Madison: University of Wisconsin Press, 1993.

——. *Consequences of Pragmatism*. Minneapolis: University of Minnesota Press, 1982.

Rosza, G., A. Szent-Györgyi, and R. Wyckoff. "The Electron Microscopy of F-Actin." *BBA* 3 (1949): 561–69.

——. "The Fine Structure of Myofibrils." *ECR* 1 (1950): 194–205.

Rotman, B. *Ad Infinitum: The Ghost in Turing's Machine*. Palo Alto, Calif.: Stanford University Press, 1993.

Ruska, E. "The Development of the Electron Microscope and Electron Micros-copy." Nobel lecture of Dec. 1986. *Bulletin of the Electron Microscopy Society of America* 18, no. 2 (1988): 53–61.

Sapolosky, H. *Science and the Navy: The History of the Office of Naval Research.* Princeton, N.J.: Princeton University Press, 1990.

Sapp, J. *Beyond the Gene: Cytoplasmic Inheritance and the Struggle for Authority in Genetics.* New York: Oxford University Press, 1987.

Schachman, H., and R. C. Williams. "The Physical Properties of Infective Parti-cles." In Burnet and Stanley, 1:223–327.

Schaefer, V., and D. Harker. "Surface Replicas for Use in the Electron Micro-scope." *JAP* 13 (1942): 427–33.

Schivelbusch, W. *The Railroad Journey.* Berkeley: University of California Press, 1977.

Schmitt, F. O. "Chairman's Prefatory Remarks." *PNAS* 42 (1956): 789–91.

——. "Chromosomes, Genes, and Macromolecular Systems." *Nature* 177 (1956): 503–5.

——. "Festvortrag: Electron Microscopy in Morphology and Molecular Biol-ogy." *Fourth International Conference on Electron Microscopy* 2 (1958): 1–16.

——. "The Fibrous Protein of the Nerve Axon." *Journal of Cellular and Compara-tive Physiology* 49 (1957): 165–74.

——. "Giant Molecules in Cells and Tissues." *Scientific American* 197 (1957): 204–16.

——. "Molecular Biology and the Physical Basis of Life Processes." In Oncley, 5–10.

——. "Morphology in Muscle and Nerve Physiology." *BBA* 4 (1950): 68–77.

——. *The Never-Ceasing Search.* Philadelphia: American Philosophical Society, 1991.

——. "Patterns of Interaction of Biological Macromolecules in Relation to Cell Function." *PNAS* 42 (1956): 806–10.

——. "The Structural Basis of Life." In Weaver, ed., 228–32.

——. "The Structure of the Axon Filaments of the Giant Nerve Fibers of *Loligo* and *Myxicola*." *Journal of Experimental Zoology* 113 (1950): 499–515.

Schmitt, F. O., R. S. Bear, C. E. Hall, and M. A. Jakus, "Electron Microscope and X-Ray Diffraction Studies of Muscle Structure." *Annals of the New York Academy of Sciences* 47 (1947): 799–812.

Schmitt, F. O., and B. Geren. "The Fibrous Structure of the Nerve Axon in Relation to the Localization of 'Neurotubules.'" *JEM* 91 (1950): 499–504.

Schmitt, F. O., C. E. Hall, and M. A. Jakus. "Electron Microscope Studies of the Structure of Collagen." *Journal of Cellular and Comparative Physiology* 20 (1942): 11–33.

Schneider, W., and G. Hogeboom. "Chemical Studies of Mammalian Tissues: The Isolation of Cell Components by Differential Centrifugation." *Cancer Research* 1 (1950): 1–22.

Schramm, G., G. Schumacher, and W. Zillig. "An Infectious Nucleoprotein from Tobacco Mosaic Virus." *Nature* 175 (1955): 549–50.

Schultz, E. W., P. R. Thomassen, and L. Marton. "Electron Microscopic Observations on *Pseudomonas aeruginosa* Bacteriophage." *PSEBM* 68 (1948): 451–55.

Schulz, H., H. Löw, L. Ernster, and F. Sjöstrand. "Elektronenmikropische Studien an Leberschnitten von Thyroxin-behandelten Ratten." In *Stockholm Conference*, 134–37.

Schwerdt, C. E., R. C. Williams, W. M. Stanley, F. L. Schaffer, and M. E. McClain. "Morphology of the Type II Poliomyelitis Virus (MEF1) as Determined by Electron Microscopy." *PSEBM* 86 (1954): 310–12.

Sedar, A., and K. R. Porter. "The Fine Structure of Cortical Components of *Paramecium multimicronucleatum*." *JCB* 1 (1955): 583–604.

Sedar, A., and M. Rudzinska. "Mitochondria of Protozoa." In *Proceedings of a Conference*, 331–34.

Seidel, R. "The Origins of the Lawrence Berkeley Laboratory." In *Big Science: The Growth of Large-Scale Research*, ed. P. Galison and B. Hevley, 21–45. Palo Alto, Calif.: Stanford University Press, 1992.

Serres, M. *Rome: The Book of Foundations*. Trans. F. McCarren. Palo Alto, Calif.: Stanford University Press, 1991.

Seymour, F., and M. Benmosche. "Magnification of Spermatozoa by Means of the Electron Microscope." *Journal of the American Medical Association* 116 (1941): 2489–90.

Shapere, D. "The Concept of Observation in Science and Philosophy." *Philosophy of Science* 49 (1982): 485–525.

Shapin, S. *A Social History of Truth*. Chicago: University of Chicago Press, 1994.

Shinohara, C., K. Fukushi, and J. Suzuki. "Mitochondrialike Structures in Ultrathin Sections of *Mycobacterium avium*." *JB* 74 (1957): 413–15.

Siekevitz, P., and P. Zamecnik. "Ribosomes and Protein Synthesis." *JCB* 91 (1981): 53s–65s.

Silverstein, A. *A History of Immunology*. New York: Academic Press, 1989.

Sjöstrand, F. "Critical Evaluation of Ultrastructural Patterns with Respect to Fixation." *Symposia of the International Society for Cell Biology* 1 (1962): 47–68.

———. "Electron Microscopic Demonstration of Membrane Structure Isolated from Nerve Tissue." *Nature* 165 (1950): 482–83.

———. "Electron-Microscopic Examination of Tissue." *Nature* 151 (1943): 725–26.

———. "Electron Microscopy of Mitochondria and Cytoplasmic Double Membranes." *Nature* 171 (1953): 30–32.

———. "Fixation and Preparation of Tissues for Electron Microscope Examination." *Nordisk Medicin* 19 (1943): 1207–12.

———. "Method for Making Ultra-Thin Tissue Sections for Electron Microscopy at High Resolution." *Nature* 168 (1951): 729–30.

——. "Molecular Structure of Mitochondrial Cristae, Solid State Biochemistry and a Simple Theory of Respiration-Phosphorylation Coupling." *Journal of Submicroscopic Cytology and Pathology* 23 (1991): 465–89.

——. "Recent Advances in the Biological Application of the Electron Microscope." In *Proceedings of the Third International Conference*, 26–37.

——. "The Ultrastructure of Cells as Revealed by the Electron Microscope." *International Reviews of Cytology* 5 (1956): 455–533.

Sjöstrand, F., and E. Andersson. "The Ultrastructure of Skeletal Muscle Myofilments." In *Stockholm Conference*, 204–8.

——. "The Ultrastructure of Skeletal Muscle Myofilments at Various Stages of Shortening." *Journal of Ultrastructure Research* 1 (1957): 74–108.

Sjöstrand, F., and V. Hanzon. "Membrane Structures of Cytoplasm and Mitochondria in Exocrine Cells of Mouse Pancreas as Revealed by High-Resolution Electron Microscopy." *ECR* 7 (1954): 393–414.

Sjöstrand, F., and J. Rhodin. "The Ultrastructure of the Proximal Convoluted Tubules of the Mouse Kidney as Revealed by High-Resolution Electron Microscopy." *ECR* 4 (1953): 426–56.

Slater, E. "A Short History of the Biochemistry of Mitochondria." In *Mitochondria and Microsomes: In Honor of Lars Ernster*, ed. C. P. Lee, G. Schatz, and G. Dallner, 15–43. London: Addison-Wesley, 1981.

"Smaller and Smaller." *Time*, Oct. 28, 1940, 50.

"Smallest Measuring Sticks." *Time*, July 18, 1949, 53.

Smith, A. K. *A Peril and a Hope*. Cambridge, Mass: MIT Press, 1971.

Smith, H. "Squid Nerves for Research." *The Rhode Islander Magazine, Providence Sunday Journal*, Aug. 1952, 3–5.

Smith, M. R., and L. Marx, eds. *Does Technology Drive History? The Dilemma of Technological Determinism*. Cambridge, Mass.: MIT Press, 1994.

Smith, T. "Reflections." *Bulletin of the Electron Microscope Society of America* 14 (1984): 19–21.

Sobel, R. *RCA*. New York: Stein & Day, 1986.

Solomon, M. "The Pragmatic Turn in Naturalistic Philosophy of Science." *Perspectives on Science* 3 (1995): 206–30.

Spath, S. "Idealizing the Microbe: C. B. Van Niel's Conception of General Microbiology." Paper presented at History of Science Society conference, New Orleans, Oct. 1994.

"Speaking of Pictures." *Life*, June 1, 1942, 8–11.

"Speaking of Pictures." *Life*, Oct. 10, 1945, 12–13.

Spencer, S. "New Magic Eye Sees Strange New Worlds." *Philadelphia Evening Bulletin*, Feb. 13, 1941, sec. D, 19.

Stanier, R. Y. "Some Singular Features of Bacteria as Dynamic Systems." In *Cellular Metabolism and Infections*, ed. E. Racker, 3–24. New York: Academic Press, 1954.

Stanley, W. M., and T. F. Anderson. "Electron Micrographs of Protein Molecules." *Journal of Biological Chemistry* 146 (1942): 25–30.

Steere, R., and R. C. Williams. "Electron Micrographic Observations of the Unit of Length of the Particles of Tobacco Mosaic Virus." *Journal of the American Chemical Society* 73 (1951): 2057–61.

——. "A Simplified Method of Purifying Tomato Bushy-Stunt Virus for Electron Microscopy." *Phytopathology* 38 (1948): 948–54.

Stockholm Conference on Electron Microscopy, Proceedings. New York: Academic Press, 1957.

Stone, R. A. "Split Subjects, Not Atoms; or How I Fell in Love with My Prosthesis." *Configurations* 2 (1994): 173–90.

Storck, R., and J. T. Wachsman. "Enzyme Localisation in *Bacillus megaterium.*" *JB* 78 (1957): 784–90.

Strick, J. "Adrianus Pijper and the Debate over Bacterial Flagella: Morphology and Electron Microscopy in Bacteriology, 1946–1956." *Isis* 87 (1996): 274–305.

Strickland, S. *Politics, Science, and Dread Disease: A Short History of United States Medical Research Policy.* Cambridge, Mass.: Harvard University Press, 1972.

——. *Scientists in Politics: The Atomic Scientists Movement, 1945–46.* Lafayette, Ind.: Purdue University Press, 1968.

——. *The Story of the NIH Grants Programs.* Lanham, Maryland: University Press of America, 1988.

Suchman, L. *Plans and Situated Action.* Cambridge, Eng.: Cambridge University Press, 1987.

Summers, W. "From Culture as Organism to Organism as Cell: Historical Origins of Bacterial Genetics." *JHB* 24 (1991): 171–90.

Süsskind, C. "Ladislaus Marton, 1901–1979." In Hawkes, 501–23.

Szent-Györgyi, A. *The Chemistry of Muscular Contraction.* New York: Academic Press, 1947.

——. "Free Energy Relations and the Contraction of Actomyosin." *Biological Bulletin* 96 (1949): 140–61.

——. "Lost in the Twentieth Century." *Annual Review of Biochemistry* 32 (1963): 1–14.

Taylor, W., and R. C. Williams. "Shadow-Casting: A Technique to Show Surface Texture in Microscopical Material." *Anatomical Record* 96 (1946): 27–39.

Temkin, O. *The Double Face of Janus and Other Essays in the History of Medicine.* Baltimore: Johns Hopkins University Press, 1976.

Theresa of Avila. *The Life of Saint Theresa.* Trans. J. M. Cohen. London: Penguin, 1957.

Thompson, E. P. *Customs in Common.* London: Merlin Press, 1991.

Tibbetts, P. "Representation and the Realist-Constructivist Controversy." In Lynch and Woolgar, 69–83.

Tiles, M. *Bachelard, Science and Objectivity.* Cambridge, Eng.: Cambridge University Press, 1984.

Toulmin, S. *The Philosophy of Science: An Introduction.* London: Hutchinson House, 1953.

Tsugita, A., D. T. Gish, J. Young, H. Fraenkel-Conrat, C. A. Knight, and W. S. Stanley. "The Complete Amino Acid Sequence of the Protein of Tobacco Mosaic Virus." *PNAS* 46 (1960): 1463–69.

Turnbull, D. *Maps Are Territories: Science Is an Atlas.* Geelong, Victoria: Deakin University Press, 1989.

Turnbull, D., and T. Stokes. "Manipulable Systems and Laboratory Strategies in a Biomedical Institute." In *Experimental Inquiries,* ed. H. Le Grand, 167–92. Dordrecht: Kluwer, 1990.

Van Helden, A., and T. Hankins, eds. *Instruments. Osiris* 9 (1994).

van Iterson, W. "Some Electron-Microscopical Observations on Bacterial Cytology." *BBA* 1 (1947): 527–48.

———. "Some Features of a Remarkable Organelle in *Bacillus subtilis.*" *JCB* 9 (1961): 183–92.

Walter, M. *Science and Cultural Crisis: An Intellectual Biography of Percy Williams Bridgman.* Palo Alto, Calif.: Stanford University Press, 1990.

Warren, S. "Atomic Energy and Medicine." In Weaver, ed., 283–87.

———. "Radioactivity in Nagasaki." *Harvey Lectures* 41 (1946): 188.

Weart, S. *Nuclear Fear: A History of Images.* Cambridge, Mass.: Harvard University Press, 1988.

Weaver, W. "Molecular Biology: Origins of the Term." *Science* 170 (1970): 581–82.

———, ed. *The Scientists Speak.* New York: Boni & Gaer, 1947.

Weibull, C., H. Beckman, and L. Bergstrom. "Localization of Enzymes in *Bacillus megaterium,* Strain M." *Journal of General Microbiology* 20 (1959): 519–31.

Westwick, P. "Abraded from Several Corners: Medical Physics at Berkeley, 1940–1950." Paper presented at History of Science Society conference, Minneapolis, Oct. 1995.

Williams, R. C. "The Electron Microscope in Biology." *Growth* 11 (1947): 205–22.

———. "High-Resolution Electron Microscopy of the Particles of Tobacco Mosaic Virus." *BBA* 8 (1952): 227–44.

———. "A Method of Freeze-Drying for Electron Microscopy." *ECR* 4 (1953): 188–201.

———. "Replication of Nucleic Acids." In Oncley, 233–41.

———. "The Shapes and Sizes of Purified Viruses as Determined by Electron Microscopy." *Cold Spring Harbor Symposia in Quantitative Biology* 18 (1953): 185–95.

Williams, R. C., and R. Backus. "Macromolecular Weights Determined by

Direct Particle Counting: I. The Weight of the Bushy Stunt Virus Particle." *Journal of the American Chemical Society* 71 (1949): 4052–57.

Williams, R. C., R. Backus, and R. Steere. "Macromolecular Weights Determined by Direct Particle Counting: II. The Weight of the Tobacco Mosaic Virus Particle." *Journal of the American Chemical Society* 73 (1951): 2062–66.

Williams, R. C., and R. Steere. "Electron Micrographic Observations of Tobacco Mosaic Virus in Crude, Undiluted Plant Juice." *Science* 109 (1949): 308–9.

Williams, R. C., and R. Wyckoff. "Electron Shadow-Micrographs of Hemocyanin Molecules." *Nature* 156 (1945): 68–70.

———. "Electron Shadow-Micrography of the Tobacco Mosaic Virus Protein." *Science* 101 (1945): 594–96.

———. "Electron Shadow-Micrography of Virus Particles." *PSEBM* 58 (1945): 265–70.

———. "Shadowed Electron Micrographs of Bacteria." *PSEBM* 59 (1945): 265–70.

———. "The Thickness of Electron Microscopic Objects." *JAP* 15 (1944): 712–16.

Williams, R. C., R. Wyckoff, and W. C. Price. "Electron Micrography of Crystalline Plant Viruses." *Science* 102 (1945): 277–78.

Wilson, D. J. "Fertile Ground: Pragmatism, Science, and Logical Positivism." In *Pragmatism: From Progressivism to Postmodernism*, ed. R. Hollinger and D. Depew, 122–41. Westport, Conn.: Praeger, 1995.

Wimsatt, W. "Robustness, Reliability, and Overdetermination." In *Scientific Inquiry and the Social Science*, ed. M. Brewer and B. Collins, 124–63. San Francisco: Jossey-Bass, 1981.

Wood, D. *The Power of Maps*. New York: Guilford, 1992.

Wyckoff, R. "The Electron Microscopy of Developing Bacteriophage: I. Plaques on Solid Media." *BBA* 2 (1948): 27–37.

———. "Reminiscences." In Hawkes, 583–87.

———. "Ultraviolet Microscopy as a Means of Studying Cells." *Cold Spring Harbor Symposia in Quantitative Biology* 2 (1934): 39–46.

Yavendetti, M. "The American Reaction to the Use of Atomic Bombs on Japan: The 1940s." *Historian* 36, no. 2 (1974): 224–47.

Yoxen, E. "Giving Life New Meaning: The Rise of the Molecular Biology Establishment." In *Scientific Establishments and Hierarchies*, ed. N. Elias, H. Martins, R. Whitley, 123–43. Dordrecht: D. Reidel, 1982.

———. "Where Does Schrödinger's 'What Is Life?' Belong in the History of Molecular Biology?" *History of Science* 17 (1979): 17–52.

Zallen, D. "The Rockefeller Foundation and Spectroscopy Research: The Programs at Chicago and Utrecht." *JHB* 25 (1992): 67–89.

Zimmerman, M. *Heidegger's Confrontation with Modernity: Technology, Politics, and Art*. Bloomington: Indiana University Press, 1990.

Zworykin, K. P., and G. Chapman. "Television Microscopy of Living Sporulating and Germinating *Bacillus cereus* Cells." *Journal of Cellular and Comparative Physiology* 48 (1946): 301–16.

Zworykin, V. K. "Electron Microscopy in Chemistry." *Electronics* 16 (1943): 64–68, 190–96.

Zworykin, V. K., G. A. Morton, E. G. Ramberg, J. Hillier, and A. W. Vance. *Electron Optics and the Electron Microscope.* New York: J. Wiley, 1945.

Index

In this index an "f" after a number indicates a separate reference on the next page, and an "ff" indicates separate references on the next two pages. A continuous discussion over two or more pages is indicated by a span of page numbers, e.g., "57–59." *Passim* is used for a cluster of references in close but not consecutive sequence.

Dorrance Building, 158f, 187
Duggar, B. M., 47

Edsall, J. T., 178
Electron Microscope Society of America
    (EMSA), 1, 61–65, 127ff, 156, 163;
    first meeting of (1942), 61f, 231–37
    passim, 244f, 276n41
Electrophoresis, 5, 9, 35, 52, 159f, 201
Electrostatic lenses, 25, 62
Embodiment, 228–31, 247–52 passim
Emerson, Robert, 292n115
Endoplasmic reticulum, 113–23, 144,
    211f, 247, 250
Engstrom, Arne, 160
Executives, biologists as, 4f, 156, 206
Eyring, Henry, 272n123

Farquar, Marylin, 147
Fawcett, Don, 146f
Federal Communications Commission
    (FCC), 31, 64
Ferguson, Thomas, 260n13
Ferry, John, 292
Feyerabend, Paul, 262n60
Flagella, 49, 79f, 86, 88, 99
Fleck, Ludwik, 68, 273n149
Flick, John, 73
Foster, Joseph, 292n115
Foucault, Michel, 249, 272n127
Fraenkel-Conrat, Heinz, 208f, 213–
    16
Franck, James, 198
Franklin, Allan, 12f, 261n46
Franklin Institute, 76, 81
Fraser, W. D., 208f
Freund, Jules, 72
Fry, William, 292n115
Fuerst, John, 293n128
Fuhr, Irving, 188f, 292
Fujimura, Joan, 283n74
Fullam, Ernest, 107–12 passim
Fussell, Paul, 269n87

Gadamer, H. J., 255
Galison, Peter, 286n145
Gamow, George, 198
Gasser, Herbert, 105, 109, 112

General Electric, 9, 31, 61f
General physiology, 8, 33, 37–40, 159,
    184ff, 199, 251
Genetic engineering, 10, 216f. See also
    Biotechnology
Gerard, Ralph, 292n115
Geren (Uzman), Betty, 160–63, 168ff
Germ warfare, Big Ben project on, 73f,
    274n14
Gessler, Albert, 106, 112
Gestalt, 68, 154, 243
Giere, R. N., 12
Glauert, Audrey, 147
Goldman, David, 292
Gots, Joseph, 73f
Gray, Irving, 292nn114, 115
Griggs, R. F., 48f
Gross, Jerome, 160
Guillemin, Victor, 292n115
Gulliver's Travels, 225

Habermas, Jürgen, 228, 300n27
Hacking, Ian, 12f, 261n40
Hall, Cecil, 1, 27, 37–40, 62, 156, 173,
    230f, 237
Hanson, E. J., 160, 176–85
Hardy, James, 292
Harrison, Ross, 60
Hartline, Haldan, 292
Hartman, P. E., 275n26
Haskins, Caryl, 47, 51, 267n69
Heelan, Patrick, 15f, 248, 253
Heidegger, Martin, 16, 228, 300n27
Heidelberger, Michael, 50ff, 268n84,
    269n90
Hermeneutics, 16f, 224, 247–56
Hessen, Boris, 260n19
Higher (i.e., intermediate) voltage, 35, 62,
    77, 276n41
Hill, A. V., 192
Hillier, James, 32–37 passim, 43, 48, 65,
    127, 175, 207–9 passim, 224f, 235,
    263n11; bacteriology research of, 73–
    80 passim, 89–95 passim, 112, 242, 246,
    297n35
Hillman, H., 23ff
Histochemical Society, 1953 meeting of,
    89, 129–32

Library of Congress Cataloging-in-Publication Data
Rasmussen, Nicolas.
   Picture control : the electron microscope and the transformation
of biology in America, 1940–1960 / Nicolas Rasmussen.
      p.   cm. — (Writing science)
   Includes bibliographical references (p.   ) and index.
ISBN 0-8047-2837-2 (cl.) : ISBN 0-8047-3850-5 (pbk.)
   1. Electron microscopes—History.   2. Electron microscopy—
History.   3. Biology—History.   I. Title.   II. Series.
QH212.E4R37   1997
570'.28'250973—dc21   97-1230
                                                             CIP

Original printing 1997

Printed in the USA
CPSIA information can be obtained
at www.ICGtesting.com
JSHW021435221024
72172JS00002B/6